David George Price

Engineering Geology
Principles and Practice

T0178403

David George Price

Engineering Geology

Principles and Practice

Edited and Compiled
by M. H. de Freitas

With 182 Figures and 54 Tables

 Springer

Author

David George Price (deceased)

Editor

Dr. Michael H. de Freitas
Department of Civil and Environmental Engineering
Imperial College London
South Kensington
London SW7 2AZ, United Kingdom

Contributors

Dr. Michael H. de Freitas (Imperial College London, United Kingdom)
Dr. H. Robert G. K. Hack (ITC, The Netherlands)
Ian E. Higginbottom (Wimpey Laboratories London, United Kingdom)
Prof. Sir John L. Knill (deceased) (Imperial College London, United Kingdom)
Dr. Michiel Maurenbrecher (TU Delft, The Netherlands)

ISBN: 978-3-642-06725-9
e-ISBN: 978-3-540-68626-2

© 2010 Springer-Verlag Berlin Heidelberg

Cover design: WMX Design GmbH, Heidelberg

Printed on acid-free paper   30/2132/AO – 5 4 3 2 1 0

springer.com

To Valerie

Cover Picture

Here we see the front end, or head, of a tunnel boring machine stripped down to its essentials and awaiting these before assembly to its tail. Its design and construction illustrates the need for engineering geology. To the left is the cutting head, carrying discs to match and overcome the strength of the rock to be encountered – in this case strong rock is the anticipated material, hence the discs. They are mounted on a strong cutting head designed to both rotate and carry the thrust required to load the discs against the rock, so they may grind and crush the rock to bore the tunnel. That head is largely unprotected for the hole it will bore will not collapse; it is predicted that the rock mass will be self supporting. Thus no tunnel lining segments will need to be placed behind the head and so there will be no structure against which the cutting head can react to generate its forward thrust. For this reason the machine has been given lateral pads (one is facing the viewer) to extend and press against the tunnel wall, so anchoring the head whilst its forward facing jacks press the cutter and its discs against the tunnel face. The rock mass is predicted to be stiff, deforming little under these lateral loads; it could not anchor the machine if it were not so. No lining also implies no ground water; this is predicted to be a dry tunnel. Space has been left in the lower third of the head; this will be occupied by a conveyor that will carry the debris from the cutting head up through the machine where it will discharge into a transporting system (another conveyor or train or trucks) to be carried out of the tunnel; a stable range of fragment sizes is predicted from the comminution at the tunnel face. At some stage in the design of this machine an engineering geologist has had to advise on the geology of the materials to be encountered, the structure, stiffness and stability of the mass in which they are contained, the presence of ground water and the likelihood of it draining to the tunnel, the response of the ground to loading by disc cutters, the response of the ground to unloading as the bore proceeds, the disposal of the cuttings and the danger to personnel that rock dust from the tunnel face, and gasses from the surrounding ground, may generate. Insurers and Health and Safety legislation may require these variables to be the subject of risk analyses. If the tunnel is to pass under urban areas then noise and vibration of tunnelling must also be predicted for the loads and advance rates that are required. Radiation of acoustic waves, the sort when strong enough that will be heard and felt at ground level, is usually anisotropic and requires ground structure to be understood. Hundreds of thousands of tons of broken rock may come from the excavation, some may be suitable for concrete aggregate and used elsewhere in the project, the bulk may be used for embankments leading to and from the tunnel itself, or for reclaiming land elsewhere; all these decisions involve input from engineering geology. Finally, the road on which the machine is parked is a small cutting; that cutting will get larger as the hillside into which the tunnel will bore is approached until a rock face is exposed into which the tunnel may start boring. This is the tunnel portal and invariably it has to penetrate landslides that cloak the surface of the slope through which it cuts without disturbing the slope or the developments that may be founded upon it. All this and more is engineering geology.

Preface

David Price had written the greater part of this book by the time he died; it has been completed by his colleagues as a tribute to the many contributions he made to the subject of engineering geology through his professional and academic life.

David graduated from the University of Wales in 1954 with the degree of Geology with Mathematics and Physics, joined the Overseas Division of the Geological Survey and was despatched to what was then British Guiana, to map economic mineral reserves and construction materials. He returned to the UK in 1958 to join the construction company George Wimpey. The post-war boom was beginning and David was engaged as an engineering geologist. In those days industry appreciated the need for research, as little was known for the tasks that had to be completed, and David joined a remarkable group of scientists and engineers at Wimpey's Central Laboratory at Hayes; the young reader can best visualise this as an "industrial university". At that time formal education and training in engineering geology did not exist and as David recalled "... *no one really knew what they were doing; we followed the principles of our subject, used common sense, learnt from what happened on site and talked to those who seemed to know more than we on the subject in hand.*" It was David's generation that established "Engineering Geology", as we now know it, in the UK and he played a full part in its foundation.

In 1975 David was appointed to the Chair of Engineering Geology at the Technical University of Delft. There he witnessed both the academic growth of the subject and the portents of its demise, for by the time he retired, in 1993, David could see that university funding would threaten the teaching of his subject. It was time to write the book that had long been his intention; a text that set out simply and clearly "... *what engineering geology is about and how it's done*" and to do so uncluttered by theory and equation, which were to be used only to illustrate the information required from the field for design and analyses; a book for those new to the subject, based on his lifetime's experience as a practicing engineering geologist, and on his view of the priorities for its practice.

David started writing the work in the summer of 1995 but died shortly before his text was finished. He agreed it should be completed by his colleagues and it has been our happy task to complete that left undone, to smooth that left rough, and to connect that left unconnected. Where possible the text has remained as written. The chapters were divided amongst those of his colleagues who could help and their names stand next to the chapters with which they were involved. *Ian Higginbottom*, a contemporary from his days at Wimpey and life-long friend whom David viewed in many ways as his mentor. *John Knill*, also of that vintage, Reader and later Professor of Engineering Geology at Imperial College London, close friend and academic ally who was influential in David's move from industry to academia. *Michiel Maurenbrecher*, long time academic colleague and loyal member of David's staff in Engineering Geology at Delft. *Robert Hack*, of the International Institute for Geo-Information Science and Earth Observation (ITC) at Delft and Enschede, a former student of David who became a close associate and shared with him much teaching, and many research projects.

David George Price, 1932–1999

Euro Ing. C.Geol., MICE, FGS
Professor Emeritus of Engineering Geology
Technical University of Delft

Acknowledgements

David wrote; *"I wish to acknowledge the contributions to my career and the assistance I have received over many years from my colleagues Ian Higginbottom, Colin Davenport, Robert Hack, Michiel Maurenbrecher, Arno Mulder, Peter Verhoef, Willem Verwaal and Niek Rengers. Royalties from the book should be lodged with the Geological Society of London and used by the Engineering Group to fund field work in Engineering Geology for students from Britain and the Netherlands."*

As editor I must record, with thanks, the invaluable support I received from Christine Butenuth, whose suggestion it was that the book should be saved by David's colleagues, and who provided urgently needed assistance with the final preparation of the text. My work was entirely dependent on that of the co-authors whom I thank for their patience and tenacity with the task they were set. It was David's wish that this should be a project involving "Delft and Imperial" and that it became. Post graduates, Stephen Gazard and Baudrey Kock, helped prepare the script for delivery on time, and Marion Schneider, our editor at Springer, has been responsible for publishing the work. Finally, I must thank Imperial College for providing the opportunity to complete this contribution to engineering geology.

M. H. de Freitas

Imperial College London
August 2007

Contents

Chapter 1

The Basis of Engineering Geology

1.1
Development of Engineering Geology

While ancient man must have had some intuitive knowledge of geology, as evidenced by the feats of mining and civil engineering performed in the distant past, the present science of geology owes much of its origin to the civil engineers working in the eighteenth century. These engineers, while constructing the major engineering works associated with the industrial revolution, had the opportunity to view and explore excavations in rocks and soils. Some, intrigued by what they saw, began to speculate on the origin and nature of rocks, and the relationships between similar rocks found in different places. Their ideas and theories, based on the practical application of their subject, formed the groundwork for the development of geology as a science. Engineers such as Lewis Evans (1700–1756) in America, William Smith (1769–1839) in England, Pierre Cordier (1777–1862) in France and many others were the 'fathers' of Geology.

Their interest in geology often stemmed from a 'need to know'. They were confronted with real engineering problems which could only be solved with the help of both a knowledge and understanding of the ground conditions with which they were confronted.

In the later nineteenth century both geology and engineering advanced, geology becoming a more-or-less respectable natural philosophy forming part of the education considered suitable for well brought up young ladies. Engineering, characterised by the canal and railway construction carried out by the 'navvy', on the other hand, remained as an eminently practical subject. The theoretical understanding of engineering was driven by practical engineering problems. The geological knowledge of the engineer, confronted by increasingly difficult engineering challenges, did not progress as rapidly as geology, advanced as a science under the leadership of geologists such as James Dana (1813–1895) in America, Albert Heim (1849–1937) in Switzerland and Sir Archibald Geikie (1835–1924) in Britain. Thus, by the end of the nineteenth century the majority of civil engineers knew relatively little about geology, and very few geologists were concerned about, or interested in, its engineering applications.

This widening division between geology and engineering was partly bridged in the nineteenth and early twentieth century by the development of soil mechanics by engineers such as Charles Coulomb and Macquorn Rankine, who developed methods of calculating the deformations of earth masses under the stresses imposed by engineering works. The great leap forward may be considered to have taken place with the publication of *"Erdbaumechanik"* by Karl Terzaghi in 1925, which brought together old knowledge, and added new theory and experience to establish soil mechanics in its own right as a discipline within the field of civil engineering. Subsequent publications by Terzaghi and others have continued to recognise a clear understanding of the fundamental importance of geological conditions in civil engineering design and construction. However, this appreciation has not proved to be universal and many engineers continued to rely on inadequate geological knowledge, or over-simplified ground models.

Failures of engineering works in particular, such as that of the Austin Dam in Texas in 1900 and the St. Francis Dam in California in 1928, showed that there was often a lack of appreciation of the importance of geological conditions in engineering design.

Such failures emphasised the need for expert assessment of geological conditions on civil engineering sites and there was, by the 1940s, a trend for civil engineers to employ geologists in an advisory capacity. However, while certain gifted individuals, such as Charles Berkey in the United States (Paige 1950) and Quido Zaruba in Czechoslovakia (Zaruba and Mencl 1976), performed this function very well it was not always a successful liaison. Few geologists had sufficient engineering knowledge to understand the requirements of the engineer and few engineers had more than the most superficial knowledge of geology.

Despite these problems the recognition that liaison was required slowly brought to the fore a new breed of earth scientist, the 'engineering geologist'. Most of the early engineering geologists were geologists who had gravitated into engineering employment, educating themselves by study and experience. Certain notable engineers, such as Robert Leggett in Canada (Legget 1939), developed their geological knowledge to achieve the complementary aim.

Eventually engineering geology became sufficiently developed for the subject to form part of university curricula. Thus, in Imperial College, London, engineering geology was taught at postgraduate level to both geologists and engineers as early as 1957 under the guidance of John Knill. Courses progressively developed elsewhere in England, Europe, America and Canada during the subsequent decades. Now there are few countries in the world where engineering geology, in some form or other, is not taught as an academic discipline.

While educational opportunities developed, the number of practising engineering geologists increased until in California in the United States they were sufficiently numerous to band together to form a professional association. This expanded in 1963 to become the Association of Engineering Geologists (AEG), covering all the United States and now with an international membership. In 1967, the International Association of Engineering Geology (IAEG) was formed. This provides, for engineering geology, the international association equivalent to the International Society for Soil Mechanics and Geotechnical Engineering for soils engineers and the International Society for Rock Mechanics for rock mechanicians. Reputable journals for engineering geology have also developed, such as "*Engineering Geology*", published by Elsevier. Both the AEG and the IAEG have their own publications. National groups also began to publish journals of high reputation, such as the "*Quarterly Journal of Engineering Geology and Hydrogeology*" published by the Engineering Group of the Geological Society of London

Hopefully, having read this short description of the origins of engineering geology, the reader is convinced that the subject exists. The next question is "What's it all about?"

1.2
Aims of Engineering Geology

Every discipline must have an aim and purpose. The Association of Engineering Geologists includes in its 2000 Annual Report and Directory the following statement:

"Engineering Geology is defined by the Association of Engineering Geologists as the discipline of applying geologic data, techniques, and principles to the study both of a) naturally occurring rock and soil materials, and surface and sub-surface fluids and b) the interaction of introduced materials and processes with the geologic environment, so that geologic factors affecting the planning, design, construction, operation and maintenance of engineering structures (fixed works) and the development, protection and remediation of ground-water resources are adequately recognised, interpreted and presented for use in engineering and related practice."

The IAEG has produced a statement on similar lines which sets out the redefinition of its mission in 1998 as The International Association for Engineering Geology and the Environment.

The exact phraseology, and interpretation, of such statements varies from country to country depending upon national and local practice. Thus many "engineering geologists" are essentially geologists who deliver basic geological data to engineers, without interpretation. At the other end of the scale some engineering geologists might design foundations and slope stabilisation, thereby spending much of their time as geotechnical engineers. Much clearly depends on the training and experience of the geologist involved, and the attitudes of the organisation in which he or she is employed.

A particular problem lies in the field of hydrogeology (or geohydrology). In some countries much of exploration for sources of potable water is carried out by engineering geologists. In other countries this is undertaken by specialised hydrogeologists who are quite separate from their engineering geological brethren. Again the national culture of science and engineering influences the trend.

Engineering geology may exist under, or be a part of, other titles, such as "*geological engineering*", "*geotechnical engineering*", "*earth science engineering*", "*environmental geology*", "*engineering geomorphology*" and so on. If there is a difference in the content of the disciplines described under these names it probably lies in the training and experience of the practitioner. Engineering geology is taught in some countries as a postgraduate (Masters) degree course following on from a first degree or other qualification. If the first degree is in geology then the product after the Masters degree will be that of an engineering geologist; if the first degree is in engineering then the product may be considered as a geotechnical engineer.

Whatever their origins and training, engineering geologists contribute to the task of providing a level of understanding of ground conditions that ensures the engineering works are constructed to estimates of time and cost. In addition, such works should not fail as the result of any misunderstanding or lack of knowledge about the nature of the ground conditions. Engineering failures may cost lives and cause injuries, will certainly cost money, and will result in consequential delay. To prevent such failures and incidents occurring, the influence of the geology of the site on the design and construction of the engineering work must be determined, understood and clearly explained. The problem is how to achieve this level of understanding or, in other words, how to attain the aims of engineering geology.

1.3
Attaining the Aims

Behind every discipline there must be a basic philosophy or a way in which that discipline approaches its problems. The philosophy of engineering geology is based on three simple premises. These are:

1. *All engineering works are built in or on the ground.*
2. *The ground will always, in some manner, react to the construction of the engineering work.*
3. *The reaction of the ground (its "engineering behaviour") to the particular engineering work must be accommodated by that work.*

The first premise would seem to be fairly obvious but it would appear that, not uncommonly, the work of designing and executing a project is sub-divided between various types of engineers, architects and planners so that no single person may have a comprehensive overview of the complete project. Thus the vital concept that *the structure is but an extension of the ground* may be lost, or even never acquired, by a particular member of the team in the course of his contribution to the work. The premise that the ground will always react to the construction of the engineering work also seems self-evident. The problem is to assess the magnitude and nature of the reaction of the ground to both the construction and the operation of the project. This ground reaction, the engineering behaviour of the ground, could be small and insignificant, or massive and perhaps disastrous, depending on the nature of site geology and the engineering work. It must, however, be known in order to fulfil the third premise, namely that the engineering work be designed so that it can be constructed and will operate within the bounds of the site geological conditions without sustaining significant damage as the result of the reaction of the ground.

To determine the engineering behaviour of the ground the engineering properties (in the broadest sense) of the ground mass and the proposed design of the engineering work must be known. These two streams of data must be brought together and processed in order to determine, by calculation, the engineering behaviour of the ground. It is of vital importance that the acquisition and processing of data is done systematically to ensure that no significant factors are omitted from the analysis. The problem is devising a system to do this. Some years ago John Knill of Imperial College and David Price of THDelft began writing a book on engineering geology together. This died still-born after a few chapters, but one of the joint products was the concept that the sequence of operations to be followed to arrive at the engineering behaviour of the ground could be expressed by three verbal equations. These were:

material properties + mass fabric = mass properties

mass properties + environment = the engineering geological matrix[1]

---

[1] The original term used was *"situation"* (Knill 1978). Matrix implies a database in which the relationship of the components is defined – a highly desirable but rarely achieved goal in engineering geology.

the engineering geological matrix + changes produced by the engineering work
= the engineering behaviour of the ground

1.4
Materials and Mass Fabric

The terms used in these equations require some explanation. *Materials* may be rocks, soils and the fluids or gases contained within them. *Material properties* are the properties which are of significance in engineering, such as density, shear strength, deformability and so forth. *Mass fabric* describes the manner in which the materials are arranged within the mass (in beds, dykes, veins, sills, etc.) and includes the discontinuities (joints, faults, etc.) which ramify through the mass. It is not possible to calculate the reaction of the ground mass to engineering construction unless it is known how all the various materials are distributed within the volume of ground stressed by the construction.

In Fig. 1.1 the building at the top left sits on compressible clay of uniform thickness overlying effectively incompressible rock; deformation of the clay and settlement of the structure into the ground is uniform. Under the building in the centre the clay is not of uniform thickness. Settlement in this case is differential, being larger over the greatest thickness of clay. The building will tilt, and may crack, but may not suffer great damage. However, where the clay thickens towards both ends of the building with rock nearest to the ground surface under the centre, as in the top right drawing in Fig. 1.1, differential settlement may produce disastrous results effectively breaking the back of the building. This simple example shows the importance of understanding the subsurface distribution of materials.

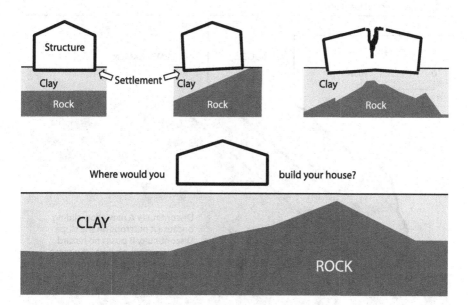

Fig. 1.1. The importance of the distribution of materials in the groundmass relative to the position of the structure

The distribution and orientation of discontinuities, such as bedding planes, faults and joints, is equally important. In the two cases in Fig. 1.2 the discontinuity A could

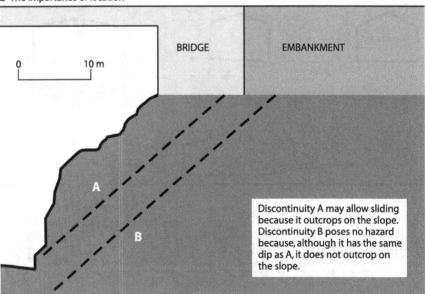

a The importance of discontinuity orientation

BRIDGE

EMBANKMENT

0   10 m

A

B

Discontinuity A may allow sliding because it outcrops on the slope. Discontinuity B poses no hazard because it does not outcrop on the slope.

b The importance of location

BRIDGE

EMBANKMENT

0   10 m

A

B

Discontinuity A may allow sliding because it outcrops on the slope. Discontinuity B poses no hazard because, although it has the same dip as A, it does not outcrop on the slope.

Fig. 1.2. The importance of location and orientation on the engineering significance of discontinuities

permit sliding of the foundation under the bridge because it daylights in (i.e. intersects) the slope face. Discontinuity B, with similar characteristics to discontinuity A, would not pose a hazard even though it has a similar orientation because it does not daylight.

1.5
Mass

It is necessary to decide what is meant by *mass*. The ground mass can be defined as that volume of ground which will be influenced by, or will influence, the engineering work. The ground influenced by the engineering work could be, for example, that volume of ground stressed by the extra load of a building, a bridge or a dam. In a tunnel the mass could consist of the volume of ground affected by the withdrawal of support caused by tunnel excavation and that volume of ground from which water has been lost by drainage into the tunnel excavation. The ground mass influenced by the engineering work is generally very much larger than the mass in direct contact with the engineering work. Thus, in a mining area, the engineering construction could be damaged by subsidence resulting from mining deep below the construction. The "mass" would thus extend down to the level of the mined ore body. Dams built in valleys may be endangered by landslides from valley sides – these may be of some antiquity and of natural origin, but new movement may be triggered by the construction process. In such circumstances the involved "mass" would extend into these landslides, which could be situated well outside the construction site.

The mass may also be that of an engineering work constructed from naturally occurring geological materials such as earth and rock fill dams, breakwaters, bunds and dykes. In such cases the material properties are those of the construction material while the mass fabric includes layers and discontinuities incorporated through engineering design and construction.

The reader may wonder why so much attention is paid to mass while, in so many text books, particularly those on soil mechanics, the emphasis is placed on material properties while mass properties appear to be if not totally neglected to be under-emphasised. Emphasis is herein given to mass because engineering works are built on ground masses, and it is the reaction of the mass which must be calculated. In soil mechanics many of the traditional methods of calculation assume that the properties of samples of material (mostly measured in the laboratory) are those of uniform, isotropic, horizontal layers which compose the mass. Under many engineering sites soils do lie more-or-less horizontally and their properties may be relatively uniform over distances commensurate with the size of most engineering works. Thus the assumption can, in many cases, be correct. The problems come when the assumption is made without examining the geology, and without confirming that this degree of uniformity is, in fact, present.

The first verbal equation given in Sect. 1.3 above provides sufficient information to make a calculation of the behaviour of the mass with regard to the proposed engineering work. This may be reasonably correct but engineering behaviour is often influenced by externally imposed factors of a more general and fundamental nature which may be grouped together under the general title of the "environment". These factors could have sufficient weight to render all such calculations on the basis of the first formula invalid.

1.6
Environmental Factors

The environment includes such features as climate, stress in the ground, and natural hazards, and can include time. The principal climatic factors are rainfall (amount, time of occurrence, intensity etc.), temperature and wind. Rainfall is of particular importance in that it relates to the moisture content of materials and mass; it is well-known that moisture content is one of the factors which determine the properties of materials and mass. Particular combinations of moisture content and temperature produce such special environments as the permafrost condition and the dry hot (or cold) deserts. Materials and masses of identical lithological and structural nature may behave quite differently under the action of an engineering process depending on the climate in which they exist.

1.6.1
Climate

Every region of the Earth's surface is subjected to a particular climate. This climate might be considered to comprise the yearly average weather, including such factors as rainfall, wind, temperature, hours of sunshine and so forth. In most parts of the Earth none of these factors is constant throughout the year. Thus, in the case of rainfall this may be defined as annual rainfall, seasonal rainfall (the amount in spring, summer, autumn and winter), monthly rainfall or even rainfall intensity. The rain, when it does fall, will not necessarily fall uniformly. In some countries a rainfall of say, 200 mm in a month, might fall in small amounts over a number of hours every day; in other areas all 200 mm could fall in one afternoon. In Papua New Guinea the author experienced a rainfall intensity of about 250 mm in one day of which 50 mm fell in one hour. Similarly, temperature can be considered as the average for the year, season or month, and also as maximum and minimum temperatures for year, season, month or day. Some regions, particularly those near the equator may have remarkably uniform temperatures, with little difference between maximum and minimum throughout the entire annual cycle. Other areas, such as in desert or mountain regions, might have variations of say, 30 °C in the course of one day.

How these climatic variations come about is a long and complex story, outside the scope of this text, but essentially the main factors influencing the nature of the climate of a particular location are its latitude (influencing the amount of warmth received from the sun), its altitude (also influencing insolation as well as rainfall), and its proximity to the sea.

The reader may ask why climate could be considered to be of significance with regard to Engineering Geology and Civil Engineering. Several important engineering geological factors are related to climatic conditions. In almost all engineering geological problems the presence of water and the depth of the groundwater table play important roles in determining the behaviour of the ground mass in response to the engineering process. Many site investigations take place in a relatively short period of time, often in the "dry" season of the year. Depending upon the amount of rainfall, and the infiltration of that rainfall into the ground, it may be found that there are significant variations in level of the water table. Knowledge of the climate of an area can give some indication as to whether it is necessary to undertake long-term observations of ground-water table level or not.

The distribution of rainfall throughout the year will indicate when rainfall is likely to prove a problem in the execution of a civil engineering contract. This is not only an issue of workers not wishing to get wet but also a question of clays softening to the point at which their liquid limit is achieved (i.e. they become viscous fluids rather than soft solids), thus affecting trafficability. Alternatively, very heavy rainfall in areas where infiltration is low, and which have recently been denuded of vegetation, can cause soil erosion leading to extensive gullying. Dry periods with strong winds can cause much dust movement with detrimental effects on workers, machinery and sensitive equipment. Very low temperatures can freeze water lines and thicken lubricating oil. Very high temperatures can thin motor oil to the point where it fails to lubricate machinery adequately and all construction work must cease.

These brief examples demonstrate the need to recognise the significance of climatic factors when dealing with site evaluation problems and the practicalities of construction work. While the "normal" climatic variations can have influence on the success of an engineering process, "abnormal" climatic events such as hurricanes, typhoons, whirlwinds, etc. may be disastrous. They are not necessarily abnormal when seen in the context of long-term climate characteristics, for they can occur almost every year in certain large regions of the Earth. However, the exact areas that they will effect will not be known in advance so that specific locations may suffer hurricanes, for example, only very infrequently. Because they are, relative to the general climate, of infrequent occurrence, of extreme intensity and may have severe consequences, such climatic events may be treated under the heading of "*natural disasters*".

1.6.2
Stress

All materials exist under certain stress conditions. The magnitude and direction of stress may strongly influence the reaction of the ground to the engineering process, and particularly in the case of underground works. Stresses encountered in the ground may result from the following causes:

Gravity. The weight of the material above any level below the ground surface will cause material at that level to compress. Vertical compression of an unconfined specimen is accompanied by lateral expansion which, in a confined situation, results in the development of horizontal stress. The ratio of horizontal stress (σ_h) to vertical stress (σ_v) is the lateral stress ratio K where:

$$K = \frac{\sigma_h}{\sigma_v} \qquad (1.1)$$

Tectonics. Tectonic stresses, which may be residual from past tectonic movements or active from present tectonic activity. These stresses appear to be mostly horizontal but are often strongly directional.

Erosion. Topographically related stresses can result from the redistribution of stresses by the erosion of valleys especially in steep mountainous areas. In Norway 'rock bursts' (explosive failures of overstressed rock) occur in tunnels driven beneath the steep slopes on the sides of fjords.

The distribution of stress around a tunnel or within a slope is much affected by the major discontinuities, such as faults, within the rock mass. The problem just quoted from Norway is particularly difficult due to the fact that tunnels tend to run parallel to fjord directions, so that bursting occurs along the entire length of the tunnel.

The stress which is imposed on a soil or rock mass, either naturally or through an engineering structure, is influenced by the presence of water which itself imposes a stress regime determined by the pressure of water. The imposed stress, from whatever source, gravity, engineering loads etc, less the water pressure acting as a water pressure in either pores or joints, is called the *"effective stress"*.

1.6.3
Natural Hazards

Many parts of the world are afflicted by recurring natural hazards. The most well known of these in engineering terms are earthquakes, but hurricanes, typhoons, sandstorms, floods, volcanism, tidal waves and snow avalanches can also be included. No engineering work can be undertaken in areas where these problems occur without recognising the significance of such natural hazards. The five most prominent natural hazards are floods, windstorms, volcanic eruptions, earthquakes and mass movements (landslides).

Floods

Of all the natural hazards it is possible that floods are the most destructive. Although not as spectacular as volcanic eruptions or as dramatic as earthquakes, they are of regular occurrence and located in river valleys where people live and work. Commonly they occur in the alluvial plains which are also rich agricultural lands. Floods may not only kill people and destroy property but also kill the animals and destroy the crops which form the food supply and underpin the economy of an area. Thus floods may not only have an immediate effect but also be a blow to short- and long-term subsistence.

The majority of floods fall into two categories. These are the 'rainstorm' flood, associated perhaps with such events as hurricanes or typhoons, and the 'coastal' flood caused by a rise in sea level brought about by storms. It follows then that the majority of floods occur in areas underlain by Quaternary deposits, for it is the recent sediments which form the flood plains and river bottoms. However, the origin of the flood may lie far outside the area which is flooded.

The river and stream beds which traverse flood plains have a limited water carrying capacity and, if the amount of water arriving from the catchment area exceeds this capacity, then floods will inevitably ensue. The excessive amount of water being discharged into the river from the catchment area is the result of infiltration into the ground being unable to absorb the intensity of rainfall in the catchment. The basic cause of low infiltration lies in the geology of the catchment area; if the rocks forming the catchment basin are permeable, such as sandstones or limestones, then infiltration can be high. But if the rocks are impermeable, such as mudstones or crystalline rocks, infiltration is low and can more easily be exceeded by the intensity of rainfall.

This means that in certain areas of the world heavy rainfall on distant mountains can give rise to floods on arid lowlands. The annual flooding of the Nile delta has been considered as a beneficial result of this phenomenon because of the introduction of a new layer of fertile silt, but to the majority of areas this excess water is very seldom a benefit. What this does imply is that no proper study of a flood problem can be undertaken without knowledge of the geology and climate of the whole river basin area.

In low lying coastal areas, dykes and sea walls give protection against sea floods. In such polder areas the dykes serve to protect land which has been reclaimed from the sea by dyke construction and land drainage. Such sea defence works must stand against the forces of waves, winds and tides, and the unfortunate combination of all three in unusual conditions may breach defence works. One of the most dramatic of these events was the flooding which occurred in the Netherlands in January 1953 when 1 490 people were drowned; the damage was estimated at about U.S.$2.5 billion. Since this event the Delta Works have been constructed to prevent this happening ever again. However, in comparison with other major floods, such as those which have occurred in China when the Huang Ho or Yangtze rivers have flooded, this incident can be considered as "minor".

A solution to these problems lies in the construction of some sort of defence works. In the case of river floods natural levees may be raised or dykes constructed, but by far the most effective technique is to control river flow by the construction of dams within the river basin. Such dams may also provide hydroelectric power (thereby supporting regional infrastructure) or may store water for irrigation in dry seasons. It is one of the ironies of nature that many areas which are damaged by floods in rainy seasons may also suffer from drought for the remainder of the year.

In urban areas the problems may be more difficult for there is less freedom to construct major works. In many cities studies are made to map flood plains at risk and, by flood routing studies, determine the effects of river control. The hazard of floods in urban areas may partly originate in land subsidence caused by natural long-term settlement, groundwater extraction (as is the case in Bangkok) or as the result of natural isostatic subsidence. In coastal zones in seismic areas flooding as a consequence of the generation of tsunamis (tidal waves) must also be considered.

Storms

The most well known types of major storms are hurricanes and typhoons which occur in regions of tropical and sub-tropical climate. Hurricanes are found mostly in the area around the Caribbean while typhoons occur in the China Sea. Modern meteorology presents explanations for these phenomena and the frequency and possible paths of these storms can be established. The path and onset of storms can be forecast, and appropriate warnings issued.

Civil engineering structures must be designed to resist wind pressures from such storms in which wind velocities may exceed 100 km h^{-1}. This will mean additional loadings upon foundations; if these are piled then some piles may be need to perform in tension. However, other effects of such storms may be somewhat more serious. In coastal areas there is sometimes a rise in sea level of some metres above the normal which may bring about extensive coastal flooding and erosion.

Such storms may also be associated with extremely heavy and intense rainfall. Apart from the flooding from overcharged rivers that will result, an increase in moisture content of sediments on and within slopes may bring about landsliding. High run-off may cause gullying of cut soil slopes.

Volcanic Eruptions

There are over 500 volcanoes which are classified as 'active' and it is estimated that, in the last 500 years, some 200 000 people have lost their lives as a direct consequence of volcanic eruptions. However, although this is a large number and while all precautions must be taken to prevent such loss of life, volcanic eruptions would appear to be of relatively minor importance in comparison with other natural disasters. Thus, for the sake of comparison, a single storm and the associated floods which occurred in Bangladesh are thought to have killed 500 000 people.

It is not proposed in this book to delve deeply into the various types of volcanic eruption for the hard fact is that there can be little done, other than by very limited civil engineering works, to mitigate these hazards. It should be pointed out that the only sure way to avoid the dangers of active volcanoes is to live somewhere else! Unfortunately, expanding populations creep closer to volcanic hazards (as is the case in Italy around Mt. Vesuvius) and, should explosive eruptions take place, future death tolls may be far greater than those which have occurred in the past.

Earthquakes

Many textbooks handle the problem of earthquakes and civil engineering exclusively from the viewpoint of the consequences to a structure when subjected to earthquake vibrations. In such an approach topics such as the point of origin of the earthquake, the frequency of earthquake events etc. are either not considered, or are given limited treatment. Similarly, some seismological texts, when considering the behaviour of the ground under the earth tremors, do not appear to consider the reaction of the ground at shallow depth. Proper handling of an earthquake problem requires the following components of knowledge:

1. An estimate of the likely strength, frequency and location of future earthquakes. This may be derived from a study of the geology of the region around the construction site and a survey of past earthquake events.
2. A study of site geology in order to assess the likely ground response to a possible future earthquake event. This would determine whether any possibility exists of phenomena such as liquefaction, land spreading, flow slides, etc. which are associated with weak, saturated Quaternary deposits.
3. An assessment of the likely response of the proposed structure to the anticipated tremors and any other ground response events associated with the earthquake.
4. An assessment of tsunami potential generated from earthquakes causing displacement of the sea floor.

Equal attention has to be given to each facet of knowledge to ensure that, at the end of the exercise, a suitably protected construction has been designed.

If the study being undertaken is concerned with the construction of new centres of industry or habitations attention must be given to the effects of earthquakes on the necessary infrastructure (roads, water, electricity, etc.) as well as on the development itself. This is required so that, in the event that a major earthquake takes place, sufficient infrastructure facilities remain to allow relief measures to be implemented.

Mass Movements

Mass movements are essentially landslides, but may also be held to include avalanches; they can occur in almost any material, rock or soil. With regard to mass movements in Quaternary sediments it is probably true to say that the very largest mass movements are commonly associated with earthquake-induced liquefaction. Otherwise most attention is given to slides in river terrace deposits, in superficial soils covering rock slopes, to slides in man-made excavations cut into Quaternary deposits for road, rail and other works, and to slides developed in man-made embankments. Recently the hazards associated with mass movements on the continental slope have been highlighted as these could send tsunami like waves radiating across the ocean surface to affect vastly greater areas of coastal development than the area of the original slope movement.

Manmade Hazards

It may also be considered that the activities of man are also part of the environment (perhaps as "*un-natural hazards*") within which a proposed engineering work will have to exist. These activities would be those not directly concerned with the proposed construction and could include such features as subsidence due to mining, oil, gas or water extraction, generation of seismic activity by pumping into deep wells or impounding reservoirs, and so forth. The effect of such activities must be taken into account when planning new works.

New works may also be affected by toxic land contamination resulting from past industrial activity. Ground-water and surface water may be polluted by leachates from contaminated land or poorly confined waste deposits. The storage and disposal of radioactive waste is a subject of major importance, much debate, and individual concern. No one wishes to live close to such a disposal site but it could be it might be better to live on top of a storage facility contained within safe geological conditions rather than some distance from one located in less suitable geological conditions.

The pressure of human activities on the environment of the Earth has reached a level such that man must now be considered to be one of the significant agencies determining the character of his own environment.

1.6.4
Dynamic Processes

It is also important to understand that the processes that modify landscape and geology are dynamic. Such water-associated landscape features as beaches, bars, sand spits, river courses, etc. can be dynamically stable representing a balance between forces operating at any given time. What may be a small change, caused by civil engineering construction, may be reflected at a distant location. Thus the building of dams on the Ebro river in Spain has reduced sediment deposition in the Ebro delta with significant consequences on coastal form, which will eventually influence harbours and agriculture. Other changes, such as the 'greenhouse effect', may be a cause of eventual sea level change, with subsequent potentially disastrous effects. It is perhaps rather arrogant to assume that mankind is the only source of such effects, far too little is known about 'normal' variations in the Earth's environment consequential on deeper seated changes. The human lifespan is too short to view such changes, and stability is often presumed, although knowledge of Quaternary geology shows that there were major climatic and sea level changes in the Pleistocene. It might be appropriate to consider that the present geological environment is but a transient phase in the continuing Pleistocene succession of ice ages.

1.6.5
Time

With regard to time it is well to remember that all materials, whether natural or man-made, are subject to weathering and decay in the progress of time. Consequently the possible change in geotechnical properties of material and mass with time must be considered when assessing engineering behaviour of the ground. The first thoughts of the engineer are generally to consider what may happen during and shortly after the construction of an engineering work. However, consideration must also be given to how the ground may react throughout the whole planned lifetime of that construction. Most engineering geologists have seen cut rock slopes that are stable for a few years after construction but become unstable once weathering has had the chance to reduce the strength of the material from which it was made and the discontinuities it contains. Time may thus be considered to be an environmental parameter.

The majority of engineering works are constructed with the intent that they should be able to operate without substantial maintenance or repair for a certain time. This engineering lifetime is usually not less than 50 years. The behaviour of construction materials (concrete, steel, brick and the like) over such a period of time in a particular environment is generally quite well known and assuming that there are no major construction faults the anticipated engineering lifetime is often an underestimate. However, the ground on which the engineering work is built, or within which the engineering work has been excavated, is also subject to decay by the process loosely described as weathering. Thus, a cut slope designed for a material of a given strength (this strength being measured before the slope is excavated) may well become unstable as the exposed material in the excavated slope becomes weathered. This is a particular problem in the argillaceous rocks.

In soils, whether of Quaternary age or older, it is now generally recognised that in the design of slopes in clay consideration has to be given to both 'short-term' and 'long-term' stability. In the latter case allowance is made, by modifying the strength parameters used in analysis, for the long-term effects of weathering and water pressure on the stability of the slope.

1.7
Analysis

All the factors leading up to the description of the engineering geological situation defined in the three equations set out in Sect. 1.3 may be established through the process of site investigation. Thereafter the engineering behaviour of the ground with respect to the proposed engineering work is determined by calculation and judgement. If the calculated ground behaviour is such that it cannot be accommodated by the construction process and would damage the completed work, make construction or maintenance uneconomic, or in any way impair the feasibility of the project, then the project must be redesigned or moved to a more suitable location. Redesign on an existing location effectively takes place by modifying the size and shape of the ground mass influenced by, or influencing, the construction. Thus if a building cannot be founded on shallow foundations because of the poor quality of the underlying soil (Fig. 1.1) the use of piled foundations could be considered as one method of redesign. This, in effect, means that the ground mass being loaded extends to greater depth, is confined and perhaps incorporates stronger strata than those present at shallow depth.

The three verbal equations describe an approach, in effect a process of thought, which should be followed to give reasonable certainty that the eventual calculation of the engineering behaviour of the ground will be accurate. The organisation of this book follows the pattern of these equations, which provide the link between successive chapters.

1.8
Essential Definitions

Both geology and engineering suffer from a confusion of terminology. Many learned societies and professional institutions have set up committees and working parties to resolve problems of terminology and the resulting reports may serve as guides through the morass of minor terminological confusion. More fundamental terminological problems are sometimes so deeply rooted that change is more or less impossible.

Thus many geologists refer to sands, silts or clays as 'soft rock', presumably working on the principle that, in geological time, they will become 'harder', eventually becoming 'hard rocks'. However, many tunnelling engineers would think of soft rock as shale, weak sandstone, mudstone, etc, rocks which can generally be excavated fairly easily. In civil engineering, naturally occurring geological materials are divided into 'soil' and 'rock'. The 'soil' is not only that of the agronomist but includes also un-cemented granular materials (such as boulders, cobbles, gravel, sand, silt) and cohesive materials (the clays). Granular soils have loose, easily separable, grains; cohesive soils are generally plastic and can be deformed and moulded without breaking.

A rock defined in engineering terms is an essentially rigid and often brittle material, far greater in strength than a soil, which cannot be moulded or bent without breaking. It should be noted that such criteria may also be fulfilled by hard clays which are desiccated, although such material would not be considered as rock but as soil in a special condition.

The distinction between rock and soil would seem to be clear enough but problems arise in practice. One major difficulty is that many excavation contracts are ar-

ranged so that there is one rate of payment for excavating 'rock' and another for excavating 'soil'. If all rock is fresh, and all soil of alluvial origin, for example, there need
be no problem. However, rock is often weathered to the degree that it has the geotechnical properties of a soil but is still geologically recognisable as a rock. To the academic geologist a weathered rock is still rock; to the geotechnical engineer it is a soil.
Weathered rock can be dug out as easily as soil but should payment be made for rock,
in the geological sense, or soil, in the geotechnical sense? Similarly, if super-size boulders are found in glacial soils, are these rock because of their size or merely extra-
large particles in soil?

Clarification of the confusion is not helped by certain attitudes which have developed over the years. Thus some soils engineers consider that certain of the weaker rocks
(such as chalks and shale), which can be sampled and tested using soils boring and testing equipment, are effectively soils. While there is a certain justice to this argument, it
could be dangerous in some aspects of geotechnology such as the slope stability of
such materials, where slope collapse could take place by sliding on discontinuities not
related to material strength which commonly determines soil slope stability.

A common, but perilous, assumption is that beneath the surface cover of soil all
materials found below *rockhead*, the uppermost boundary between rock and soil, will
be rock in the engineering sense. This is not so. In many geological formations, perhaps hundreds of millions of years old, layers may be found that have the geotechnical
characteristics of soils. Thus in Europe un-cemented sands are found in the Triassic
and Permian, and clays occur beneath coals in the Carboniferous Coal Measures. On
the other hand 'engineering rocks' may be found in alluvial soils of quite recent origin. Thus cemented sandstones and limestones may be found within recently deposited sediments, mostly in tropical areas, and a hard, cemented layer, a "*duricrust*", may
be developed in near-surface dry environments.

The contractual and engineering problems that may originate in the subtleties of
distinction between rock and soil can mostly be solved by the application of common
sense. However, it is much better for the geotechnologist to develop the habit of thinking of both soils and rocks as simply 'materials' whose behaviour will be determined
by their geotechnical properties.

1.9
Training and Professional Development in Engineering Geology

The preceding pages have reviewed the facets of engineering and natural sciences that
enter into the content of the discipline of Engineering Geology. This review provides
a guide to the content of the training that is required to develop a fully rounded engineering geologist. Some of the basic subjects taught in any training in engineering
geology are presented in Fig. 1.3.

Within the limited time available for any university training none of the subjects
listed in Fig. 1.3 can be taught to any great depth so, for example, a properly qualified
and able soil mechanics engineer must know more about soil mechanics than an engineering geologist. The engineering geologist might then be unreasonably accused
by any professionals specialised and expert in one of the subjects listed in Fig. 1.3 as
being 'a jack of all trades and master of none'. The recognition and definition of the
problems that may come from the interaction of engineering and geology in fact re-

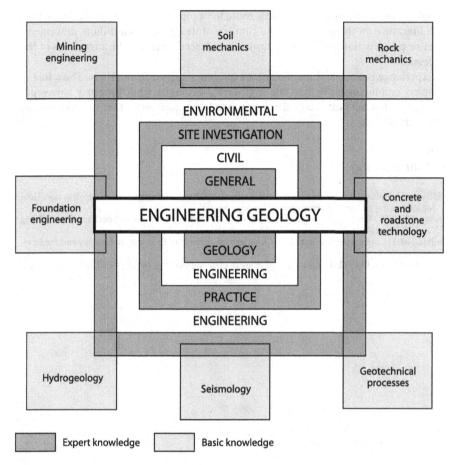

Fig. 1.3. Subjects included in engineering geological training courses

quires such broad knowledge. It has been often found, in the author's experience, that a problem recognised and defined is a problem solved. This solution may not necessarily be provided by the engineering geologist but by the expertise of some other professional; it is not important who solves a problem but that the problem is resolved.

Most engineering geologists, once employed, find themselves drifting into one or another field of specialisation depending upon the character of the company employing them. Thus an engineering geologist employed by a dredging company is unlikely to gain much experience in the construction of nuclear power station foundations and vice versa. Perhaps first employment is best gained on a major project such as the construction of a hydroelectric scheme, or motorway to view as many facets of civil engineering as possible and thereby gain a general feel for the processes and difficulties of engineering.

The engineering geologist must be prepared to learn continually. It is very possible that the young engineering geologist may find him or herself at a day's notice on an aeroplane, off to investigate a possible route for a pipeline through the desert, armed with literature on the geology of the country of destination, sand dune movement, pipeline construction and 'how to survive in a desert'. Such are the attractions of the profession.

Experience counts, and this includes experience gained from others. Thus, in any company employing more than one engineering geologist, while they may have separate duties, they are best retained as a group to learn from each other's experiences as well as their own.

1.10
Further Reading

Eddleston M, Walthall S, Cripps J, Culshaw MG (eds) (1995) Engineering geology of construction. Geological Society of London (Engineering Geology Special Publication No. 10)
Fookes PG (1997) Geology for engineers; the geological model, prediction and performance. Q J Eng Geol 30:293–424
Knill JL (2003) Core values: the first Hans-Cloos lecture. Bulletin of Engineering Geology and the Environment 62:1–34
Legget RF, Karrow P (1982) Handbook of geology in civil engineering. McGraw-Hill, New York

Part I
Investigating the Ground

Chapter 2

Geological Materials

2.1
Important Characteristics of Geological Materials

Engineers work with large volumes of soil and rock which will contain variable amounts of fluid in their pores and fractures. It is helpful to distinguish the material from which these volumes are made from the mass which they form. Sediments are made from particles, big and small, and "rocks" are made from rock!

2.1.1
Sediments

The coarsest sediments are those produced by landsliding and glaciation which may transport fragments of rock so large that an examination at close range may fail to recognise that they have been displaced. More commonly, rock fragments found below eroding cliffs may be many tonnes in weight. Such very large fragments may be further eroded during river transportation to gravel and boulder size. These fragments are recognisably rock but, as they disintegrate to yet smaller and sand-size grains, the grains tend to be largely of single minerals. The type of mineral of which they are composed will depend upon the source rock and the degree of abrasion suffered during transportation. Thus the most common sand-forming mineral is quartz, but in limestone areas the grains may be predominantly calcareous. If there are local sources of less erosion-resistant minerals, such as mica, these may be found mixed with more resistant minerals transported from distant sources. Grains of all sizes will be, to some degree, rounded by abrasion during transportation and the degree of roundness achieved is of geotechnical significance, for angular grains tend to interlock and give greater shear strength than more rounded grains. The distribution of grain sizes in sediment may vary (Fig. 2.1).

Uniformly graded sediments comprise more or less equally distributed representatives of many grain sizes. Well graded sediments are mostly of one grain size while gap graded sediments lack a range of grain sizes. Well graded sediments tend to have greater porosity and thus greater permeability than uniformly graded sediments because there are fewer finer particles to fill pore spaces between larger particles. At silt size, the particles and the pore spaces between them are very small so that permeability is very low and movement of water slow. For this reason, attempts to compact water saturated silt may result in raised pore water pressures and the subsequent liquefaction of the deposit.

All of the sediments so far described are granular and the grains do not adhere to each other. The yet finer grained sediments, clays, are formed of particles less than

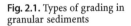

Fig. 2.1. Types of grading in granular sediments

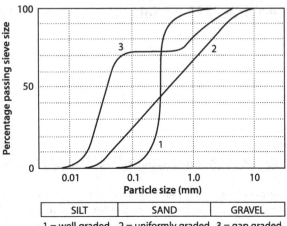

1 = well graded   2 = uniformly graded   3 = gap graded

0.002 mm in diameter and often much smaller, and are commonly very small plates of clay minerals bonded together by electro-chemical forces. This bonding gives the clay cohesion allowing the material to be moulded. Clay minerals result mainly from the weathering of other rock forming minerals. Thus kaolinite results from the weathering of feldspars in granitic rocks. Clay micas are a common constituent of clays and range, with increasing degradation, from the primary mineral muscovite to the secondary minerals sericite and, at the end of the series, illite. Muscovite has a low content of combined and adsorbed water but this, together with interstitial water, increases to the end of the series. The montmorillonite group of clay minerals, now known by the earlier term '*smectites*', are renowned for their capacity to expand on taking in water and give rise to extreme swelling. Members of the chlorite group of clay minerals also have a capacity for swelling.

Organic deposits of plant remains may be laid down in swampy environments to form highly compressible peat deposits. Plant remains may also be mixed intimately with clays to form organic clays and have a significant influence on the behaviour of these materials.

2.1.2
Intact Rock Materials

'Intact' rock is commonly taken to mean a piece of rock about the size of a laboratory test specimen (usually a cylinder of core no larger than about 100 mm long and 50 mm diameter) without obvious cracks or breaks. Most rocks are formed of mineral grains or other rock fragments bonded together in some way. The amount of pore space present, the size of the pores and the nature and quantity of the cement has a major effect on the mechanical properties of the intact rock material. In general terms, the greater the porosity, the weaker the rock and, of course, the weaker and less abundant the cementing mineral, the weaker the rock also.

No rock can be stronger than the minerals of which it is composed and the nature of the rock forming minerals has a dominating influence on the behaviour of some rocks. Thus rocks formed from soluble minerals such as calcite and gypsum, pose problems from past solution and future solubility. Evaporites are effectively monomineralic rocks whose formation is associated with the evaporation of mineral charged water. They include such minerals as halite (rock salt), gypsum and anhydrite and potash salts (such as a carnallite and polyhalite). *Evaporites*, particularly rock salt, will flow under pressure and may be found in salt domes pushed, and still moving from their original position, upward into overlying strata. The susceptibility of these rocks to creep gives problems in mines and tunnels.

2.1.3
Fluids and Gasses

The main fluids of importance in engineering geology are water and oil. It is almost impossible to over-emphasise the importance of water in determining the engineering behaviour of geological materials and masses. Water is almost incompressible and when present in the pore spaces of a material, if only in small amounts, will modify the behaviour of that material under stress. The behaviour of clays, in particular, is very much dependent on moisture content. Water is seldom pure and contains dissolved minerals, such as sulphates, which may react with engineering materials. Salinity and acidity limits its use in such processes as the manufacture of concrete. On freezing, water expands and ground heave and ice wedging are important mechanisms causing ground disruption and slope instability.

Oil as a fluid in the ground is not often of direct importance in civil engineering unless there are projects proposed at great depth in which oil may be inadvertently encountered. However, it is a common cause of contamination in industrial sites and becomes a component of sediments that has to be considered when such sites are re-developed.

Many different gasses can occur underground particularly in volcanic areas but, apart from air, in the pores of ground above the water table the most important commonly encountered gas is methane. Methane is explosive and is associated with underground works in carbonaceous rocks, but may be also be derived from peaty organic matter contained within unconsolidated clays and silts. It is non-toxic but may be asphyxiating by reducing the oxygen content in the air. The author has had an unfortunate experience with hydrogen sulphide (the rotten egg gas) which is highly toxic and comes from the decay of organic materials and the action of acid waters on pyrites. It was released during a water well pumping test in the Middle East; at first the smell warned of its presence but it then seemed to disperse. It was however the author's sense of smell that was deadened and who, with continued inhalation, turned slightly blue and was carried off unconscious.

2.2
Description of Materials

The identification and description of soil and rock in samples or in the mass is commonly one of the tasks with which the young geotechnologist is entrusted in his or her first employment. It thus may sometimes be thought to be something of little importance, to which scant attention may be given by more experienced personnel. This is not so. In most projects the design of the project is based upon an evaluation of ground conditions which is established from the description of samples and natural or excavated outcrops, and thence the testing of chosen materials. Most testing is undertaken on materials which represent the various bodies identified by visual inspection. Thus correct identification is of vital importance and is the firm basis on which all geotechnical models are founded. Many significant engineering mishaps have originated in poor description, and the inaccurate description of ground conditions is a common cause for contract claims.

Identification and description are a basis for classification. Features or properties of the material are identified or measured and then described using standard phraseology. Features may be colour, mineralogy, structure, texture, presence of weathering etc.; properties may be strength, deformability, permeability etc. Classification using these features and properties assigns a name to the material, grades properties and so forth. The aim of this procedure is to give a means of reliable communication between geotechnologists so that no engineering errors occur as a consequence of misunderstanding.

The name to the soil is given by its grain size. Many grain size classifications exist, each with minor variations reflecting problems in the particular country of their origin. Terminology is not uniform throughout the world. A simple grain size classification for soils, following BS EN ISO 14688-1, is given in Table 2.1. A full description may include such physical properties as relative density (for sands) or strength (for clays) and a description of geological structure. Other features such as colour may be added to help distinguish one stratum of material from another. A description of sand might thus be:

"*Dense thinly bedded grey fine SAND,*" while the description of clay might be: "*Stiff thinly laminated brown CLAY*"

Table 2.1. Grain size classification for engineering soils

Soil name	Qualifier	Grain size (mm)
Boulders		> 200
Cobbles		63 – 200
Gravel	Coarse	20 – 63
	Medium	6.3 – 20
	Fine	2 – 6.3
Sand	Coarse	0.63 – 2
	Medium	0.2 – 0.63
	Fine	0.063 – 0.02
Silt	Coarse	0.02 – 0.063
	Medium	0.006 – 0.02
	Fine	0.002 – 0.006
Clay		< 0.002

Many soils are mixtures of two or more grain sizes to give descriptions such as 'silty SAND', 'sandy CLAY' and 'silty sandy GRAVEL'. Detailed guidance as to the percentage of subordinate grain sizes required to give a qualified remark (such as 'slightly sandy', 'sandy' or 'very sandy') is given in BS EN ISO 14688-1. It is important to include accurate assessments of the proportions of subordinate grain sizes in a description because their presence may significantly modify the behaviour of the main body of material.

Strength is one of the most important parameters for engineering purposes and so scales of strength have been devised. One such scale is given in Table 2.2. It should be stated whether the rock is fresh (un-weathered) and thus at its greatest strength or has been weakened by weathering to, say, a decomposed condition. A typical description for a fragment of rock material might be "*Decomposed thinly bedded red coarse micaceous SANDSTONE, weak*".

Geotechnology deals mostly with the mechanics of earth materials and masses, and the reader may wonder why it is necessary to give a geological name to a rock material if its mechanical properties are known. Rock names are based on mineralogy, texture, grain size, genesis, geographical location and other parameters which may have no direct bearing on mechanical properties. This is in contrast to soil names which, since they depend on grain size, tell something about their properties such as porosity and permeability. However, rocks must be named *because of the association of the rock type with other phenomena which may be of significance in engineering*.

Table 2.2. Strength scale for clays and rocks

Strength category		Description	Field definition
Guide to shear strength of clays (kPa)	20	Very soft	Extrudes between fingers when squeezed
	40	Soft	Very easily moulded with fingers
	75	Firm	Moderate finger pressure required to mould
	150	Stiff	Moulded only by strong finger pressure
	300	Very stiff	Can be indented by thumbnail
		Hard	Can be scratched by thumbnail
Guide to unconfined compressive strength of rocks (MPa)	1.0	Extremely weak	Crumbles in hand
	5	Very weak	Thin slabs break easily under hand presssure
	25	Weak	Thin slabs break under heavy hand presssure
	50	Medium strong	Lumps or core broken by light hammer blows
	100	Strong	Lumps or core broken by heavy hammer blows
	250	Very strong	Lumps only chip with heavy hammer blows. Dull ringing sound
		Extremely strong	Rocks ring on hammer blows. Sparks fly

For example, *Limestone* if associated with underground caverns, would suggest the possible presence of solution cavities, high permeability, and water resources. *Quartzite* is associated with high strength and abrasiveness. Basalt, if *extrusive* as lava flows, is associated with scoriaceous and rubble deposits between flows, very irregular thickness, interlayers of weathered material, buried soil horizons and poorly defined jointing, except in the centre of thick flows. If *intrusive* as thin dykes or cylindrical plugs, jointing may be well defined. Shale and schist have very closely spaced bedding or foliation discontinuities, giving many opportunities for failure and anisotropic properties.

Exact identification is not the task of the engineer but, because the engineer on site may be the first person to see new excavations, he or she should be able to identify approximately any rock exposed, for this identification might lead to the early recognition of a problem. This identification need not be as accurate as a geologist should give, but should be not too far from the truth. Figure 2.2 gives an 'Aid to the identification of rocks for engineering purposes' which may be used by engineers. Engineering geologists are, however, required to give the correct name to a rock for this might lead to associations with more complex problems than those mentioned.

The distinction between igneous, sedimentary and metamorphic rocks is often best made by a simple visual examination of the outcrop.

2.3
Properties and their Measurement

The following factors will determine the properties of a dry material:

- the types of minerals to be found in the grains and the physical properties of those minerals,
- the size and shape of the grains,
- the density of the packing of the grains,
- the nature and strength of the bonding between the grains,
- the distribution of these grains within the material, whether random, layered, orientated and so forth.

These factors are clearly of importance and there have been attempts to correlate basic characteristics with material properties. As an illustration of one of the early works that of Koerner (1970) on sands may be cited. Following undertaking many triaxial tests he concluded that:

$$\varphi d = \varphi f + \varphi \delta \tag{2.1}$$

where φd is the friction angle measured in drained triaxial shear, φf is the frictional component and $\varphi \delta$ is a dilational component. A value of φf was proposed based on the mineral grain properties of the saturated sands tested such that:

$$\varphi f = 36° + \Delta\varphi_1 + \Delta\varphi_2 + \Delta\varphi_3 + \Delta\varphi_4 + \Delta\varphi_5 \tag{2.2}$$

where $\Delta\varphi_1$ = a particle shape correction; $\Delta\varphi_2$ = a particle size correction; $\Delta\varphi_3$ = a correction for grading; $\Delta\varphi_4$ = a correction for relative density, and $\Delta\varphi_5$ = a correction for

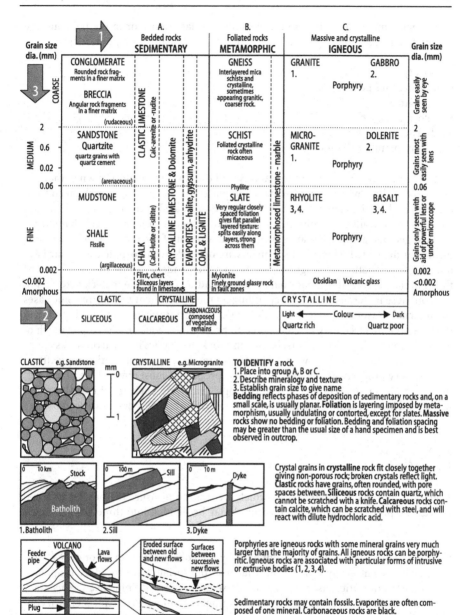

Fig. 2.2. Aid to the identification of rocks for engineering purposes (based on a similar table in BS5930 (1999) and proposals by the author to International Standards Organisation)

mineralogy. Most of the five factors considered to determine material properties are incorporated in Koerner's formula.

Presently, material testing is undertaken in the laboratory or in the field and the parameters measured are generally related to the following basic properties:

- density
- porosity and permeability ‡
- strength ‡
- deformability ‡
- abrasivity ‡
- environmental reactivity

The last property relates to the stability and durability of materials when they are taken out of their 'natural' environment and placed in another. Thus, rock taken from a quarry and used as concrete aggregate may, when placed in the new environment of concrete, chemically react with the cement and change its mineralogical character.

Most geological materials are anisotropic as a result of the way in which they were formed or deposited. Thus most sediments are bedded, metamorphic rocks may have lineations or foliations and igneous rocks may be banded, so that the properties of the materials vary with the internal structure and texture of the material. Results of tests on these materials will vary depending on the direction of testing. In some cases the internal anisotropy may be so slight as to be insignificant and for all practical purposes the material may be considered to be homogeneous and isotropic; much of the background theory of soil mechanics and rock mechanics is based on the assumption that the materials dealt with are isotropic and homogeneous. Anisotropy affects almost every property of a material but those properties most affected are marked '‡' in the list above. In certain clearly anisotropic materials, such as schists, slates and shales, and layered sediments, variations in material properties due to anisotropy may be of vital importance in certain projects.

2.3.1
Types of Test

There are very many different tests which are used to measure the properties listed above. The properties which are measured may be divided into three groups. These are:

- *Material properties* – which are obtained by such observations or laboratory testing as give a scientifically valid result whatever appropriate test procedure is applied. The instrumentation used to measure these properties will influence the accuracy of the result. Examples of such properties might be mineral content, chemical content and density.
- *Test properties* – which are measured by observing the reaction of the test specimen to a given test procedure. The property measured is recognised as a property in materials science but the result obtained depends very much on test procedure. Examples of such properties are permeability, porosity, strength, modulus of deformation, cohesion and so forth.
- *Empirical properties* – which are the result of performing the test in a standard way using particular apparatus on a particular style of test specimen. These tests are of-

ten associated with parameters which are not recognised in physical science but are useful with regard to a particular engineering process. Examples might be such 'properties' as ASTM Toughness, the Slake Durability Index, the Los Angeles abrasion value, and the Polished Stone Value.

Many tests are difficult and expensive to undertake and in the beginning of an investigation it is sometimes sufficient to have an approximate idea of the value of certain properties. The results of some easily and cheaply executed tests may have a relationship to the results of more complex and expensive tests. If this relationship is known then the cheaper test may be undertaken to obtain an approximate idea of the value of an otherwise more expensively obtained property. Such tests are called *index* tests.

2.3.2
Limitations of Testing

The reaction of a ground mass to an engineering process depends partly on the engineering behaviour of the materials of which the ground mass is composed. Material behaviour is determined by testing in which the intent is to subject samples of materials, in the condition that they are found underground, to the changes in conditions which will arise as a consequence of engineering construction. This intent is seldom realised firstly because it is almost impossible to extract a sample from the ground in a truly undisturbed condition and secondly because, for practical reasons, testing must be undertaken in a much shorter time than that in which the changes consequent to engineering construction take place. Clearly the condition of the sample tested is also of significance to the accuracy of the test result, which would be influenced by any flaws developed as a result of sample extraction and preparation. In particular, differences in moisture content between the material in situ and as a laboratory sample have major influence on test results. As well as these factors the way in which the tests are performed is of importance.

2.3.3
Size and Shape of Sample

In the simple uniaxial compressive strength test, the test specimen is usually a solid cylinder. The distribution of stress within the specimen is such that failure under uniform load equal to that exerted by the test machine is not expected for samples whose length to diameter ratio is less than 2:1. In soil mechanics this ratio is generally accepted and also mostly in rock mechanics. However, in rock mechanics there is no generally accepted standard for testing, although some recommendations (such as the "*Suggested Methods*" of the International Society for Rock Mechanics, Brown 1981) are widely adopted.

Special shapes of sample have been devised to give more reliable results for certain tests. The '*dog bone*' sample in Fig. 2.5 is intended to give tensile and compressive strength values free of the complications caused by stress distribution in the sample, but the cost of preparation of such shapes is usually prohibitive.

It is obvious also that the size of the sample must be considerably greater than the grain size of the material. Thus, for example, a 50 mm diameter, 100 mm long specimen of basalt would be acceptable for a strength test, but this size of sample would be inappropriate for a coarse conglomerate in which the size of an individual 'grain' might be a significant proportion of the size of the specimen. Test specimens are mostly specimens without fractures and are thence described as specimens of intact rock.

2.3.4
Rate of Loading, Testing Machine and Platens

The rate of application of load on the test specimen has a significant effect on the value of failure strength measured (Price and Knill 1966). Almost all rates of loading applied in the laboratory are far higher than the rates likely to be applied in engineering construction, so that to some extent, all values are false. Usually, the rate which is adopted is the slowest which can be used in the circumstances. Sometimes testing standards or recommendations suggest rates to be used.

Testing machines may be regarded as either "*soft*", i.e. "flexible" or "*stiff*". The test frames of the former deforming significantly under load so that, at the first crack in the test specimen the release of strain energy from the machine causes explosive failure of the specimen. Stiff machines were built so massively that there was little deformation of the frame. Some differences in strength results have been observed between the different machine types, but they are not very significant unless the tests are high load tests in rock mechanics or, for the particular project, post failure behaviour must be examined. Nowadays the 'stiffness' is created by servo-controlled systems that allow the deformation of the sample to control the rate at which it is loaded, so avoiding explosive failures.

In rock mechanics, platen conditions are important. Results may be affected by differences in metal or other material used for these plates. A more basic problem arises from the constraints on sample diametral expansion imposed by friction developing between platen and sample when their stiffness is not the same. This has been partially solved by the design of platens incorporating devices to allow the sample to dilate radially but such devices are too complex for standard use.

2.3.5
Standards

Results of tests on materials will vary depending upon the test procedure applied and this variation may be of engineering significance. To be assured that results of tests from different laboratories are comparable tests are undertaken following established *standards* or *norms* set up by national or international bodies. Examples of such organisations are the British Standards Institution (BSI), the American Society of Testing and Materials (ASTM), the International Standards Organisation (ISO), the International Society of Rock Mechanics (ISRM), the Deutsche Industrie-Norm (DIN) and the Nederlands Normalisatie Instituut (NNI). Not all tests are standardised and standards for particular tests vary, usually but slightly, from country to country. All routine tests should be performed to a standard and that standard recorded in reports and scientific papers.

2.4
Density and Unit Weight

Most rocks and all soils can be regarded as containing voids between grains. These voids may be filled with water, air or both. Thus, three conditions are possible – dry, partially saturated and saturated. The in situ or 'bulk' density of the material is determined by the relative proportions of mineral particles, water, and air in a given volume of material, and the specific gravity of the mineral particles.

Density (ρ) is the amount of mass in a given volume and the units are Mg m^{-3}. Unit weight (γ) is the weight of a unit volume, and the units are N m^{-3}. Usually, because the Newton is a small force, unit weight is expressed in kN m^{-3}. Formulae appropriate to the calculation of various parameters associated with density and unit weight are given in Table 2.3.

Table 2.3. Formulae in common use for parameters related to density and unit weight

Density (ρ)	$\rho = \dfrac{\text{Mass }(M)}{\text{Volume }(V)}$	(2.1)
Unit weight (γ)	$\gamma = \dfrac{\text{Weight }(W)}{\text{Volume }(V)}$	(2.2)
Particle specific gravity (G)	$G = \dfrac{W_s}{V_s \gamma_w}$	(2.3)
Void ratio (e)	$e = \dfrac{\text{Volume of voids}}{\text{Volume of solids}} = \dfrac{V_v}{V_s}$	(2.4)
Moisture content (m)	$m = \dfrac{\text{Weight of water}}{\text{Weight of solids}} \cdot 100\%$	(2.5)

Note: Because of the way in which moisture content is defined, it is possible to have materials with $m > 100\%$.

Degree of saturation (s)	$s = \dfrac{\text{Volume of water}}{\text{Volume of voids}} = \dfrac{V_w}{V_v} \cdot 100\%$	(2.6)
Bulk unit weight (γ_b)	$\gamma_b = \dfrac{\text{Total weight}}{\text{Total volume}} = \dfrac{W_s + W_w}{V_s + V_v}$	(2.7)
Saturated unit weight (γ_{sat})	$\gamma_{sat} = \dfrac{\text{Saturated weight}}{\text{Total volume}}$	(2.8)
Dry unit weight (γ_d)	$\gamma_d = \dfrac{\text{Dry weight}}{\text{Total volume}}$	(2.9)

V_s = volume of solids; V_v = volume of voids; V_w = volume of water; W_s = weight of solids; W_w = weight of water; γ_w = unit weight of water.

Most granular soils have a wide range of unit weight within which the soil may occur in situ. Figure 2.3 shows a 'theoretical' soil composed of perfectly spherical uniform diameter grains at minimum bulk density with, for this soil, the largest possible spaces between the particles, so that the void ratio is at maximum while the bulk density is minimal. If the soil is compacted to maximum bulk density by vibration and/or weight of material above, the particles will close together, giving minimum size pore spaces and void ratio. The in situ density lies usually somewhere between maximum and minimum; the relative density expresses where the in situ density lies between minimum and maximum values for a particular soil and is a measure of the state of compaction which the soil has reached.

Relative Density (*R.D.*) may be therefore expressed in terms of void ratio (*e*) or unit weight (*γ*):

$$R.D. = \frac{e_{max} - e}{e_{max} - e_{min}} = \frac{\gamma_{max}}{\gamma_{in\,situ}} \left[\frac{\gamma_{in\,situ} - \gamma_{min}}{\gamma_{max} - \gamma_{min}} \right] 100\% \qquad (2.3)$$

where, for a particular sample, γ_{min} = minimum possible unit weigh, γ_{max} = maximum possible unit weight and $\gamma_{in\,situ}$ = in situ unit weight.

Relative density assumes that the maximum density is achieved without alteration to the size and shape of the grains, i.e. the grains are not fractured and broken into smaller pieces by compaction and that the grains are solid, i.e. they do not contain pores as do the grains of pumice and chalk. The degree of compaction achieved by a natural deposit may be classified by relation to the relative density. Terminology is given in Table 2.4.

Since relative density suggests the degree of compaction a granular soil has achieved relative to the maximum it might achieve it indicates the settlement that can take place under extra load, such as that of a foundation. In Fig. 2.3 there are, in two dimensions, four point contacts between the grains at minimum density and six contacts at maximum density, indicating an increase in shear resistance of granular materials at higher degrees of compaction (Table 2.4). Because relative density relates to both settlement

Fig. 2.3. In situ density related to maximum and minimum densities

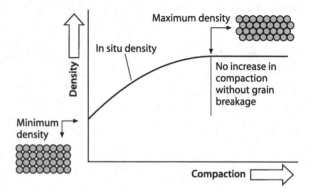

Table 2.4. Terminology for relative density and approximate relationships to other properties of common granular soils such as sand, silt and gravel

Relative density (%)	Dry unit weight (kN m^{-3})	Descriptive term	ϕ (°)	SPT 'N' value
0 – 15	< 14	Very loose	< 28	0 – 4
15 – 35	14 – 16	Loose	28 – 30	4 – 10
35 – 65		Medium dense	30 – 36	10 – 30
65 – 85	16 – 17.5	Dense	36 – 41	30 – 50
85 – 100		Very dense	> 41	> 50

on and strength of foundation materials it is of importance in foundation design. Generally, relative density is measured using the Standard Penetration Test (SPT) in boreholes and the SPT-'N'-value is much used in foundation design. In trial pits it may be assessed approximately by the resistance of the soil to excavation. It is also an important parameter in the assessment of the potential for liquefaction of sands during earthquakes.

2.5
Porosity and Permeability

Porosity is a measure of the voids in sediment and rock, and the masses they form. Permeability is a measure of the extent to which these voids are connected.

2.5.1
Porosity

Porosity is a measure of the volume of voids in a material or mass. In materials porosity depends upon the space between grains; in masses it would also include any space provided by open fissures and joints.

$$\text{Porosity}(n) = \frac{\text{Volume of voids}}{\text{Total volume}} \tag{2.4}$$

Intergranular porosity is often determined in the laboratory by comparison between dry and saturated weights of the sample. To become fully saturated, voids must be in contact, and both interconnections and voids must be large enough to allow the flow of water under reasonable pressures. One or both of these conditions is often not satisfied, particularly for fine grained materials, and the results of such tests are properly described as apparent porosity. Note that the execution of the porosity test may damage the sample. There are some porous rocks, such as vesicular basalts, that are porous but whose porosity cannot be measured by conventional means because the voids are not in contact i.e. the material is impermeable.

2.5.2
Permeability

Permeability is a general term which describes the ability of a porous medium to allow the flow of fluid through it. For flow to take place through a saturated material there must be a difference in total head. In engineering geology the usual concern is with the flow of water through a granular soil or rock mass. The head of water h is the height of the water level above the point at which the head is measured. Thus, the pressure of water at that point $p = h\gamma_w$, where γ_w = unit weight of water.

In 1856, Darcy conducted laboratory experiments to analyse the flow of water through sand, and showed that, if Q is the quantity of water passing through a sand tube of cross-sectional area A in time t, then the specific discharge q of water flow is given by:

$$q = \frac{Q}{A} \tag{2.5}$$

In this experiment the flow was intergranular and non-turbulent; these are conditions most likely to be encountered in engineering geology except where flow is fast as in gravel, open rough fractures in rock or through some clays where the clay minerals may interfere with free flow.

Darcy also showed that the specific discharge was directly proportional to the hydraulic gradient (i), which is the loss of total head per unit length of flow (Fig. 2.4). Thus $q \propto i$ and $q = Ki$, where the constant of proportionality K (the *coefficient of permeability*) is dependent upon both the properties of the porous medium and the fluid. Since the fluid is water, K is also called the *hydraulic conductivity*. The properties of the fluid which effect K are fluid density and viscosity, which change with temperature. K is therefore temperature dependent. Thus

Fig. 2.4. The hydraulic gradient between A and B is $(h_2 - h_1) / dl = dh / dl = i$

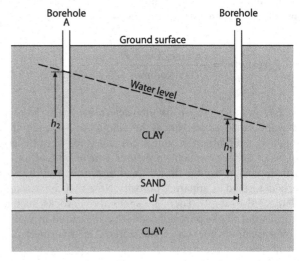

$$Q = AKi \tag{2.6}$$

and

$$v = \frac{Q}{A} = Ki \tag{2.7}$$

where v = a theoretical velocity of the flow across the area A. The coefficient K has units of velocity.

In soil mechanics and rock mechanics the usual unit for K is m s^{-1}, but in other applications common units are mm s^{-1}, cm s^{-1}, m year^{-1}, ft day^{-1}. The constant of proportionality, K is determined by the properties of both fluid and medium. Guide values of hydraulic conductivity for rocks and soils are given in Table 2.5. For the medium, the properties which determine the rate of flow are mean grain diameter (d), distri-

Table 2.5. Guideline values for hydraulic conductivity (permeability) of rock and soil

	Type	Hydraulic conductivity k (m s^{-1})
Rock materials	Granite	$10^{-13} - 10^{-12}$
	Slate	$10^{-13} - 10^{-12}$
	Limestone	$10^{-12} - 10^{-9}$
	Dolomite	$10^{-11} - 10^{-10}$
	Mudstone	$10^{-9} - 10^{-10}$
	Schists	10^{-6}
	Sandstone	$10^{-5} - 10^{-9}$
Rock masses	Gneiss	10^{-5}
	Granite	10^{-5}
	Lignite	10^{-4}
	Sandstone	10^{-4}
	Mudstone	10^{-6}
	Limestone	$10^{-4} - 10^{-6}$
Soil	Gravel	$1 - 10^{-3}$
	Sand	$10^{-2} - 10^{-5}$
	Silty sand	$10^{-5} - 10^{-7}$
	Silt	$10^{-7} - 10^{-9}$
	Clay	$< 10^{-9}$

Note: These values are taken from the literature.
They serve as a guide to relative permeabilities.

bution of grain sizes, nature of grain packing and degree of interconnection of the pore spaces. For a particular medium, these properties will not change and therefore will be the same regardless of which fluid is flowing. This is a property which is exclusively that of the medium, and is termed the *specific* or *intrinsic permeability*. The constant of proportionality K can now be re-defined, thus:

$$K = \frac{Cd^2 \rho g}{\mu} \qquad \qquad (2.8)$$

where density ρ and viscosity μ are properties of the fluid, and Cd^2 is a function of the medium alone. C is another constant of proportionality and is a function of the medium properties other than mean grain diameter which determines d^2. Consequently, the intrinsic permeability = Cd^2. Confusion arises, because the designated symbol for intrinsic permeability is k, which is the symbol used in soil and rock mechanics for coefficient of permeability, which more properly should be called hydraulic conductivity.

Intrinsic permeability is used mostly in the petroleum industry, where the fluid under consideration may be petroleum, water or even methane or a mixture of all. The units of intrinsic permeability are m^2, but since values in this unit are very small, petroleum engineers have adopted the darcy as the unit (1 darcy = 0.987 μm^2 and 1 $\mu m^2 = 10^{-12}\ m^2$).

2.6
Strength

All geological materials have some ability to resist failure under the action of stresses; this is their strength. Most values quoted as the 'strength of' a certain material are the stresses at failure, the ultimate failure strength. Usually, testing is done on small samples in the laboratory. Strengths measured are:

- uniaxial (or unconfined) compressive strength, which is the stress at failure of a sample under compression;
- uniaxial (or unconfined) tensile strength which is the stress at failure of a sample under tension;
- triaxial strength, which is the stress at failure of a sample that is confined. This is usually accomplished by placing the sample under compression while it is restrained laterally by a minor horizontal stress.

The units of strength are force/area, for example $kgf\ cm^{-2}$, $N\ mm^{-2}$, $kN\ m^{-2}$, $MN\ m^{-2}$. Recently it has become fashionable to use pascals, particularly for compressive strength (1 pascal (Pa) = 1 $N\ m^{-2}$).

2.6.1
Water Content and Drainage

The effects of water content and drainage conditions on test results are best understood in relation to intergranular stresses (effective stresses). At a point at depth d below the surface which is overlain by saturated material of unit weight γ_{sat}, the total stress at $d = d\gamma_{sat}$. The pressure of water in the pores between grains of material is $d\gamma_{water}$. Total pressures are reduced by water pressures to give intergranular pressures, so that intergranular pressure = $d\gamma_{sat} - d\gamma_{water}$.

When a saturated sample is loaded in a test so that no water can drain from the sample (an undrained test), the test load is taken partly by the intergranular contacts and partly by the pore water. The pore water is virtually incompressible; increasing strain on and shortening of the sample cannot now produce reduction of pore size and greater compaction of the sample, but can only cause pore water to push between grains, lower intergranular friction and cause failure at lower stress levels than would otherwise be the case.

All undrained tests on fully saturated soils whose particles are not bound together, give values of friction equal to zero. The fact that some undrained triaxial tests on these materials give values of φ above zero is because they are not tested in a truly fully saturated condition and some water movement into unsaturated pore spaces is possible. For the same reasons, saturated rock gives lower strength values than dry rock.

2.6.2
Strength Tests

Test arrangements and formulae to calculate strength are given in Fig. 2.5.

Compressive Strength

This is the stress (load/area) at which a sample of material (cylinders in the case of soil or rock, cubes in the case of concrete) fails under compressive stress.

Tensile Strength

This is the stress at which a sample of material fails under tensile stress. Tensile strength may be measured directly, but in practical terms it is difficult to do. In rocks it may be measured indirectly by the *Brazilian test* or the pierced disc test. A further variation of these tests is the *Point Load test*, which is now used as a field method for the determination of rock strength.

The Brazilian test attempts to measure tensile strength by developing tension across the diameter of a rock disc which is being subjected to compression through a vertical load. Because the tensile strength of rock is between one quarter and one tenth of the compressive strength, the tensile stresses being developed horizontally as a result of vertical compressive stress should be enough to cause tensile failure before failure in compression can occur. In fact, intense contact compressive stresses between sample and platen may cause local fracturing which then proceeds from the circumference towards the tensioned centre. The values of tensile strength measured are thus incorrect. A modified Brazilian test using curved platens to overcome stress concentrations has been tried (Mellor and Hawkes 1971).

Shear Strength

Material loaded under both major and minor stresses fails in shear. The shear strength of a material under shear load depends upon confining stress and is not a unique value.

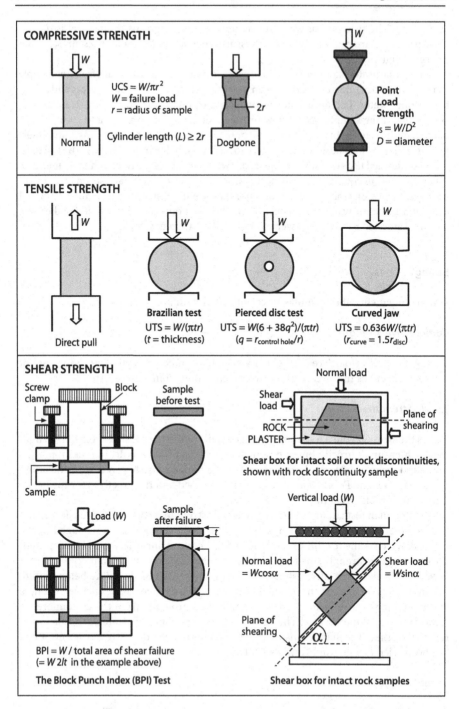

Fig. 2.5. Simple laboratory tests on rock samples. *UCS* and *UTS* = unconfined compressive strength and unconfined tensile strength

Unconfined, or uniaxial, strengths may be considered as shear strengths under no confining stress, a situation which rarely occurs in nature. Shear strength is described in terms of parameters c (cohesion) and φ (angle of shearing resistance).

Consider a block of weight W resting on a horizontal plane (Fig. 2.6). The weight W produces an equal and opposite reaction R, without any tendency for the block to move. If a horizontal force S is applied and increased until the block is just about to slide, the reaction, or resultant force R, will be inclined at an angle α to the vertical, so the horizontal component of R is $S = R \sin \alpha$ and the vertical component of R is $W = R \cos \alpha$. The angle α will have increased to a limiting value φ. At this point the horizontal force resisting sliding, $S = W \tan \varphi$. The parameter $\tan \varphi$ is known as the coefficient of friction.

Shear strength can be seen to be related to frictional resistance within a material. However, materials whose grains are bonded together may have additional strength produced by the adhesion or cementing between grains. Coulomb's law states that:

$$\sigma_\tau = c + \sigma_n (\tan \varphi) \tag{2.9}$$

where: σ_τ = shear stress at failure (shear strength) = S / A; c = cohesion (adhesion or cementing); σ_n = normal stress on sliding plane = W / A, and φ = angle of frictional resistance.

According to these properties, three types of geological material may exist. These are:

- $c = 0$ (materials exhibiting no cohesion, such as dry sand)
- c and φ materials (soils and rocks with both cohesion and internal friction)
- $\varphi = 0$ (materials exhibiting no internal friction)

In the *shear box* test (Fig. 2.5), shear force and normal force are applied directly, and it is only necessary to plot graphs of shear stress against normal stress, and shear strength (Fig. 2.7) against shear displacement to find c and φ and then peak and residual shear strengths. Peak strength may be defined as the maximum resistance of the sample to shear force while the residual strength is the resistance of the failed shear surface after considerable movement.

Shear boxes may be used to determine c and φ on intact material samples (Fig. 2.5), or along existing discontinuities which must be aligned along the plane of shearing in the box.

Fig. 2.6. The concept of shear strength

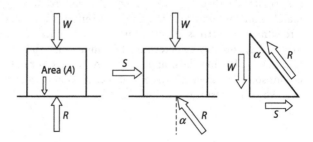

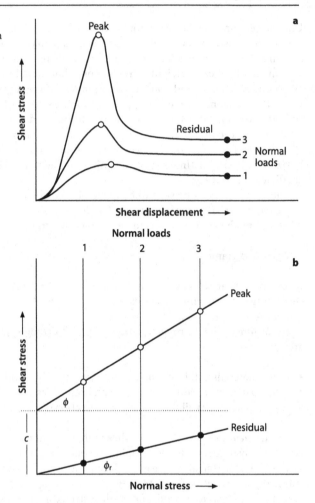

Fig. 2.7. a Stress/displacement graphs; **b** Graphical calculation of peak and residual strength parameters

Shear strength parameters of intact rock are difficult to measure, because loads and strengths are generally high, at least compared to soils, but it can be done (Fig. 2.5). Most shear testing on rock is done on discontinuities such as joints, which requires the sample to be cast into a form which fits the box. To determine the shear strength at failure, shear displacement must be plotted against shear stress. A peak failure load is normally clearly evident (Fig. 2.7a). After failure shear stress can fall to an approximately constant value. This residual strength value may be plotted on the shear strength/normal stress graph to find residual values of c and φ, which are designated to c_r and φ_r and may be determined for intact material or for existing discontinuities (Fig. 2.7b). The ring shear apparatus may be used for determination of residual values for soils. The shear displacement is circular and limited only by the length of time over which the test is conducted.

Triaxial Strength

To try to determine the effect that stresses caused by surface construction would have on underlying soils or rocks it would seem sensible to take samples from various depths below the construction, subject them to the appropriate confining overburden pressure and thence impose the additional vertical construction stresses appropriate to the sample depths. The *triaxial cell* has been designed to do this.

In the *triaxial test* (Fig. 2.8) the cylindrical sample is placed in a test cell which is constructed so as to allow the application of an all round hydraulic pressure on the sample and also the application of a load which will bring the sample to failure. The test sample is surrounded by an impervious membrane to prevent contact with the cell fluid. The hydraulic pressure is kept constant for each test and the deviator stress $(\sigma_1 - \sigma_3)$ increased until failure occurs. The Mohr circle diagram is used to plot the shear and normal stresses.

There are very many possible variations of the triaxial test. A truly accurate test would impose the construction load at the rate with which construction takes place and allow for drainage of the sample. Such testing is expensive, so a variety of tests are possible approaching reality with an increasing degree of complexity and expense. In the simplest, the *quick undrained test*, samples are tested without drainage, as this gives a measure of their strength when weakest, and without the measurement of their pore water pressures (which takes time). The triaxial test is not confined to soils and increasing strength of the apparatus allows stronger materials to be tested (Hoek and Franklin 1968).

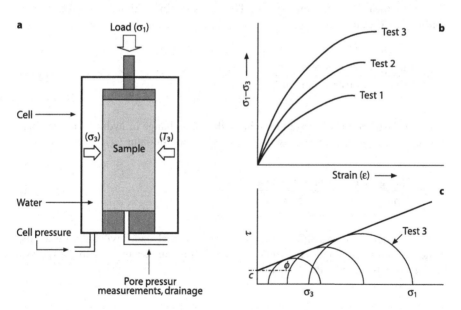

Fig. 2.8. a the triaxial test apparatus; **b** plots of deviator stress vs. strain for three cell pressures; and **c** Mohr circle plots to calculate c and ϕ

2.7
Deformation

If a sample of material is placed under load, it deforms; that is, it changes in shape and perhaps also in volume. If, when the load is removed, the deformation instantly and completely disappears, the material may be said to be elastic. In this case the relationship between stress and strain is often linear. If a cylinder of length l is loaded by a stress σ applied to the top of the cylinder, the length may be shortened by an amount dl. The strain ε is expressed by:

$$\varepsilon = \frac{dl}{l} = \frac{\text{Length decrease}}{\text{Original length}} \tag{2.10}$$

For elastic materials, strain is related to stress by Hooke's law, so that:

$$E = \frac{\sigma}{\varepsilon} \tag{2.11}$$

E is commonly referred to as Young's Modulus.

If an elastic cube is deformed by shear stress, the cube changes its shape to become a rhomb. The small angle $d\theta$ which lies between the sides of the cube and that of the rhomb is a measure of the shear strain. The stress/strain relationship is given by the Modulus of Rigidity (G).

$$G = \frac{\text{Change in shear stress}}{\text{Change in shear strain}} = \frac{\tau}{d\theta} \tag{2.12}$$

If the elastic solid is changed in volume due to a change in hydrostatic pressure, then the stress/strain relationship is given by the Bulk Modulus (K).

$$K = \frac{\text{Change in hydrostatic pressure}}{\text{Change in volumetric strain}} = \frac{d\sigma}{d\varepsilon_{\text{vol}}} \tag{2.13}$$

If a cylinder of Young's Modulus E is shortened by axial load, it will bulge outwards. If the shortening of length is dl, and the increase in diameter is dd, then the ratio of these strains is given by Poisson's ratio v:

$$\frac{dd}{dl} = v \tag{2.14}$$

2.7.1
Types of Rock Deformation

Few, if any, materials behave in a truly elastic manner, and various styles of deformation may be observed from the stress/strain curve plotted during a load test. If the plot is more or less linear, then the material may be described as elastic. If not, the material may be described as 'plastic' or some mixture of elastic-plastic behaviour (Fig. 2.9). It is then prudent to speak of the *Modulus of Deformation* rather than a Young's modulus. The moduli chosen for use in deformation calculations depend mostly on the levels of stress expected in the engineering situation (Fig. 2.9). See also Table 2.6.

Creep. If a specimen is loaded to failure in a testing machine, the test is often completed in a very short time, and the strength value measured may be described as the *instantaneous* or short-term strength. If the same specimen were instead to be subjected to a lower stress for a long period of time, it may be found that the rock fails by a process known as *creep*, at a stress sometimes very much lower than that which gave the instantaneous strength. Values of *long-term strength* may be as low as 25% of instantaneous strength. In general terms the greater the initial stress the shorter the time to creep failure. Recognition of the effects of creep on rock strength is important in the design of mine pillars, particularly if the magnitude of constant overburden load stress is a high proportion of the instantaneous strength. In civil engineering, the loads applied to rock materials are usually but a small fraction of their instantaneous failure strength so unless the rock is very weak and/or the loads are unusually high, there is little need to be concerned about the effects of creep.

2.7.2
Consolidation of Soils

Application of load to soils produces 'consolidation'. In the case of clayey soils, the load is carried initially by both the solid particles of the soil and the water within the pore spaces. In time, with continual application of the load, the water will slowly be driven out of the material. This is a slow process because of the low hydraulic conductivity (permeability) of the clay, but as the water is driven out, there will be an increase in

Fig. 2.9. Types of rock deformation under stress

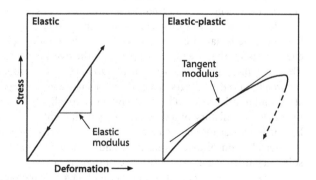

Table 2.6. Illustrative values of elastic moduli for some fresh rocks

Rock type	Unconfined compressive strength (dry) (MPa)	Elastic (tangent at 50% UCS value) modulus (dry) (GPa)
Augite olivine basalt[a]	398	75.6
Dolerite[a]	261	52.7
Dunite[a]	133	97.6
Granite[a]	201	66.3
Quartzite[a]	229	52.1
Granite biotite gneiss[a]	308	71.7
Mica gneiss[a]	154	41.3
Crystalline limestone[b]	190	70
Calcarenite[b]	3.5	1.2
Chalk[b]	22	11

[a] From Verhoef et al. 1984; [b] from Swart 1987.

effective stress as the load is carried more and more by the soil particles. The result will be a decrease in volume of the soil i.e. consolidation, the magnitude of which is related to the load. Engineers are concerned to know by how much a clay will consolidate, and in what time, so that the amount of settlement of a structure supported by the soil can be determined. The amount of consolidation may be expressed by the coefficient of volume change, m_v:

$$m_v = \frac{\text{Volumetric change in vertical direction}}{\text{Unit of pressure increase in vertical direction}} \qquad (2.15)$$

The higher the value of m_v the more the soil will consolidate (Table 2.7).

Consolidation is also a natural process. In an uninterrupted sequence of deposition any soil will consolidate under the load of a younger bed laid down on top of it. Such soils are said to be *normally consolidated*. These are mostly younger soils. Beds of older clays may, after their deposition, have been subject to erosion so that some of the upper beds have been removed. The beds remaining then display consolidation characteristics that are not appropriate to the overburden that presently covers them. They are then said to be *over-consolidated*. Over-consolidation can also result from the weight of the ice sheets that covered glacial clays during the Pleistocene and melted at the end of this period. The differences in consolidation characteristics depend upon the geological history of a deposit. Thus settlement of structures may be expected to be much greater on younger normally consolidated soils than on the older over- or geologically consolidated soils. It should be noted that consolidation, as described in soil mechanics, is but the early stage of diagenesis.

To predict the effect that settlement will have on an engineering structure it is also necessary to know how quickly this will take place so that the structure may be de-

Table 2.7. Typical values of m_v (from Smith 1968)

Soil	m_v (m^2 MN^{-1})
Peat	10 – 2
Normally consolidated clay	2 – 0.25
Stiff clay	0.25 – 0.125
Hard clay (boulder clay)	0.125 – 0.0625

signed to accommodate these movements. Tests are thus conducted to determine both the coefficient of volume change, m_v, to deduce the magnitude of settlement, and the coefficient of consolidation, C_v, (expressed in units of area/time) to deduce the rates of settlement.

2.8
Abrasiveness

Quartz is a common rock forming mineral and, because of its resistance to weathering, one of the commonest components in soils. It is also one of the hardest minerals, being number 7 on Moh's scale of mineral hardness, implying that it will scratch most other common minerals. It will also scratch steel. There are a number of engineering processes which involve cutting rock using hard picks, teeth, bits etc. The economy of the operation may depend upon the amount of wear on the cutting tools, which have to be replaced when they become blunt. This wear will depend upon the quantity, grain size and shape of the quartz in the rock or soil, how firmly the quartz is cemented into the matrix (expressed by the rock or soil strength), the properties of the cutting device and the mechanics of cutting.

A number of tests have been developed to assess abrasiveness of a rock. Most involve pressing a steel pin against the rock with uniform force, scratching the rock for a specified distance and observing the loss of steel from the pin. The Cerchar test involves moving a specially shaped pin a short distance over a prepared rock sample and thence measuring weight loss. In the pin-on-disc test a specially prepared pin is pressed against a rotating disc of specimen rock for a given number of revolutions. In both of these empirical tests the quality of the steel making up the pin is specified. Relative abrasiveness may be assessed from knowledge of rock mineral content and strength. Schimazek and Knatz (1970), using a pin-on-disc test, developed the 'F' value (a wear factor), using the formula:

$$F = \frac{Qtz.Eq. \times d \times BTS}{100} \ (\text{N mm}^{-1})$$
(2.16)

where $Qtz.Eq.$ = quartz equivalent or content (% by volume), d = mean diameter of quartz grains (mm), and BTS = Brazilian tensile strength (MPa).

The higher the value of F the greater the weight loss of the pin and, it may be presumed, the greater the wear on cutting tools. The reader is warned that the subject of abrasion is one of great complexity relating to the science of tribology, which deals with friction, wear and lubrication and is thus clearly of importance in mechanical engineering. Verhoef (1997) has reviewed the subject of wear on cutting tools and the form of site investigations necessary to gather appropriate data to attack these problems while Deketh (1995) has reviewed abrasion tests.

2.9
Environmental Reactivity

Rocks and soils underground exist within their environment, which may be defined in terms of stress, temperature, moisture content etc. When ground materials are exposed the environment will change and the ground materials may themselves change in response. Stress release may induce rebound and cracking but also the minerals forming the materials may change. Some rocks, for example shales and mudstones, may on exposure rapidly weather and decay and begin to revert to their original muds.

Slaking tests, in which small samples of rock are subjected to repeated wetting and drying, may demonstrate susceptibility to weathering if the sample breaks up into smaller pieces after a number of wet/dry cycles. The slake durability test (Franklin and Chandra 1972) subjects samples of rock to both wetting, drying and some degree of abrasion.

Many minerals contained in materials become unsuitable if subjected to changes in their environment as the consequence of engineering. Thus, for example, pyrite and marcasite will suffer expansion as the result of oxidation on exposure to air, moisture and perhaps bacterial action and higher temperatures. If these minerals are present in rock (and they are quite common in some Carboniferous shales and mudstones) they will expand when exposed. This could result in upward heave of foundations built on these rocks, changes in the stability of tunnel walls etc. Other minerals such as chalcedony and chert will react to the cement in concrete if they are present in rocks used for concrete aggregate.

2.10
Index Tests

Certain tests require little in the way of sample preparation and are so simple to conduct that the characteristics of soils and rocks, regardless of how complex these are, have been linked to them. Thus a simple test can indicate a range of material characters and the behaviour to expect from them, such as strength, permeability, and likely change in the presence of water. These tests are called *Index tests*.

2.10.1
Soils

Grading

Granular materials are described with reference to their particle size and grading curves. Grading curves are obtained by passing the material through a series of sieves down to fine sand size. Silts and clays are graded by observing the velocity of fall of the grains through water. The result of particle size analysis is a grading curve (Fig. 2.1). Reference may be made to the effective size (D_{10}) of a material. This is the largest diameter in the lower 10% of the material. The uniformity coefficient is D_{60}/D_{10}. The value of the grading curve as an index is that correctly naming the soil leads to associations with soil behaviour which have been learnt by general experience.

Atterberg Limits (also called Consistency Limits)

An important test for clayey soils is the Atterberg limit or consistency test (BS1377). This examines the effect of changes of moisture content on the plasticity of clay soils. In the worst condition, clay could be just a suspension of clay particles in water, which would then act as a rather dense fluid. If the water is allowed to dry out the point will be reached where the material is a very soft solid and exhibits a small shear strength. This limiting value of moisture content is the *liquid limit (LL)*. As further moisture dries out from the material it becomes stronger. Above a certain moisture content the soil will behave plastically, that is it will deform under stress but not regain its shape when the stress is removed. If the moisture content is less than this *plastic limit (PL)* the soil, when loaded, may break with a brittle fracture. The *plasticity index (PI)* is the range of moisture content within which the soil acts as a plastic material. Thus:

$$PI = LL - PL \qquad\qquad (2.17)$$

2.10.2
Rocks

Point Load Test

Engineering in rock is usually concerned with deformation of rock mass rather than strength of rock material, and material strength tests serve only as indicators of quality. The Point Load test (Broch and Franklin 1972) has been used since 1972 and has become the most popular of the simpler techniques. It is used in laboratory and field for logging mechanical strength of rock specimens obtained from boreholes sunk by rotary core drilling. A core is loaded between two 'points' which are steel cones subtending an angle of 60° and terminating in a hemisphere of 5 mm radius. The core is usually loaded across a diameter. The test specimen should be at least 1.5 times as long as the diameter. The strength value obtained from the Point Load test varies according to core diameter. In the standard test, results are converted to a standard 50 mm diameter core by the use of a conversion chart.

Correlations between Point Load strength (I_S) and unconfined compressive strength have been derived from experimental evidence, but unfortunately the multiplying factor appears to vary from one rock type to another. Bowden et al. (1998) record authors giving correlation factors (UCS/I_S) ranging from 4 to 50, with the majority falling in the range 10 to 30. For stronger rocks correlation factors of about 24 seem reasonable. For weaker rocks, into which the points may sink before the rock breaks, about half this value may be appropriate. The Point Load test is useful but it has a number of problems (Brook 1985; Price et al. 1978). It may also be noted that, in sedimentary rocks, much Point Load testing takes place along the bedding, the direction of greatest weakness and thus good correlation with the unconfined compressive strength, measured at right angles to the bedding, seems unlikely.

Enthusiastic and unrestrained Point Load testing on cores may destroy core which could have been used for more complex testing requiring the 2:1 height/diameter ratio. The combination of these problems suggests that other rock strength index tests are to be preferred and the use of the Point Load test on core be discontinued.

Block Punch Test

The Block Punch test is a test, developed for use as an index test, in which a thin plate or disc of rock is sheared by a punch device (Fig. 2.5). The Block Punch Index (*BPI*) is given by:

$$BPI = \frac{W}{t} \times (L_1 + L_2)$$
(2.18)

where W = failure load; t = disc or plate thickness and L_1 and L_2 are the lengths of the shear planes.

Extensive testing (van der Schrier 1988) has given correlations with unconfined compressive strength ($UCS = \pm 6 \times BPI$). The accuracy of the test seems to be as good as the Point Load strength test as an index for unconfined compressive strength and it has the advantage that thin plates or discs, about 10 mm thick, can be cut by diamond saw from small pieces or cores of rock. Thus cores too fragmented to fulfil the Point Load test specimen dimension criteria can be tested while longer cores can be tested without destroying those big enough for more complex testing.

Schmidt Hammer Test

The Schmidt hammer is a device designed to estimate the strength of concrete (but which can be used on rock) by measuring the rebound of a spring-impelled hammer striking on the surface of the mass. Several varieties of the hammer differ in the energy imparted by the spring – the L-type is generally the most satisfactory for rocks. The rebound number is converted to unconfined compressive strength by means of graphs, which are based on laboratory testing evidence. Correlation between rebound number and *UCS* is not absolute, and rather like the use of Point Load Strength index, it might be better used for comparison between rocks if the link to *UCS* were avoided. The value of rebound is affected by the attitude of the hammer, whether vertically pointing down or up, horizontal or inclined.

Velocity Tests

The velocity of waves through a material can give some indication of other properties of the material, such as strength, density and elastic moduli. Measurements may be made on intact specimens in the laboratory, or on rock masses in the field. A comparison of laboratory and field velocities may give an indication of rock mass properties.

The velocities of waves through a cylindrical sample of material can be measured accurately. The pulse generated at one end of the specimen is picked up at the other end, and recorded. For a known density the elastic constants can be calculated:

$$E = 2\rho V_s^2 (1 + v) \qquad (2.19)$$

$$v = \frac{0.5 \left(V_p / V_s \right)^2 - 1}{\left(V_p / V_s \right)^2 - 1} \qquad (2.20)$$

$$K = \rho \left(V_p^2 - \frac{4}{3} V_s^2 \right) \qquad (2.21a)$$

$$G = \rho V_S^2 \qquad (2.21b)$$

where V_P and V_S are compressional and shear wave velocities, ρ is density, E is Young's modulus, v is Poisson's ratio, G is shear modolus and K is bulk modulus.

Slaking and Slake Durability

Slaking tests consist of cyclically saturating and drying a fragment of rock and observing its loss of weight after a number of cycles. The slake durability test (Franklin and Chandra 1972) includes an element of erosion for rock fragments are placed in a rotating drum. The test is suitable only for weak or weathered rocks there being generally too little effect on strong fresh rock.

Combined Parameter Indices

An index may be established using more than one property. An important index is the ductility number (sometimes referred to as the brittleness index) which is the ratio UCS/UTS (Fig. 2.5). Values of ductility number of more than 15 indicate a rock which will shatter when impacted by cutter tools. Values below about 9 indicate that the rock will crush under tool impact. The brittle/ductile behaviour boundary is of great importance in rock cutting.

2.11
Range of Values for Material Properties

Although the mechanical values for a rock or a soil can vary considerably, they do exhibit a characteristic range of values and this range becomes the sort of value that "would be expected".

2.11.1
Soils

Some typical values for the properties of granular soils are given in Tables 2.8 and 2.9. Ranges of shear strengths for clays are given in Table 2.2.

The properties of clay soils are very dependent upon their moisture content and they can be much more complex, mineralogically, than more granular soils. Consequently the range of values of properties is much greater (Table 2.10). There are some special 'problem' soils encountered in the world, which give particular engineering difficulties. In many cases the properties are due to the mode of origin and/or subsequent geological history, in which environment often plays a leading role. Some examples are given below.

Loess

Loess is a yellow, calcareous, porous, wind blown silt, characteristic of Pleistocene sequences in several large areas on every continent. Generally more than 80% of the loess particles consist of quartz, 10 to 20% of feldspar and up to 20% of carbonate. The percentage of clay size particles may range up to 15%. The structure of the material is dominated by the fine quartz particles, between twenty and fifty micrometres in length. The open structure of the quartz particles is often strengthened by clay minerals, and cemented by calcium carbonate. The whole fabric is essentially delicate, which makes the structure liable to sudden collapse. Loess soils cause great difficulties for civil engineers. They tend to collapse on saturation under load, or at least when the moisture content is changed they consolidate very rapidly – 95% of settlement can occur in ten minutes. The effect is local, which can lead to very large differential settlements of structures founded on them. Similar loss of strength may occur after disturbance by other means, for example by large loads or earthquakes.

Sensitive Clays

The shear strength of an undisturbed clay is generally greater than the shear strength of the same remoulded clay at the same moisture content. There is a loss of strength on remoulding, and the ratio of the undisturbed to remoulded strength is called the

Table 2.8. Some properties of granular soils

Property	Gravel	Sand
Specific gravity	2.5 – 2.8	2.6 – 2.7
Bulk unit weight (kN m^{-3})	14.1 – 22.6	13.7 – 21.6
Dry unit weight (kN m^{-3})	13.7 – 20.6	13.0 – 18.6
Angle of friction (°)	33 – 45	27 – 46
Porosity (%)	25 – 40	25 – 50
Shear strength (kPa)	180 – 550	100 – 400
Permeability (m s^{-1})	$10^{-1} - 10^{-4}$	$10^{-3} - 10^{-6}$

Table 2.9. Some properties of Dutch sands

Unit weight (kN m^{-3})	Relative density (%)	Natural moisture content (%)	ϕ (°)	Location
14.8	21	4	30	Eems
15.5	53	8	33	Eems
16.3	67	8	35	Eems
16.0	70	4	40	Eems
17.1	95	8	41	Eems
16.9	91	4	42	Eems
17.1	95	12	43	Eems
18.7	90	12	44	Echteld
20.0	100	8	46	Berg en Terblijt

Table 2.10. Some properties of silt and clay soils

Property	Silt	Clay
Specific gravity	2.63 – 2.67	2.55 – 2.76
Bulk unit weight (kN m^{-3})	17.65 – 21.2	14.5 – 21.2
Dry unit weight (kN m^{-3})	14.2 – 19.2	11.6 – 21.2
Void ratio	0.34 – 0.82	0.42 – 0.95
Liquid limit (%)	24 – 36	> 25
Plastic limit (%)	14 – 25	> 20
Permeability (m s^{-1})	10^{-6} – 10^{-9}	10^{-9} – 10^{-12}
Cohesion (kN m^{-2})	< 70	15 – 200
Angle of friction (°)	25 – 35	4 – 17
Coefficient of consolidation (m^2 yr^{-1})	12	5 – 20

sensitivity (S_t) of the clay. The measure of sensitivity proposed by Terzaghi is the ratio of the peak undisturbed strength (C) to the remoulded strength (C_r). Thus:

$$\text{Sensitivity}\,(S_t) = \frac{C}{C_r} \qquad (2.22)$$

All clays show some degree of sensitivity; the term 'quick' is applied to clays which have a sensitivity of over 16 (Skempton and Northey 1952). The more sensitive the clay, the more difficult the engineering problems posed. The term 'quick clay' is sometimes used to denote clays of such sensitivity that they behave as viscous liquids when in a remoulded state. Small earthquakes may trigger of a mass movement which may be responsible for great loss of life and damage to property.

Sabkha

The term 'sabkha' is used in Arabic speaking countries to mean 'salt encrusted flat surface', of which there are many in coastal and inland areas of Arabia, particularly along the southern shore of the Persian Gulf. In engineering geology and civil engineering, sabkha is now used as a description of geological material originating in particular environments of deposition, and consisting of sandy or silty sediment which is at least partly cemented by salts, usually calcium carbonate and calcium sulphate, but often with a variety of other salts present also. Most geologists apply the term only to the coastal environment. At times of rain, flash flood or extra high tidal level this crust may soften very rapidly.

The geotechnical difficulties may be summarised as follows:

- very variable bearing capacity, vertically;
- very variable deformability (compressibility);
- rapid change due to growth or change, for example hydration, of new or existing minerals;
- effect of salt on concrete and steel.

Duricrust

If, in a semi-arid climate, the ground-water table lies not far below surface, water drawn to the surface will evaporate and deposit any minerals held in solution within the near surface materials, resulting in their cementation. If the material is already rock then pore spaces become filled and make the rock stronger. The minerals in solution may be leached from materials far below surface so duricrust layers may be underlain by leached material, weaker than it was in its original condition.

The cementing material is most commonly calcite so that form of duricrust is termed *calcrete*. However, a silicious cement gives *silcrete*, a ferruginous cement *ferrocrete* and so forth, while their are many local names for these materials. They are widespread in many parts of the world and much attention has been drawn to their occurrence in the Middle East (Fookes 1978). Thicknesses of duricrust may range from a few centimetres to several metres with considerable changes in thickness in but a short horizontal distance.

Duricrusts are formed near surface and commonly near the sea but there have been considerable variations in climate and sea level in Pleistocene and Holocene times. Because of this duricrusts may now be found underwater (Dennis 1978) or above sea level along shore lines. It is thus important, in a site investigation, to establish the recent geological history of the area under investigation to determine whether or not duricrust is likely to occur.

Glacial Soils

There is a very large range of glacial soils whose nature and properties vary according to the material available, the environment of transport, i.e. on top of, within, or near the base of the glacier, and the environment of deposition. Thus, tills are soils deposited directly by the ice and fluvio-glacial soils are deposited by meltwater either close to the ice front, i.e. *periglacial,* or at greater distance, i.e. *proglacial.*

The term *till* is used in preference to the more commonly used term 'boulder clay' to which it is synonymous. Boulder clay is a generic term rather than a lithological description, but it is generally misunderstood by engineers who may complain with justification that the material described as boulder clay on the map has neither boulders nor clay as constituents. The character of till is determined by a number of factors:

- the position of transport – on or in the glacier
- the nature of the bedrock over which the glacier is passing
- the nature of the rock through which contributory glaciers have passed
- the mode of deposition of the material

Some tills originate from the 'plastering' of sediment transported at or very near the base of the glacier and dropped on to the bedrock. Such material would be over-ridden by the ice and if clayey would be over-consolidated as a result. The material is likely to consist of finely divided 'rock flour' produced by the grinding action of the ice. If the bedrock was shale or mudstone, the proportion of clay mineral in the till will be high, and the till will be plastic. If the bedrock is arenaceous or igneous, the clay mineral fraction may be very small, but the primary mineral particles are likely to be of clay size. However, plasticity and moisture content are very different in the two materials, but in both cases there is likely to be a scarcity of boulders. A till originating in this way is described as lodgement till.

Material which is carried along on the surface or within the glacier will be dropped as the ice melts, a process which is gradual. Accumulations of such material are known as ablation tills, and are likely to have a much higher proportion of boulders than lodgement tills. The boulders and gravel sized fractions are angular, and it is common to find clasts of rock which must have been transported long distances. All tills are variable in their nature and consequently in their properties. In general they consist of an assortment of rock debris ranging in size from rock flour to boulders, but all sizes are not usually present. Gap grading is usual.

Glacial debris may be transported by meltwater and thereby become sorted and stratified; these are *fluvio-glacial deposits.* Many such deposits consist of gravel and sand and these frequently have low relative densities. Such pockets of low density are difficult to locate, and the problem is usually anticipated and dealt with by assuming their presence and using compaction equipment before construction begins. Coarse grained fluvio-glacial deposits are often important sources for concrete aggregate.

Of the great variety of fluvio-glacial deposits, *varved* clays are perhaps the most difficult, from a geotechnical point of view. These are deposited in areas away from the glacier front, in quiet water conditions but where the sediment is contributed on an annual basis. Fine material reaches the depositional area in late winter and early spring, when the meltwater current is slow: in late spring and summer the faster current transports larger sized material, so that in a complete year a varve consists of clay sized particles giving way to silt sized. This variation in particle size produces variation in material properties. Thus, a range of liquid limits from 30 to 80, and plastic limits from 15 to 30, has been reported from a Canadian varved clay, which also has a sensitivity of about 4.

Organic Soils

Organic matter in the form of partly decomposed vegetable or animal remains, even when present in very small amounts, may significantly affect the physical and chemical properties of soil. Small amounts of colloidal organic matter in a clay, for instance, will significantly increase both liquid and plastic limit and increase compressibility. The presence of organic matter inhibits the setting of cement.

Peat is an accumulation of partially decomposed vegetable material. Complete decomposition is prevented if the vegetation accumulates in water-logged anaerobic conditions, where a series of complex biochemical and chemical changes may interact to produce humus and thereafter peat. There are many varieties of peat depending upon the nature of the vegetation which contributed and the details of the conditions and reactions. However, two main varieties may be distinguished for geotechnical purposes – the fibrous, mainly woody, peat, and the amorphous, granular variety. The difference may be summarised in the table of some geotechnical properties (Table 2.11).

Peats formed in early post-glacial times may occur at depth, buried beneath more recent sediments. Frequently, therefore, the geotechnical problems concern the bearing capacity and consolidation characteristics of the material. For peat, these are difficult to define. Calculations of settlement are difficult because peat does not conform to normal consolidation theory. Amorphous granular peat behaves in a similar manner to clay, but fibrous peat has rather different consolidation characteristics because of the variation in size of the fibrous material and consequently the size of the voids between the fibres.

Where possible, peat is usually removed and replaced before construction. Alternatively, piles may be used to take foundation loads below the peat. If sufficient time is available, the peat may be pre-consolidated by 'surcharging'; i.e. by loading before

Table 2.11. Some geotechnical properties of peat

Property	Amorphous granular peat	Fibrous peat
Void ratio (%)	< 10	< 25
Moisture content (%)	> 500	< 3 000
Bulk unit weight (kN m^{-3})	< 12	5 – 10
Dry unit weight (kN m^{-3})	6.5 – 13	6.5 – 13

the foundation is constructed. With sufficient load, peat may compress to 25% of its original volume.

2.11.2
Rocks

The range of strengths for rocks is much greater than for soils. Table 2.12 gives a general range of values for common groups of rocks. Geological names for rocks are given to indicate genesis or mineral content but may have no direct link with engineering properties. Thus 'sandstone' is a stone composed of sand grains – if there is little cementation and compaction of the grains the sandstone may be very weak. If the grains are densely packed and the cementing mineral is strong (such as quartz) and there are few pore spaces, the sandstone would be extremely strong. A further complication in rocks is that all rocks near the surface are weathered, that is, they have lost some strength as the result of decay of the minerals of which the rock is composed. Thus fresh granite is stronger than a highly weathered granite. Table 3.6' in Chap. 3 shows how the material properties of granite and dolerite vary with weathering.

Correlations between Rock Properties

There are rough relationships between the various rock properties that are commonly measured. Thus higher strength rocks tend to have larger moduli values and sonic velocities while more porous and less dense rocks tend to be weaker. These relationships are vague because of the varied mineralogy of the rocks cited but become clearer if almost monomineralic rocks are examined. Swart (1987) conducted a series of tests on limestones; the rocks tested, briefly described in Table 2.13, were all limestones in the petrological sense.

The results of tests on oven dried samples of these rocks are portrayed in Figs. 2.10 to 2.13. The plots are linked by bulk density on the horizontal axis. Figure 2.10 shows

Table 2.12. General range of strengths for common rock types (based on the experience of the author)

Very weak	Weak	Moderately weak	Moderately strong	Strong	Very strong	Extremely strong	
MPa	1.25	5	12.5	50	100	200	500
	◀——— Clastic ——— Calcarenite	Chalk	**Limestones**	◀——— Crystalline —————▶ Siliceous			
			◀—— High porosity: ——▶ weak cement ◀—— **Sandstones**	Low porosity: ——▶ strong cement	Quartzite: quartz grains with ——▶ quartz cement		
	◀————————————	**Shales** ◀—▶ Fissile	**Mudstones** ——————▶ Bedded				
			◀— **Slates** ——— Very anisotropic	**Schists** Anisotropic	**Gneisses** —————▶ Less significantly anisotropic on scale of test specimen		
				Igneous rocks (excluding flow volcanics) ◀— Coarse	Fine —▶		

Table 2.13. Age, location and other characteristics of the limestones whose properties are displayed in Figs. 2.10 to 2.13

Sample No.	Age	Location	Grain size	CaCO₃ [MgCO₃] content (%)	Clastic (Cl) or Crystalline (Crys)
1	Magnesian Limestone (Permian)	Yorkshire (UK)	Medium	62 [38]	Cl
2	Chalk (Upper Cretaceous)	near Hull (UK)	Fine	98 [1]	Cl
3	Lower Oolitic Limestone (Jurassic)	Lincolnshire (UK)	Medium	93 [1]	Cl
4	Upper Oolitic Limestone (Jurassic)	Yorkshire (UK)	Medium	84 [1]	Cl
5	Wenlock Limestone (Silurian)	Shropshire (UK)	Coarse	98 [1]	Crys
6	Wooldale Limestone (Carboniferous)	Derbyshire (UK)	Medium	98 [1]	Cl/Crys
7	Vinalmont Limestone (Carboniferous)	Hainault (B)	Medium	99 [1]	Cl/Crys
8	Muschelkalk (Triassic)	Gelderland (NL)	Fine	82 [4]	Cl/Crys
9	'Fossil Limestone' (Carboniferous)	Dinant (B)	Medium	97 [2]	Cl/Crys
10	'Red Reef Marble' (Upper Devonian)	Neuville (B)	Fine	91 [2]	Crys
11	'Grey Marble' (Middle Devonian)	Aywaille (B)	Medium	97 [2]	Crys
12	'Marl' (calcarenite, Maastrichtian, Upper Cretaceous)	Sibbe (NL)	Medium	97 [2]	Cl

density vs. unconfined compressive strength and it can readily be seen that strength increases with density, reaching a maximum as rock density approaches that of calcite. However, if the clastic limestones are separated from the clastic/crystalline and the crystalline, the plot can be presented as two curves, A and B. The clastic rocks are largely homogeneous and fall on a smooth curve (A). However, the remainder have been visually assessed to be variously inhomogeneous, containing textural discontinuities (from, for example, recrystallised fossils) representing potential flaws which, when stressed, may bring the sample to failure. The straight line B indicates the range of strengths of the inhomogeneous limestones suggesting that the chance inclusion and location of textural flaws is the strength determining factor. On this

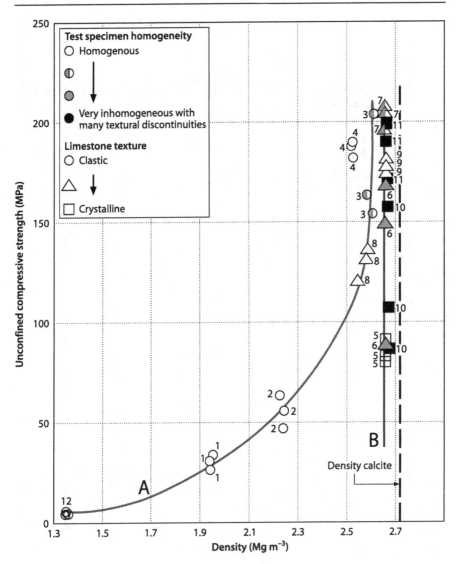

Fig. 2.10. Limestones – unconfined compressive strength vs. dry bulk density (see text)

evidence it would seem unlikely that any rock classification can be based accurately on compressive strength.

Figure 2.11 shows average tangent modulus vs. density; and here almost all the rock types now fit reasonably on a smooth curve, suggesting that for classification purposes elasticity is to be preferred to strength. Figure 2.12 depicts porosity vs. density *and* sonic velocity vs. density. The relationship between porosity and

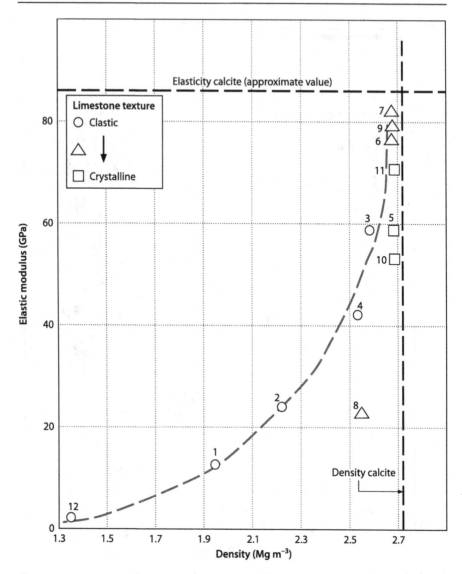

Fig. 2.11. Limestones – tangent modulus vs. dry bulk density (see text)

density is striking while it can be seen clearly that as rock density approaches that of calcite so does rock velocity approach that of calcite. Figure 2.13 combines strength, porosity and velocity plotted against density showing how, for these particular rocks, each of these properties might be used as an approximate guide to strength.

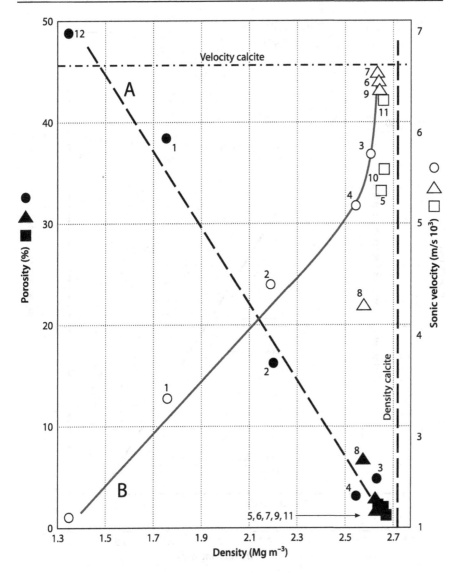

Fig. 2.12. Limestones – total porosity and velocity vs. dry bulk density (see text)

The Choice of Test for Rocks

While the results of just about all forms of soil testing are used in some form of calculation, the results of some forms of rock testing are seldom so used. Much testing is undertaken simply to gauge the quality of rock material and the investigator must ask

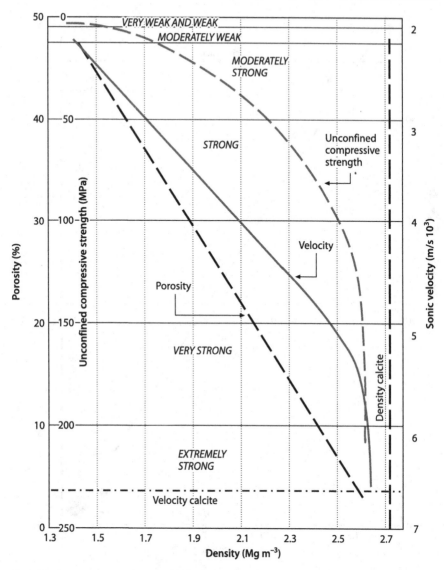

Fig. 2.13. Limestones – correlations with strength (see text)

if the task at hand requires the application of expensive testing or whether a simple quality test, such as hitting the rock with a hammer, would be sufficient to the purpose. However, if such tests are done, they must be well done, to standards. Another difficulty is the choice and availability of material for testing. In examining and testing cores the concern is to identify those weak rocks which could become overstressed by some engineering process. Unfortunately weak layers are those least likely to produce cores sufficiently intact to give test specimens of appropriate dimensions for accurate testing. Accordingly it may well be wiser to place emphasis on those index tests which may be performed on small test specimens, such as the Equotip test (Chap. 3) and the Block Punch test. Such tests could be undertaken in large numbers to gain a better appreciation of rock material variability.

2.12
Further Reading

Bell FG (1999) Engineering properties of soils and rocks, 4th edn. Blackwell Scientific Ltd., Oxford
Fookes PG (ed) (1997) Tropical residual soils. The Geological Society of London (A Geological Society Engineering Group Working Party revised report)
Forster A, Culshaw MG, Cripps JC, Little JA, Moon CF (eds) (1991) Quaternary engineering geology. The Geological Society of London (Geological Society Engineering Geology Special Publication 7)
Legget RF, Karrow PF (1982) Handbook of geology in civil engineering. McGraw-Hill, New York
Pusch R (1995) Rock mechanics on a geological base. Elsevier, Amsterdam

Chapter 3

Geological Masses

3.1
Mass Fabric

The word '*mass*' is already defined in Chap. 1 as '*the volume of ground that will be influenced by or will influence the engineering work*'. All rock and many soil masses have discontinuities and their presence in the rock or soil mass is of the utmost importance to all engineering works in rock or soil. Much research has been undertaken into the characteristics and behaviour of discontinuities. Whole conferences, such as that on 'Rock Joints' held in 1990 (Barton and Stephenson 1990) have examined the subject in great detail. However, there is a major gap between acquiring such knowledge by research and the application of that knowledge to practice. *The problem is that the investigator must forecast the character of the discontinuities in the mass, and thence their likely behaviour, from the evidence available from outcrops or boreholes. Adequate data is seldom available.*

3.1.1
Discontinuities

The properties of discontinuities of greatest importance to a volume of ground are: shear strength, influencing, for example, the stability of slopes, underground excavations, and foundations, and stiffness, influencing the deformation and permeability of the ground.

Types of Discontinuities

Discontinuities must be classified in relation to their type. Two basic types of discontinuities may be distinguished. (1) *Integral discontinuities* – which are discontinuities that have yet to be opened by movement or weathering; they have tensile strength and, hence, a true cohesion. Intact bedding planes, foliation planes, and strongly cemented joints are integral discontinuities. (2) *Mechanical discontinuities* – discontinuities which have been opened as a response to stress or weathering; they have little or no tensile strength but do generate shear strength. They may be divided into:

- *Bedding, schistosity or foliation planes* – these are formed by changes of material or mineral arrangement in the rock.
- *Joints* – which result from strains of tectonic or diagenetic origin and often fall into well-defined sets whose members are oriented essentially parallel to each other.

- *Fractures* – which result from strain due to man-made stresses (blasting etc.) or geomorphological strains (land sliding, creep etc.). They do not necessarily fall into well-defined sets.
- *Faults and shears* – which result from tectonic, geomorphological and man-made strains, with shear movement on either side of a shear surface.

Bedding, schistosity and foliation planes, joints and fractures are normally a regular occurring feature in the mass at approximately regular distances, with similar orientation, and with more or less similar characteristics (Fig. 3.1). These are normally grouped in a "*set*" or family, e.g. bedding plane family and joint set, etc. Both faults and shears may fall into well-defined sets but may occur on a scale greater than that of most engineering works.

3.1.2
Shear Strength

The main concern regarding discontinuities in engineering geology is their resistance to shear stress. This is described by Coulomb's law, which is:

$$\tau = c + \sigma_n \tan\varphi \qquad\qquad (3.1)$$

where τ = shear strength (N m^{-2}), c = cohesion (N m^{-2}), σ_n = normal stress on the sliding plane, and φ = angle of sliding resistance.

The cohesion may be caused by 'true cohesion', e.g. there is tensile strength between the two discontinuity surfaces, or may be 'apparent cohesion' caused by irregularities

Fig. 3.1. Bedding planes as mechanical discontinuity planes causing failure of a slope in limestone

along the discontinuity surfaces. The surfaces of mechanical discontinuities may be smooth and planar or exhibit varying degrees of roughness. For a particular rock the greater the roughness of the surface the greater the resistance of the discontinuity to shearing. If normal stresses are relatively low then shear movement along a rough surfaced discontinuity must be accompanied by vertical dilation (Fig. 3.2). Patton (1966) analysed the situation by proposing that the *angle* of shearing resistance of a surface containing asperities inclined at angle i, is ϕ_b, the basic friction angle, plus i, so that the shear strength of the discontinuity with asperities is:

$$\tau = \sigma_n \tan(\phi_b + i) \tag{3.2}$$

One way to determine ϕ_b is to use a shearbox test in which vertical and horizontal displacements are measured. $dv/dh = \tan i$ and with Eq. 3.2, ϕ_b can be determined. The ϕ_b has been measured in the laboratory by many researchers to give values ranging from about 25° to 35°, the lower values generally pertaining to sedimentary rocks and the higher to igneous rocks. If $\phi_b + i = 90°$ sliding is not possible and breaking of the asperities has to occur before shear displacement is possible. The asperities fail also if normal and shear stresses are high relative to the strength of the asperities. Asperity failure of a discontinuity will give an '*apparent cohesion*' (Fig. 3.3). The residual angle of friction, ϕ_r is the angle of friction obtained after a discontinuity has been displaced. ϕ_r may equal ϕ_b, however it may not if asperities have broken, because rolling particles of debris influence the measured friction in a displaced discontinuity and the material remaining on the discontinuity wall after breaking of the asperities may also be of better quality, as it is probably less weathered than the outside of the asperities.

Scale of Roughness and Anisotropy

Roughness may be seen on different scales. Various authors have described roughness scales and descriptions for roughness (Fig. 3.4) show the relations as suggested by Hack (1998). Up to mm scale the roughness is normally formed by grains or crystals in the rock, as, say, might be found in shrinkage joints in granite. On the 0.01 to 1 m scale, roughness is normally due to depositional features in sedimentary rock or foliation undulations in metamorphic rock. At greater scales discontinuity surface undulations may

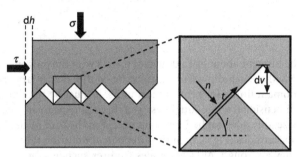

Fig. 3.2. Dilation (*dv*) following sliding on asperities under horizontal applied stress (τ) and vertical applied stress (σ)

n is the normal and t the shear stress on the contact plane, resulting from the applied stresses σ and τ

Fig. 3.3. The behaviour of rough and smooth discontinuities with varied shear and normal stresses

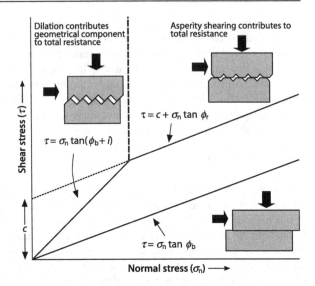

Fig. 3.4. Amplitude versus wavelength for discontinuity roughness (after Hack 1998). For small amplitudes and wavelengths, the roughness is of a triangular form whereas with larger amplitudes and wavelengths the roughness changes to a more sinusoidal form. Lustre is not included in the boundary Non-visible to Visible roughness. The boundaries in the graph are *dashed*, as these are not exact

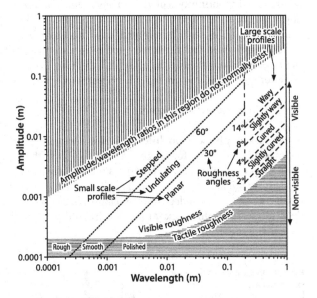

be brought about by folding and it rests with the judgement of the observer to decide whether such dip variations may be considered as large scale roughness or to mark domain limits for geotechnical units. It must be remembered that profiles are two-dimensional while roughness is a three dimensional property. On the scale of grain size, this may not be important except in the case of the lineation features known as slickensides (Fig. 3.5); in the direction of lineation these are smooth but, normal to it, they are rough. At larger scales features such as ripple marks may be similarly distinctly anisotropic.

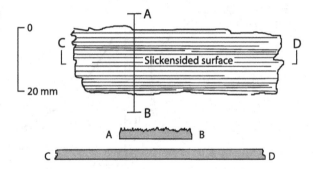

Fig. 3.5. Roughness dependent on direction for slickensided surfaces

Fitting versus Non-Fitting Roughness

Figure 3.2 shows a regular saw tooth profile as roughness profile. In reality, roughness profiles of discontinuities are not so regular and are three-dimensional. In general, only one perfect matching fit exists between the walls of the discontinuities. If the walls are displaced relative to each other, the fit will become less. If the displacement is large enough the roughness profiles become completely non-fitting. The effect on shear strength is that the contribution of the roughness angle i gradually diminishes with larger displacement. The effect has been described by Rengers; the so-called 'Rengers envelop' (Fig. 3.6). This effect makes it very important that the description of a discontinuity includes whether the discontinuity walls are fitting or displaced. In the later case an estimate should be given as to how much the i-angle has been reduced.

Discontinuity Wall Strength

A full evaluation of discontinuity shear strength requires not only assessment of roughness but also of wall strength. Accordingly, it is necessary to measure wall rock strength. This is almost impossible by conventional laboratory testing because of the extremely small size of the samples but recourse may be had to impact testers applied to the discontinuity surface. The Schmidt hammer gives too strong a blow to be of use (for the N-value reflects both near surface and deeper rock properties) but the Equotip tester (Verwaal and Mulder 1993) gives a much lighter blow which rebound reflects rock properties at shallow depth. Its use to assess wall strength has been described by Hack et al. (1993) (Fig. 3.7). If it is necessary to measure wall strength, it also follows that for some rocks that are particularly susceptible to weathering, estimates should be made of the *rate* of strength reduction by weathering of the discontinuities in any new construction.

Aperture and Discontinuity Infill

Openness of discontinuities in outcrops may have arisen from slope relaxation, near surface movement or plant root wedging and thus (except perhaps in the case of solution cavities in limestone) may in no way reflect internal conditions. If discontinuities are open, they may be either wholly or partially filled by material from either weath-

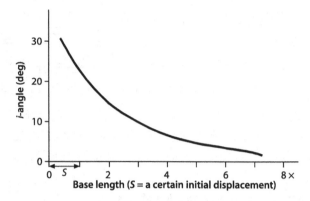

Fig. 3.6. Rengers envelope (after Fecker and Rengers 1971)

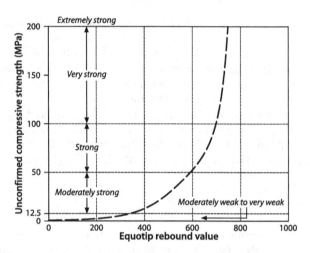

Fig. 3.7. Equotip rebound values vs. unconfined compressive strength (after Verwaal and Mulder 1993)

ering in situ or from outside. Thus bedding planes opened by weathering may be filled with clay or limestone solution cavities filled by washed in debris. Such infilling will clearly affect the shear strength of the discontinuity. If the infill is weak, the discontinuity shear strength may be less than that of the discontinuity with walls in contact. If the infill results from mineralisation as, for example, quartz or calcite, the infill may be stronger than, and closely bonded to, the rock; the discontinuity may then be described as 'healed'. Table 3.1 (BS EN ISO 14689-1) gives descriptive terms for discontinuities that have a continuous aperture or an infill of continuous thickness in all directions. However, discontinuities often do not have a continuous aperture or infill thickness. For such discontinuities, Table 3.1 looses its merits and other description criteria should be used, such as, descriptions based on the ratio of volume enclosed by the discontinuity to the total rock mass volume.

Field Estimate of Discontinuity Friction Angle

Determining the shear strength of a discontinuity is one of the most difficult tasks while it is probably the most important property to measure. A simple and often ad-

Table 3.1. Descriptive terminology for aperture

Class	Descriptor	Numeric value (mm)		
Closed	Very tight		<	0.1
	Tight	0.1	–	0.25
	Partly open	0.25	–	0.5
Gapped	Open	0.5	–	2.5
	Moderately wide	2.5	–	10
	Wide	10	–	100
Open	Very wide	100	–	1 000
	Extremely wide		>	1 000

equate assessment can be made with the tilt test. Two pieces of rock including the discontinuity are tilted while the angle of the discontinuity with the horizontal is measured (Fig. 3.8) with, for example, the inclinometer included in a geological compass. The angle measured at the moment the top block moves is the 'tilt-angle'. If no infill is present and the two blocks had fitting discontinuity surfaces, the tilt angle equals the small-scale roughness angle plus the angle of friction of the surface material. If the discontinuity roughness is completely non-fitting, the tilt-angle equals the material friction only. Undisturbed infill material will seldom be present in this test, but if it is, the tilt-angle includes the influence of the undisturbed infill. If it is not possible to obtain a sample with undisturbed infill material, it is sometimes possible to scrape infill material from another still in situ discontinuity and place this between the blocks of the tilt test sample. The thickness of the so-formed infill layer should be the same as in situ. The measured tilt-angle is then including the influence of remoulded infill material. Cohesion is not measured in the tilt test separately. If present, either real or apparent, it will be included in the tilt-angle. Steps on discontinuity planes causing hanging of the blocks, for example, result in a very high tilt-angle (which may be up to 90°). Whether this is realistic for the friction of the in situ discontinuity has to be judged on the strength of the cohesion or asperities in relation to the stresses in the in situ rock mass. Stresses in the rock mass will generally be far higher than those used during the tilt test and may cause either shearing of real cohesion or shearing of asperities along the in situ discontinuity. Clearly, the tilt-angle is only representative for small-scale roughness and low normal stresses as tilting meter-scale rock blocks is normally not possible for the average engineering geologist!

A more sophisticated methodology is to use classification systems to estimate friction or shear strength along a discontinuity. An example is the relation given in Eq. 3.3 (Barton 1971):

$$\tau = \sigma_n \tan\left[JRC \log_{10}\left(\frac{JCS}{\sigma_n}\right) + \varphi_r\right] \tag{3.3}$$

Fig. 3.8. Tilt test

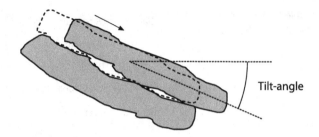

Tilt-angle

JRC is the joint roughness coefficient (a number, low for smooth planar surfaces rising with increasing roughness, estimated by visual comparison of the discontinuity surface to standard roughness graphs), *JCS* is the joint wall condition strength, σ_n is the normal stress on the discontinuity, and ϕ_r is the residual friction angle. If no other test is available, it is possible to use the tilt-angle of a non-fitting discontinuity without infill material as residual friction angle.

Another example of a classification system for estimating discontinuity shear strength is the 'sliding criterion'. Based on back analyses of slope stability a sliding criterion was developed to easily estimate the shear strength of a discontinuity (Hack and Price 1995; Hack et al. 2002). The discontinuity is characterised following Table 3.2. The roughness is characterised by visually estimating large (Fig. 3.9) and small-scale (Fig. 3.10) roughness by comparing to standard profiles and by establishing tactile roughness, infill material, and presence of karst. The different factors for the different characteristics are multiplied and divided by an empirically established factor. This results in the so-called '*sliding angle*':

$$\varphi_{\text{sliding angle}}(\text{degrees}) = \frac{Rl \times Rs \times Im \times Ka}{0.0113} \tag{3.4}$$

The '*sliding angle*' is comparable to the tilt test idea but on a larger scale. The sliding angle gives the maximum angle under which a block on a slope is stable. The 'sliding criterion' has been developed on slopes between 2 and 25 m high. The 'sliding criterion' applies for stresses that would occur in such slopes, hence, in the order of maximum 0.6 MPa.

3.1.3
Persistence (Continuity)

Discontinuities may be persistent for long distances, for example, bedding planes, they may be persistent for a certain length and end in intact rock, or they may abut against other discontinuities. The shear strength along the discontinuity is dependent on the persistence. Intact rock has to be broken before displacement can take place if the discontinuity ends in intact rock. Boundary blocks have to move before abutting discontinuities can be displaced. If possible, persistence (sometimes called continuity) of discontinuities should be measured but opportunities to take such measure-

Table 3.2. Discontinuity characterisation for 'sliding criterion' (after Hack and Price 1995)

Condition of discontinuities			Factor
Roughness, large scale (*Rl*) (visual area > 0.2 × 0.2 and < 1 × 1 m²)	Wavy		1.00
	Slightly wavy		0.95
	Curved		0.85
	Slightly curved		0.80
	Straight		0.75
Roughness, small scale (*Rs*) (tactile and visual on an area of 20 × 20 cm²)	Rough stepped/irregular		0.95
	Smooth stepped		0.90
	Polished stepped		0.85
	Rough undulating		0.80
	Smooth undulating		0.75
	Polished undulating		0.70
	Rough planar		0.65
	Smooth planar		0.60
	Polished planar		0.55
Infill material (*Im*)	Cemented/cemented infill		1.07
	no infill-surface staining		1.00
	Non softening and sheared material, e.g. free of clay, talc, etc.	Coarse	0.95
		Medium	0.90
		Fine	0.85
	Soft sheared material, e.g. clay, talc, etc.	Coarse	0.75
		Medium	0.65
		Fine	0.55
	Gouge < irregularities		0.42
	Gouge > irregularities		0.17
	Flowing material		0.05
Karst (*Ka*)	None		1.00
	Karst		0.92

ments are limited to outcrops. Persistence cannot be measured from cores. Persistence is usually quoted as a simple one-dimensional measurement; the terminology for its description is given in Table 3.3. However, in reality persistence exists in two dimensions. Thus in Fig. 3.11a joint *y* seen in an outcrop perpendicular to the direction of strike may have but small persistence across a bed (e.g. abutting) while, seen on the bedding plane, it may have long persistence in the direction of strike. In Fig. 3.11b, joint *y* has limited persistence in the direction of strike. The difference in two-dimensional persistence between joints *y* in Fig. 3.11a and b could have great significance in slope stability. Sandstone blocks could part along *y* and slide down slope over the shale in Fig. 3.11a, but in Fig. 3.11b parting and down slope block sliding would be inhibited by the limited persistence of joint *y*. It is suggested that persistence in banked (see Fig. 3.12) and bedded rocks be recorded as a proportion of bank thickness. Thus, for example, in Fig. 3.11a the persistence of *y* could be given as >20 *t* in strike direction, where *t* is bank thickness, and in Fig. 3.11b the persistence of *y* could be given as ¼ to 1½ *t* in strike direction.

Fig. 3.9. Large scale roughness profiles (after Hack and Price 1995)

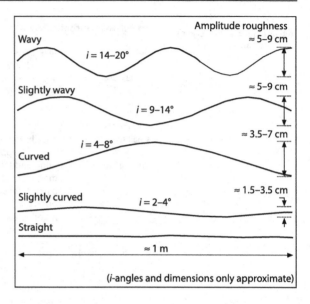

Fig. 3.10. Small scale roughness profiles (after Hack and Price 1995)

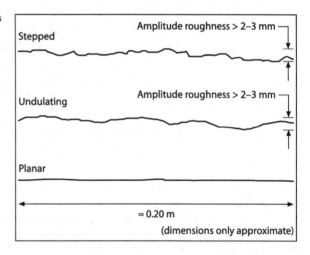

3.1.4
Orientation

The importance of discontinuities in any particular project depends partly on their orientation relative to directions of imposed stresses. The geologist records the orientation of discontinuities (which may be geological surfaces such as bedding planes giving an indication of geological structure) using a geological compass/clinometer usually placed on the surface being measured at some readily accessible outcrop. In the much more detailed surface mapping required for engineering geological purposes dip and strike measurements of discontinuities have

Table 3.3. Terms for the description of one-dimensional persistence (after BS EN ISO 14689-1)

Term	Numerical value (m)
Very low	< 1
Low	1 – 3
Medium	3 – 10
High	10 – 20
Very high	> 20

Fig. 3.11. Persistence

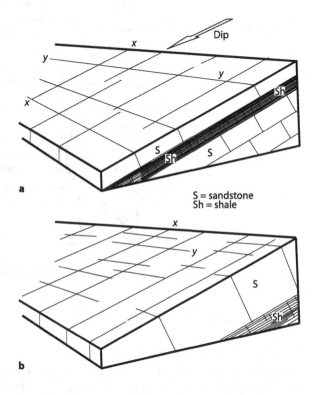

S = sandstone
Sh = shale

to be taken regardless of whether they are easy to reach or not. For work in tunnels, the compass/clinometer must have built-in illumination. Clinometers should have the capacity to sight-in the inclination of a discontinuity that cannot be reached for a contact measurement, say, for example, in a tunnel roof. Discontinuity orientation may also be measured from terrestrial stereo-photographs or laser scans (Slob et al. 2002).

There are various ways of expressing orientation. For engineering geological purposes, the best way is to record fully dip and strike data, leaving no possibility for misinterpretation of the record. Check the method you propose to use with another person – do they understand it!

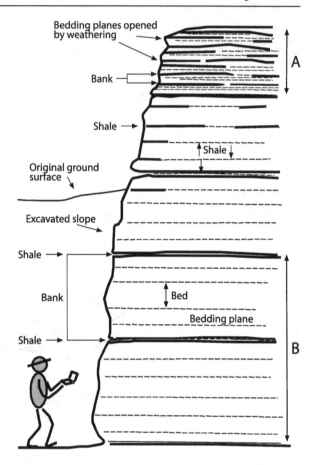

Fig. 3.12. Banks and beds

3.1.5
Spacing

It is quite clear that in describing a ground mass mention must be made of the spacing of discontinuities, e.g. the perpendicular distance between two discontinuities from the same set. This spacing is relevant to problems of slope stability, tunnel stability, excavation, groundwater flow and foundation bearing capacity. The terminology, which is standard for the description of discontinuity spacing, is given in Table 3.4. However, the use of this terminology does pose some problems when applied to bedded rocks. Assume the rock slope illustrated in Fig. 3.12 is in sandstone. The upper part of the slope is old, the lower part newly excavated, but because some, mostly higher, bedding planes have been opened by weathering and some assemblies of beds are separated by thin shaley units, the spacing of the bedding planes does not reflect the thickness of the geotechnical layers which would be of significance

Table 3.4. Terminology for bedding and discontinuity spacing (after BS EN ISO 14689-1)

Integral discontinuities	Spacing	Mechanical discontinuities
Very thickly bedded	> 6 m	Extremely widely spaced
Very thickly bedded	2 – 6 m	Very widely spaced
Thickly bedded	0.6 – 2 m	Widely spaced
Medium bedded	0.2 – 0.6 m	Medium spaced
Thinly bedded	60 mm – 0.2 m	Close spaced
Very thinly bedded	20 –60 mm	Very close spaced
Thickly laminated	6 –20 mm	Extremely close spaced
Thinly laminated	< 6 mm	Extremely close spaced

Note: For metamorphic rocks 'bedded' becomes 'foliated', 'thickly laminated' becomes 'closely foliated' and 'thinly laminated' becomes 'very closely foliated'. According to this standard bedding (integral discontinuities) is described using the numbers 2 and 63; e.g. Thickly laminated, 6.3–20 mm, and so on. A practical compromise is shown here.

Table 3.5. Rock block dimensions and description (after BS EN ISO 14689-1)

First term	Maximum dimension
Very large	> 2 m
Large	0.6 – 2 m
Medium	0.2 – 0.6 m
Small	60 – 200 mm
Very small	< 60 mm

Second term	Shape of block
Blocky or cubic	Equi-dimensional
Tabular	Thickness much less than length or width
Columnar	Height much greater than cross section

in, say, slope stability. To overcome this problem a further descriptive terms is needed. The author suggests the use of the term 'bank' to indicate a layer of geotechnical significance. Thus the sandstone in zone A in the slope would be described as 'thinly banked thinly bedded' while that in zone B would be 'thickly banked medium bedded'.

The shape and size of rock blocks depends upon the spacing of the discontinuities bounding them. The terminology proposed by British Standards (1999) is given in Table 3.5 and is designed to indicate the size of rock blocks that may come out of a quarry or be encountered in a tunnel.

3.2
Weathering

Most civil engineering works occur close to the surface and the process of 'weathering' has affected most groundmasses at shallow depth. Because of this, weathering of both engineering soils and rocks is one of the most important problems with which the engineering geologist has to contend. Weathering implies decay and change in state from an original condition to a new condition as a result of external processes. A review of weathering processes has been given by various authors (Anon 1995; Price 1995).

A most obvious sign of weathering is brown staining by the oxidation of iron bearing minerals. While solution is an agency involved in most forms of chemical weathering, it is most prominent in limestones and rocks containing halite and gypsum. In limestones, such as chalk and calcarenites, near vertical solution pipes may form extending from the limestone/overburden surface to depths of perhaps as much as 40 m. These are usually wholly but sometimes partially infilled by overburden deposits which have flowed or been washed in from above. Crystalline limestones are often strong enough to support the development of underground cavern systems through which rivers may flow. The limestones closely adjacent to such systems may be entirely unaffected by the solution weathering and be as strong as the fresh material.

Weathering takes place in all environments but is most intense in hot, wet climates where weathering may be expected to extend to great depths. While weathering may reach great depths in limestones, and rocks containing halite and gypsum, it is slow to do so and the style of weathering may change if climatic conditions change. In the northern hemisphere, where large areas have been subjected to phases of glaciation with intervening warmer periods the weathered nature of the groundmass as presently seen may reflect these changes. Thus, in the most recently glaciated areas all weathered materials may have been carved away by ice and almost fresh rock exposed. Beyond the boundaries of glaciation, weathering may reflect periglacial conditions (Higginbottom and Fookes 1971). It is generally thought that, at the end of the last glaciation, sea level rose by about 100 m. This implies that in seas less than 100 m deep, sea bed materials have been exposed to sub-aerial weathering for a substantial period of time and their properties may reflect this. Indeed, if at any time in geological history, any material has been exposed above surface, it would have been subject to weathering. Thus in Western Australia, some of the laterites exposed (and perhaps covered by more recent deposits) are thought to be of Tertiary age (Geological Survey of Western Australia 1974).

3.2.1
Influence of Weathering on Rock Mass Properties

Weathering weakens rocks. Table 3.6 shows how rock material and rock mass properties can change with weathering. The table in which the weathering is given by grades (I = fresh to VI = residual soil) shows the considerable difference in material properties that are a consequence of weathering. An ordinary geological map showing the granodiorite or the dolerite in the table would not give any indication of the

Table 3.6. Examples of variations in engineering properties of dolerite and granodiorite as a consequence of material and mass weathering. Columns 2 to 9 inclusive refer to material properties and weathering; columns 10 and 11 refer to the mass. Note: not all rocks and rock masses may weather this way

Grade	Density γ (kN m^{-3})	Porosity n (%)	Unconfined compressive strength UCS (MPa)	Unconfined tensile strength UTS (MPa)	Static deformation modulus Es (GPa)	Seismic velocity		Schmidt hammer number H^c	Rock mass friction (deg)	Rock mass cohesion (KPa)
						Longitudinal wave, Vp (m s^{-1})	Shear wave Vs (m s^{-1})			
Dolerite[a]										
I–II	28.04	0.4	160 – 180	42–48	16.5	4000–5000		64 (60–75)		
III	27.64	0.5	83 – 160	15–42	3.3	2500–4000		53 (50–60)		
IV	26.96	1	58 – 83	11–15		1800–2500		45 (35–50)		
V	26.18	3.2	24 – 58	2–11		1400–1800		25 (20–35)		
Granodiorite[b]										
I	25.6–26.96	2.6–0.4	111 – 165		31 – 34	3749–4968	2520–2883		47	17
II	25.7–26	3.3–5.9	60 – 97		14.5 – 15.3	1737–2377	1545–1840		46	16
III	25.1–25.7	1.5–2.3	33 – 48		9.4 – 11.5	1545–1840	1082–1139		38	14
IV	22.9–25.1	5.2–6.1	8 – 24		3.9 – 5.9	499–1447			17	8
V	19.8	24	0.1		0.002 – 0.013				6	3
VI	14.7	44								

[a] Dolerite data from author's own files; dolerite once exposed at Stirling Castle, Scotland.

[b] Granodiorite data from Krank K.D. and Watters R.J. (1983), except rock mass friction and cohesion. Granodiorite rock mass friction and cohesion from slope back analysis in Granodiorite in the Falset area, Spain, from Hack H.R.G.K. (1998). Grade scales follow the classification given in Table 3.7.

[c] Guide to values that might be expected in granite using an N hammer.

weathered condition of the rock, yet clearly this has great influence on the likely engineering behaviour of that rock material and mass. Because of this, it is customary to describe the weathered condition of the rock in all engineering geological descriptions of material or mass.

3.2.2
Susceptibility to Weathering

The lifetimes of most structures for civil engineering and infrastructure are in the order of 50 to 100 years. To guarantee the safe and sound design for the whole lifetime it is important to know what the geotechnical properties of the soil or rock mass are going to be at the end of this time span, i.e. the susceptibility to weathering of the soil or rock mass. If, for example, a slope is excavated in a sandstone in which the cement between the grains consists of gypsum it can be expected that in a moderate climate the gypsum will dissolve and the sandstone rock mass changes into a soil mass of loose sand grains within a few years. A slope made in fresh granite is not expected to undergo any major changes within 100 years in a moderate climate. The influence of weathering on intact rock used as building or grave-stones has been studied and also rates for weathering have been established. For most soils and rocks it is also reasonably well known how they deteriorate over long (geological) periods. However, for 50 to 100 year time spans very little is known. Research has been done to the weathering rates in underground excavations and its influence on rock mass classification ratings (Laubscher 1990) (see Chap. 4). For the extremes given in the slope example above, it is generally not difficult to make an estimate of the changes of properties due to weathering. For many other soil or rock masses, for which it is not so clear, the engineer has to estimate the susceptibility to weathering based on other exposures of known excavation date. If these do not exist, engineering judgement has to be used.

3.2.3
Standard Weathering Description Systems

Weathering can be described following a relatively simple scheme, such as in Table 3.7 which is compiled from recommendations given in BS 5930 (1981) and the report of the Engineering Group Working Party on Core Logging (Anon 1970), or from more elaborate schemes of which BS5930 (1999) is an example (Fig. 3.13, Table 3.8). While the simple scheme is easy to use, it often does not fit a particular state of weathering for a particular material in a particular environment. The more elaborate schemes have more flexibility to fit all sorts of materials and environments, but have the potential of being over complex and not understood.

The factors entering into the description of weathering are the condition of the discontinuities (joints, bedding planes, foliations etc.) and the condition of the material between the discontinuities. Weathering begins on the discontinuities that transmit water. Increasing weathering affects more and more discontinuities and eventually starts to affect the rock material. Usually rock mass weathering is assessed from a number of boreholes, and natural or artificial outcrops. Boundaries between weathering grades can never be more than approximate.

Table 3.7. Standard terminology for description of weathering of rock cores, outcrops and material

Weathering description	Grade No.	Rock core grades[a]	Rock outcrop grades[b]	Rock material descriptive terms[b]
Fresh	I (A)	No visible sign of weathering.	No visible sign of rock material weathering, perhaps slight discolouration on major discontinuity surfaces.	Rock material weathering can be described by using terms such as:
Faintly weathered[c]	I (B)	Weathering limited to the surface of major discontinuities.		
Slightly weathered	II	Weathering penetrates through most discontinuities, but only slight weathering of rock material.	Discolouration indicates weathering of rock material and discontinuity surfaces. All the rock material may be discoloured by weathering.	Discoloured: The colour of the original fresh rock material is changed and is evidence of weathering.
Moderately weathered	III	Weathering extends throughout the rock mass but the rock material is not friable.	Less than half the rock material decomposed or disintegrated to a soil. Fresh or discoloured rock is present as a continuous framework or as corestones.	Decomposed: The rock is weathered to the condition of a soil in which the original material fabric is still intact, but some or all of the mineral grains are decomposed.
Highly weathered	IV	Weathering through discontinuities and the rock material is partly friable.	More than half the rock material decomposed or disintegrated to a soil. Fresh or discoloured rock is present as a discontinuous framework or as corestones	Disintegrated: The rock is weathered to the condition of a soil in which the original material fabric is still intact. The rock is friable but the mineral grains are not decomposed.
Completely weathered	V	Rock is wholly decomposed and in a friable condition but the rock texture and structure are preserved.	All rock material is decomposed and/or disintegrated to soil. The original rock mass structure is still largely intact.	The stages above may be qualified by using terms such as 'partially', 'slightly', 'wholly'.
Residual soil	VI	A soil material with the original texture, structure, and mineralogy of the rock completely destroyed.	All rock material is converted to soil. The mass structure and material fabric are destroyed. There is a large change in volume but the soil has not been significantly transported.	

[a] After Geol. Soc. Enging. Group Working Party on 'The logging of cores for engineering purposes' (Anon. 1970). [b] After BS 5930 (1981). [c] Faintly weathered is seldom found in descriptions and may be considered more-or-less obsolete. Grade I only becomes I (A) if faintly weathered is used.

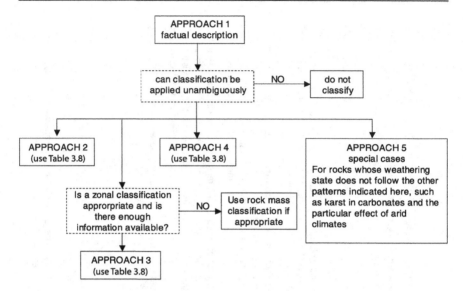

Fig. 3.13. Scheme for description of weathering (after BS5930 1999). See also Table 3.8. Note: BS EN ISO 14689-1 does not, at present, recommend Approaches 2 and 3

3.2.4
Weathering Description and Zonation

Weathering is a gradational feature and to deal with such features the usual approach is to impose boundary conditions within them so that they are divided into various grades defined by a range of characteristics. In reviewing the applicability of the grades of weathering proposed in the various systems mentioned above it is well to remember the purpose of describing weathering i.e. is to assess its significance with regard to engineering projects to be conducted in or upon the weathered rock mass. This significance depends upon two factors, first the change in engineering properties of the rock material and second, the volumetric regularity or irregularity of this reduction within the rock mass. Regarding the second, the writer distinguishes four basic types of mass weathering. These are:

1. *Uniform weathering.* A gradual decrease of weathering grade and intensity with depth, in thick strata of homogeneous lithology.
2. *Complex weathering.* An irregular weathering profile in layered lithologies that have different susceptibilities to weathering. It may mean that more weathered strata lie under less weathered strata, particularly if the strata dip from surface outcrop.
3. *Corestone weathering.* In many, mostly coarser grained, igneous rocks rounded 'corestones' of almost fresh rock may be surrounded by very decomposed highly friable material, similar to a compact sand. The corestones become larger with depth.
4. *Solution weathering.* Carbonate and most salt rocks weather by dissolution. Joints and bedding planes become open and underground caverns may develop. In strong crystalline limestones, karstic conditions may result. In the weaker calcarenites,

Table 3.8. Description state of weathering (after BS5930 1999). This scheme no longer accords with BS EN ISO 14689-1 which is considered by some to be inferior to the old BS 5930 1999

Grade	Classifier	Description
Approach 2: Rock is moderately strong or stronger in fresh state (not in BS EN ISO 14689-1) *Uniform materials*		
I	Fresh	Unchanged from original state
II	Slightly weathered	Slight discolouration; slight weakening
III	Moderately weathered	Considerable weakened, penetrative discolouration; large pieces cannot be broken by hand
IV	Highly weathered	Large pieces can be broken by hand; does not readily disintegrate (slake) when dry sample immersed in water
V	Completely weathered	Considerably weakened; slakes in water; original texture apparent
VI	Residual soil	Soil derived by in-situ weathering but having lost retaining original texture and fabric

Zone	Description (2)	Typical characteristics
Approach 3: Heterogeneous masses (mixture of relatively strong and weak material) (not in BS EN ISO 14689-1) *Heterogeneous masses*		
1	100% grades I–III	Behaves as rock; apply rock mechanics principles to mass assessment and design
2	>90% grades III <10% grades IV–VI	Weak materials along discontinuities; shear strength stiffness and permeability affected
3	50 to 90% grades I–III 10 to 50% grades IV–VI	Rock framework still locked and controls strength and stiffness; matrix controls permeability
4	30 to 50% grades I–III 50 to 70% grades IV–VI	Rock framework contributes to strength; matrix or weathering product control stiffness and permeability
5	<30% grades I–III 70–100% grades IV–VI	Weak grades will control behaviour. Corestones may be significant for investigation and construction
6	100% grades IV–VI	May behave as soil although relict fabric may still be significant

Class	Classifier	Description
Approach 4: Moderately weak or weaker in fresh state (permitted by BS EN ISO 14689-1) *Classification incorporates material and mass features*		
A	Unweathered	Original strength, colour, fracture spacing
B	Partially weathered	Slightly reduced strength, slightly closer fracture spacing, weathering penetrating in from fractures, brown oxidation
C	Distinctly weathered	Further weakened, much closer fracture spacing, grey reduction
D	De-structured	Greatly weakened, mottled, lithorelicts in matrix becoming weakened and disordered, bedding disturbed
E	Residual or reworked	Matrix with occasional altered random or apparent lithorelicts, bedding destroyed. Classed as reworked when foreign inclusions are present as a result of transportation

calcisiltites and calcilutites (chalk), solution pipes, often infilled with materials from above, may have penetrated deeply into the strata.

While a complete volume of rock mass might be described using the terms given above within that mass the weathered nature must be described in greater detail in order to more closely target an engineering problem. This is perhaps best done with regard to the engineering significance of the mass weathering observed. The following terms are proposed:

i   *Effectively unweathered.* The weathering of the rock mass is such that any engineering work may be constructed on or in it without regard to the weathered condition as found.

ii *Significantly weathered.* The weathering of the rock mass is such that some regard must be taken of the weathered condition of the rock mass in the design and construction of some particular engineering works. For example, weathering of discontinuities could imply an impairment of shear strength if the work involves the construction of rock slopes, but such weathering would have less effect on the design of foundations on the rock mass.

iii *Severely weathered.* The weathering of the rock mass is such that the weathered condition of the mass dominates the design and construction of any engineering work to be constructed on or in it. This implies both weathering of discontinuities and materials.

iv *Residual soil.* Sufficient of the rock material is decayed to the geotechnical condition of a soil to make the mass behave as a mass of soil. It may be important, because of their influence on, for example, slope stability, to distinguish between those residual soils that are structureless and those that have relict discontinuities.

The boundary between (*iii*) and (*iv*) is difficult to establish. A rock mass in condition (*iv*) may contain relict rock blocks which, if in continuous contact, may present the engineering behaviour of a rather weak rock mass. If the relict blocks are not in contact, the total mass may behave as an engineering soil. If the volume of soil material exceeds about 50% of the mass then the behaviour of the total mass should be that of a soil and be categorised as residual soil (*iv*); if the volume of soil is less than about 30% then behaviour should be that of a severely weathered rock mass (*iii*). It will be difficult to estimate such percentages in this transition zone between soil and rock mechanics and estimates are most likely to be subjective.

The terms given above could be used to describe zones of weathering within a rock mass. Such zones might be observed in outcrop and boundaries evaluated. The author has suggested a ratings system to aid establishing such boundaries (Price 1993). It also provides a numerical estimate of weathering that may form part of a rock mass classification system and has been so used by Laughton and Nelson (1996) in considering the rock mass parameters that are necessary to predict tunnel boring machine performance. The data required to zone a rock mass into categories (*i*) to (*iv*) may come from natural or artificial surface outcrops but is derived most commonly from boreholes. Thus, in core descriptions note must be made of the weathered condition of discontinuities and the extent of decay of rock materials.

In rock mechanics the condition of discontinuities is of major importance and it would seem logical to describe discontinuity weathering to provide a link to assessments of discontinuity strength. In both integral and mechanical discontinuities, three main conditions may be considered to exist. These are:

- *fresh:* no sign of weathering
- *surface stained:* no or little penetration of weathering into the wall rock (so that the discontinuity asperities have the strength of fresh rock)
- *surface weathered:* weathering penetrates to a depth greater than the roughness or undulations of the discontinuity surface (so that the discontinuity asperities have a strength less than that of fresh rock)

In the case of rock materials similar simple terms may be applied, namely:

- *fresh:* no sign of weathering
- *stained:* the material is weathered but without obvious loss in strength
- *decayed:* deeply stained and with obvious loss of strength, perhaps friable

In the case of solution weathering an additional term, "*absent by solution*", might be applied.

Achieving a weathering zonation of a rock mass investigated largely by boreholes is greatly aided if the style of weathering in outcrops of similar rock can be studied, even if these are not located close to the site. This may help the investigator appreciate the relationship of the linear data given by boreholes to the volumetric reality of the rock mass.

3.3
Ground Mass Description

A volume of ground may have some exposure that can be studied but in most cases it has to be investigated and described with the aid of core recovered from boreholes drilled for that purpose. The borehole core provides a description of the vertical profile but at some stage a general description of the volume is needed.

3.3.1
General Mass Description

Most ground masses on or in which engineering works are to be performed can be considered composed of a series of *geotechnical units* each distinguished by particular mass properties. The range of properties defining a unit is assigned by the investigator, commonly with regard to the particular type of engineering work to be performed. Thus a ground mass which is to be excavated to form a large road cutting might be divided into units relative to ease of excavation, to reuse of the excavated materials as embankment fill, or with regard to slope stability. The units related to one engineering process are not necessarily the same as those for another process for their boundaries would be defined by different parameters.

Whatever the units, the description of the mass must include descriptions of the units, their boundaries, their discontinuities, their water content and the level of groundwater within them. A common way of obtaining this information is to drill holes into the ground and describe the samples (cores) recovered from them. The process of describing such cores is known as core logging.

3.3.2
Core Logging

Poor or inappropriate core logging can lead to major civil engineering claims. Accordingly, accurate core logging is one of the most important activities for the engineering geologist. Cores taken as part of a site investigation for an engineering project should be logged by an engineering geologist rather than a geologist for the style, content and emphasis of the descriptions will depend on the recognition of the importance of the geological factors revealed in the cores to the particular engineering project. If the logging of a specific geology feature, such as structural geology, fossil content, etc., is required, a geologist specialised in the required feature should rather do this additionally.

Basic Requirements for Good Core Logging

The contractor and consultant must ensure that the cores are placed in good, well-labelled core boxes and are stored so that their condition as recovered from the borehole is maintained. They should not be allowed to dry out, be eroded by rain, frozen, etc. A well-lit large table in a clean, dry, covered store should be provided as a logging area. Help should be available to move core boxes.

The Process of Core Logging

Ideally, all the cores should be logged in one exercise and the logger must be fully aware of and understand the nature of the engineering work so that the log gives the information necessary for the design of that work. Cores may be logged as they come out of the borehole for their description may influence the further development of the investigation but this provisional log should not be the final log.

For the final log the logger must look at all the soils and rocks and decide provisionally what they should be called and in what detail they should be described. A published system of description should be used and its selection justified. The way in which the cores are described will depend upon the geology, the engineering work, and the nature of the problem that the geology poses for that work. Photos of the core boxes against a scale should be made for later reference.

The cores should be sorted out so that they fit together as well as possible to allow *Total Core Recovery* (*TCR*) and, if required, *Solid Core Recovery* (*SCR*) and *Rock Quality Designation* (*RQD*) to be measured. *TCR* is the length of core recovered divided by the length of hole drilled (the core run) expressed as a percentage (Fig. 3.14). *SCR* is the total length of solid core cylinders as a percentage of the core run. Care has to be taken in defining what is meant by '*solid*' (Norbury et al. 1986). *RQD* is the total length of core sticks longer than 4 inches (now commonly converted to 100 mm) expressed

Fig. 3.14. Measurements that may be made on rock cores

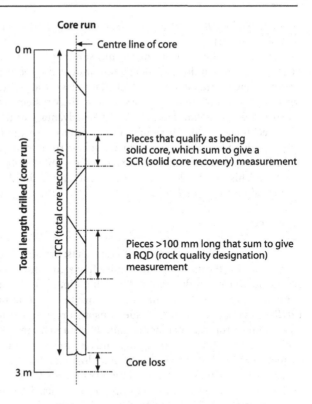

Core run

0 m

← Centre line of core

Total length drilled (core run)

TCR (total core recovery)

Pieces that qualify as being solid core, which sum to give a SCR (solid core recovery) measurement

Pieces >100 mm long that sum to give a RQD (rock quality designation) measurement

Core loss

3 m

as a percentage of the core run. According to Deere (1968) " *'The Rock Quality Desig- nation' (RQD) is based on a modified core recovery procedure which, in turn, is based indirectly on the number of fractures and the amount of softening or alteration in the rock mass as observed in the rock cores in a borehole.*" Fractures induced by handling or the drilling process should not be counted (the pieces broken by such fractures should be fitted together and their total length measured) and the pieces counted should be 'hard and sound'. If all the sticks of core measured are less than 100 mm long, *RQD* is zero and if there are no breaks in the core *RQD* is 100%. This raises the possibility that if all sticks were 99 mm long, *RQD* would be 0% and if all were 101 mm long *RQD* would be 100%. Clearly measuring mechanical discontinuity spacing against a fixed baseline has some disadvantages. Some workers have measured *RQD* against several baseline lengths to give a guide to rock block size; Edmond and Graham (1977) did this to assess block size in a rock mass through which a tunnel was to be driven.

Some rock materials are composed of thinly interlaminated sandy and clayey bands. They may come out of the core barrel as solid sticks of rock but on drying out sepa- rate into thin discs, splitting along the clayey layers. *RQD* measured on the dried out cores would be very much lower than on those just recovered from the borehole. The logger is then faced with the difficult choice between including or ignoring the dry- ing-out breaks. Inclusion gives a quality assessment of the rock material; exclusion gives an *RQD* related only to genetically, tectonically, or geomorphologically induced discontinuities. Such discing may also be a consequence of stress relief. Whether or not included in RQD measurements such discing would merit description and discussion in the report.

The uncertainties concerning the measurement and meaning of RQD suggest that RQD should be abandoned and the author does not recommend its use. It has, however, been incorporated in some Rock Mass Classification systems, and because of this will probably continued to be measured until these systems are modified or abandoned. The best reason to abandon RQD is to take advantage of the opportunities that computer technology offers. It should be appreciated that the RQD calculated from core applies only to the direction in which the hole from which the core is recovered, was drilled. An alternative to RQD is to record the number of fractures per given length of borehole. This may be done for metres depth, for each lithology or each engineering geological unit distinguished.

Recording

The logging process falls mostly into two parts; first describing the materials and second, describing the discontinuities. It is often a good idea to do the discontinuity measurements (stick length, joint and bedding dip, etc.) first. It is better to measure the depth and dip of every mechanical discontinuity so that thereafter any calculations (RQD, frequency etc.) can be made. If using a rock mass classification system somewhere in the project remember to gain the data to fill in the ratings. This may involve noting joint roughness, infilling etc., and also means that the rock mass classification system must be chosen before beginning logging.

The process of making the discontinuity measurements gives another chance to look at the rocks and consider the descriptions to be used. Geologically this is quite easy, but engineering geologically rather difficult for the observer seeks to identify geological features of significance to that particular engineering work. While it is quite easy to become absorbed in the engineering geological and rock mechanical side of logging, it must not be forgotten that a proper description of the geology must also be made. This is necessary for correlation between boreholes and with surface data. In writing the descriptions, the word order should be consistent, following a recognised system such as that offered in BS5930 (1999). Any departures from that system should also be consistent and the reason for them explained in the report.

Features in the cores resulting from drilling, have to be identified. Thus, ends of core runs tend to be more broken while core springs may produce scratch marks simulating bedding. These should be noted on the record. Because the core spring is situated above the cutting face of the drilling bit extraction of the core may leave a core stub at the bottom of the borehole. If this is recovered in the next run then the core recovery for that run could exceed 100%. If a stub is left at the bottom of the hole and the next core run completely fills the core barrel then cores may be crushed.

Logging the strength of the rock cores by some form of simple test may be necessary. This may be done using such devices as the Point Load Tester, but these break the core and may thus reduce the amount of core available for accurate laboratory testing. The author does not recommend this test. The ISRM 'Suggested Methods' (Brown 1981) includes a method of using the Schmidt hammer on cores in the laboratory, but the author prefers cross diameter acoustic logging or the use of the Equotip (a kind of mini-Schmidt hammer).

If samples are extracted then something should be put in their place (ideally a block of wood cut to length and labelled) and noted. Core boxes should have a logging record pinned to the inside of the lid on which the logger notes who and when the cores were logged and also records any samples taken.

In the initial stages of logging, errors of the driller in placing the cores in the box may be encountered. It is useful to be familiar with top and bottom criteria to identify cores that may have been put in upside down. Before logging rock it is useful to wash the cores and look at them wet, for wet cores show more than dry ones. In soils a little drying often helps reveal delicate sedimentary structures such as lamination. Cores may have to be broken to determine the rock type. If this is done note the locations of breaks made; an indelible felt pen is useful for this. If possible, at the end of the logging of a number of boreholes, all logs should be examined again to check consistency of description.

Discontinuity Orientation

The dip of discontinuities may readily be measured on cores from a vertical borehole, however, the direction of dip can only be measured with special, and usually expensive, equipment. If the direction of dip of the dominant integral discontinuity (bedding or foliation) is known, it is possible to measure the direction of dip of other discontinuities relative to it.

Daily Logging Rate

A good core logger, giving a full engineering geological description for a complex project, can usually manage to log about 30 m of core a day before mental indigestion sets in. It is important to be comfortable to do the work; logging in the mud in a mosquito-infested swamp usually produces poor logs. It is much better to do the work in a well-lit, air-conditioned hut, sitting in a comfortable chair with a pot of coffee at hand.

3.3.3
A Theoretical Example

Figure 3.15 is a drawing of three core boxes to illustrate some of the points made above. The Equotip rebound values have to be converted to unconfined compressive strength with the help of Fig. 3.7. The drawing Fig. 3.15 is two dimensional and limited in what it shows by the constraints of the graphics programme used for its construction. Details are as follows:

1. Three core boxes are illustrated containing cores obtained by rotary core drilling from 15.00 to 38.06 m depth. The upper two boxes contain 160 mm diameter core, the lowest 120 mm diameter core. The driller has marked the beginning and end of each core run, which are measured from the drill rods, and has also estimated the depth

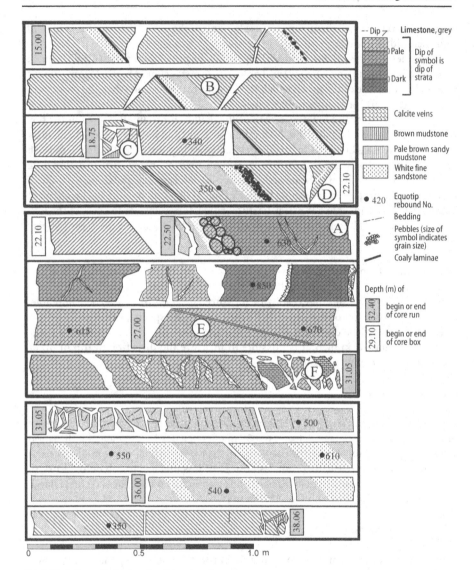

Fig. 3.15. A theoretical example of a series of core boxes

of the end or beginning of the core in each core box. If there has been any core loss then these latter depths may be inaccurate.

2. The cores have been placed in the core boxes to more-or-less fill them without breaking the cores more than is necessary. However, there will be some 'fitting' breaks and one of the first tasks in logging is to identify them. One such break is to be found at point 'A'; there are others. They would not be counted as natural discontinuities for *RQD* measurements etc.

3. The cores in the first box show a 'fining upwards' succession of rocks (conglomerate ⇒ sandstone ⇒ sandy mudstone ⇒ mudstone ⇒ coal). The stick at 'B'
   shows this succession inverted and has thus been placed upside down in the
   box.
4. The rocks at 'C', just below the 18.75 m core run marker are rather broken and may
   represent an over-drilled core stub, left in the hole at the end of the previous run
   and disturbed at the beginning of the next run.
5. The fragment of rock at 'D' would not appear to be in the correct place and may come
   from the bottom of the run at 22.50 m.
6. The joint at 'E' may be open. If, when its two surfaces are placed in contact, the thickness of the core is less than the diameter drilled then the joint is open and the aperture can be calculated.
7. At 'F', the core is very broken and at this point, the driller decided to reduce core
   diameter, presumably as a result of encountering drilling difficulties. Such difficulties would be recorded in the daily drilling log. The implication is thus that the breaks
   in the core above and below 31.05 m are a consequence of natural phenomena and
   not the drilling itself. The same remark could apply to the broken core just above
   38.06 m but here it might also be the core cracking in the vicinity of the core spring
   on breaking off the core to pull it to surface.

Describing Lithologies

There are clear differences between the strata above about 22.70 m, from 22.70 to about
31.00 m and below this last depth. In the case of the strata above 22.70 m there would
appear to be cyclic deposition in fining up sequences. There are various ways of describing the sequence. If the project concerns putting foundations on the rock mass it
would probably be sufficient to record the number of cycles of deposition and to indicate their nature. If the rocks are likely to be exposed in a slope or tunnel wall it would
be prudent to emphasise the presence of the coaly laminae which could squeeze or
which could form planes on which sliding might take place. Alternatively, a more factual approach could be taken and the rocks closely described layer by layer with measurements of depth and thickness.

At about 22.75 m a conglomerate is found composed of limestone pebbles and may
well mark an unconformity. A check on the relative ages of the limestone and overlying cyclic deposits would be of value for the unconformity surface could be that of a
very irregularly eroded landscape. The limestones show fractures, perhaps solution
opened, and healed by calcite veining. The end of the core shows solution features, there
is significant core loss, and the driller may have recorded encountering cavities as the
hole was drilled.

The limestones are not all the same. They are generally pale grey but two pieces,
perhaps once joined by a calcite vein, are dark grey and show different bedding dips.
This could be a consequence of faulting. Below about 30.90 m the mudstones and sandstones are again encountered although lacking indications of cyclic deposition. However, while their bedding below about 32 m is regular and thin between 31 and 32 m it
becomes variable with indications of folding. This can be syn-sedimentary, but is probably faulting.

Structure and Strength

Measurements of bedding dip would be taken as frequently as the dip varied. The relative orientation of joints could also be measured. Strength may be assessed by the Equotip rebound number or some other method but indicated in the log by the terms used in the scale of strength for rocks, e.g. moderately strong. The description of the cores shown in Fig. 3.15 is given as in Chap. 7.

3.4
Further Reading

Eddleston M, Walthall S, Cripps JC, Culshaw MG (1995) Engineering geology of construction. Geological Society of London (Engineering Geology Special Publication 10)
Geological Society of America (1957–1978) Engineering geology case histories 1–11. Geological Society of America
Harris C (ed) (1996) Engineering geology of the Channel tunnel. Thomas Telford, London

Chapter 4

Maps

4.1
Maps

An engineering work either influences, or is influenced by, the groundmass in its vicinity. The distribution of materials and discontinuities in the mass may be shown on maps. Most investigations include the study and/or construction of a geological map. Published maps are studied to gain an idea of site conditions before designing an investigation. Maps produced as a consequence of investigation display appropriate geological information as an aid to design and construction of the works.

4.1.1
Published Geological Maps

The majority of the earth's surface has been mapped geologically. Often the scale of the geological maps is rather small, perhaps only 1:1 000 000 but in a number of countries maps are published on scales of from 1:100 000 to 1:50 000 and in some cases at scales as large as 1:10 000. Geological maps are constructed to show:

- the *stratigraphic names*, and thus the relative age, of the geological formations in the area mapped,
- the *distribution of these formations* within the area mapped,
- the *geological structure* of the area mapped.

They may also show:

- the *lithology of the formations*,
- *geomorphological features* in the area, such as major landslides,
- features resulting from the *activities of man*, such as mines and quarries.

Because geological maps show the stratigraphy and relative age of formations, it is possible to deduce whether strata seen on the map are older or younger than their neighbouring formations. If this is known, then, by looking at the pattern of outcrops in relation to the physiography of the landscape, it is possible to gain insight into the general structure of the geology of the area. This may enable the geologist to draw cross-sections through areas to show the underground geology along that line of section. This has the immediate practical engineering geological purpose of showing, for example, the likely geology to be encountered along the line of a proposed tunnel or trench excavation.

Often these maps seem to be of little value to the engineer and may seem to some to be quite misleading. Many formation names incorporate lithological terms, such as limestone or sandstone, which were given at a time when it seemed to the geologist discovering the formation that the dominant rock was limestone or sandstone. Since the original discovery the formation name, incorporating the lithological term, may have been applied to strata of similar age but with a different lithology. Geologists recognise these problems, for this is part of their professional expertise, and they understand that to read a geological map correctly one must also study the geological memoir or guide that accompanies the map. Much of this extra information is not readily available and may be difficult to obtain quickly. It must also be understood that the majority of geological maps were made with the scientific aim of explaining the geology rather than for use in engineering practice. Accordingly, the majority of the older maps do not incorporate much, if any, detail of superficial geomorphological features, such as landslides.

4.1.2
Published Engineering Geological Maps

In recent years, there has been considerable activity in the production of engineering geological maps. These maps are specifically designed for engineering purposes and give specific attention to such features as:

- the lithology of the strata,
- the thickness of the layers and their depth below surface,
- the depth to the water table,
- the location of significant geomorphological features, such as the presence of karst features in limestone, landslides, active faults etc.,
- hazards remaining as the result of past industry, such as abandoned mines, old quarries, areas of toxic fill etc.,
- location of construction materials.

There are so many features of interest for engineers that could go on a map that it is almost impossible to put everything on one map. Also, depending on the nature of the feature, it is best displayed at a particular scale. An engineer concerned with the location of old mine workings will find that a map showing these on a scale of 1:50 000 would be of little practical value although this would be a useful scale for someone wishing to view the general hydrogeology of a region.

4.2
Geological Map Making

Before dealing with engineering geological maps it is of importance to consider the value and problems of construction of ordinary geological maps, for often standard geological maps form the basis of engineering geological maps. Traditional geological mapping was (and is) undertaken by one geologist who, with the aid of an existing topographical map (and often aerial photographs and remotely sensed imagery), traverses the countryside to search for natural or artificial outcrops of the strata. Quite frequently mapping begins adjacent to rivers and streams and along coastlines where outcrops exposed by erosion may be expected to be found. These outcrops are measured and described, the materials found placed in their stratigraphic position and the location of the observation is plotted on the map. It must not be forgotten that this location is plotted by a geologist, aided, for older maps, only by a magnetic compass, on a map, which may show few features that can be recognised and can help establish the exact location. It is not surprising therefore that the mapped locations of outcrops may be found to be somewhat erroneous if, later, their position is surveyed accurately. Newer maps may not suffer this problem as position is determined with modern methods as GPS (Global Positioning System), high resolution aerial photographs, etc. (see below).

4.2.1
Linking the Data

Having plotted outcrops, the geologist then has the task of linking the outcrops together to show boundaries between formations by studying the form of the landscape in which breaks in the slope may indicate the presence, under the cover of soil and vegetation, of harder or softer layers. Quaternary deposits may be contained within landforms, such as terraces, drumlins, or barchans that can be recognised by their shape. Thus, the geomorphology of the landscape is a tool with which the boundaries of geological formations may be detected. Geologists experienced in particular areas may also be able to look at the type of soil present and the nature of the vegetation growing to get some idea of the underlying geology. Soil type and vegetation type boundaries may also be geological boundaries. The actual plotting of these boundaries on the map is subject to the same inaccuracies as the plotting of the location of outcrops.

Today both the plotting of outcrops and boundaries is greatly aided by the use of aerial photographs and Global Positioning systems (GPS). Aerial photographs show features, such as trees, bushes, irregularities in slope etc. that are not presented on ordinary topographical maps. Geological mapping with the aid of aerial photographs is, in general, much more accurate than mapping without, depending on the scale and age of the photographs. Thus plotting boundaries on a 1:1 250 topographical map may not be significantly aided by 1:40 000 scale aerial photographs. Moreover, it must not be forgotten that aerial photographs help to evaluate accurately and locate the boundaries of features; it does not follow automatically that these boundaries reflect underground geology. However, it is fair to say that the modern geological map is probably much more accurate in terms of location of boundaries than the older map. It is thus of some importance when reviewing geological maps, to look at the date of publication and the date of mapping. These two dates may be very different, because some years may pass between finishing mapping and publication of the map. Maps first prepared many, many years ago may be several times revised. Geological surveys mostly have geologists who are responsible for the geology of particular areas who will examine any new excavations that may be made, samples taken from boreholes etc. and when sufficient information is available will revise the map.

It is obvious that geological maps are only as good as the quality and quantity of information used in their compilation. Often, for example in urban areas, there might be a large amount of geological data available gathered from boreholes, wells, road excavations, sewer excavations and so forth. In rural areas, devoid of artificial exposures and perhaps with few natural outcrops, the quantity of data available may be small. Many of the larger scale maps show locations of borings while observations on the map (such as dips and strikes, fossil locations etc.) indicate where the outcrops were seen. It is possible by looking for such observations to judge whether or not the map was prepared with the help of much data or with very little. This may also be different on one map. Often exposures are only found along roads and rivers, but few in the rest of the area. Particular attention should be taken if a map shows many measurements along easy accessible areas such as roads and around towns despite there being plenty of exposures at other locations more difficult to access. This may be an indication of low quality mapping, possibly prepared by geologists with little time and less enthusiasm (often aptly named a '*hotel or bar-map*').

National geological surveys generally retain the original field maps from which the published maps were prepared. Commonly the field mapping was done on a larger scale (say 1:25 000 or 1:10 000) than the published map (at say, 1:250 000 or 1:50 000). There may be observations on the field maps for which there was no room on the published map. It is generally a good idea to examine the field map held by the geological survey to judge the accuracy of the published map and to see if the field map contains any observations which did not get on to the published map. Great accuracy is expected in engineering geological maps for exact knowledge of ground conditions is required for engineering purposes. This degree of accuracy is seldom attained in published geological maps. Many investigations must begin with boring and trial excavations designed to confirm and thence to improve the existing geological information. However, engineers should not be too critical of geological surveys when they find inaccuracies in geological maps; the accuracy of the map is only so good as the data available to prepare it.

4.3
Understanding Geological Maps

Geological maps must be interpreted in order to understand the geological structure of an area. All geological maps show the real or interpreted position of the various stratigraphic units below the surface. Boundaries, which have been observed, are usually shown as solid lines, inferred, or assumed boundaries as dashed lines. It should be noted that there is no common definition for '*observed*' and '*inferred*'. One geologist may mark a boundary that has not been observed, but is certain in his or her opinion, as 'observed' while another geologist may mark the same boundary as 'inferred'. Other symbols on the maps relate to geological structure, the most common being those indicating the *dip* and *strike* of the strata (Fig. 4.1).

Geological surfaces that are inclined are said to dip, the angle of dip being the maximum inclination of the surface measured, by convention, from the horizontal. The strike of a surface is the direction of a horizontal line drawn on that surface, usually expressed relative to north. A geological surface could be the top or bottom of a bed of sedimentary rocks, one side or the other of an intrusive dyke, a fault or a joint; it is two-dimensional. A geological body is bounded by geological surfaces and has thickness.

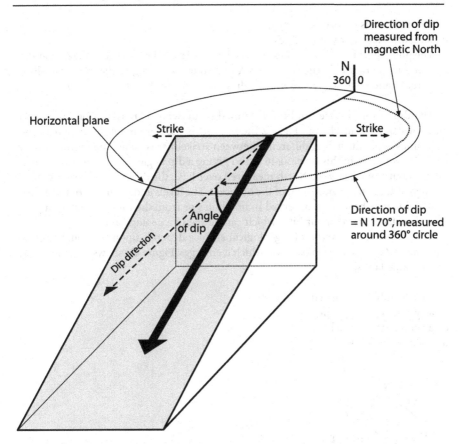

Direction of dip measured from magnetic North

N
360 | 0

Horizontal plane

Strike

Strike

Direction of dip = N 170°, measured around 360° circle

Dip direction

Angle of dip

Fig. 4.1. Dip and strike

Geological maps on larger scales are usually plotted on topographical maps which show contour lines, roads, field boundaries, buildings and other land use features and mostly have a grid co-ordinate network with a magnetic and true north symbol as part of the general information.

If a map is to be interpreted for engineering purposes scales of 1:10 000 or larger are required and the map must have contours. If the geological map shows outcrops and contour lines from which the levels of points on the outcrops can be measured then it is possible to interpret the map to indicate the levels and positions of the layers underground. To do this the interpreter must have points of known elevation on boundary surfaces between geological units. Just as points of elevation above mean sea level (the datum plane) on the landscape may be interpreted to give contour lines, points of elevation on geological surfaces may be interpreted to give *strikelines*. At least three elevations are required to draw the strikelines on any dipping plane surface. If that surface is plane and uniformly dipping, strikelines will be parallel and uniformly spaced. In Fig. 4.2 points A, B and C on a geological surface are at levels of 30 m, 10 m, and 0 m above datum respectively. To find the dip of the plane:

i   Join the points to form a triangle.
ii Measure the lengths of the lines.
iii The point at which the plane lies at elevation 20 m must lie halfway between A and B
    at X and 1/3 of the distance AC from A at Y. Join X to Y to give the 20 m strikeline.
    Zero, 10 and 30 m strikelines may be drawn parallel to the 20 m strikeline.

The dip of the surface can be determined by drawing a cross-section but is also
done by calculation. Thus in Fig. 4.2 the line CD is the horizontal distance between
strikelines while the height difference between strikelines is 30 m. The *tan(dip angle)*
is 30/CD, which is, in this case, 30/46 = 0.65 giving a dip angle of 33°. The direction of
dip is about 150° relative to North measured over East, the direction of strike 240° or
60°, depending on the viewpoint of the observer. The levels and positions of the points
A, B and C may have been observed in boreholes or from the junction between geological surfaces and contour lines which can be seen on geological maps.

The shape of an outcrop of a geological surface depends upon the complexity of
both the surface and the landscape at which it emerges. Figure 4.3 shows outcrop forms
on the same landscape for:

a   a uniformly dipping bed,
b   a symmetrical syncline,
c   a symmetrical anticline,
d   a plunging syncline.

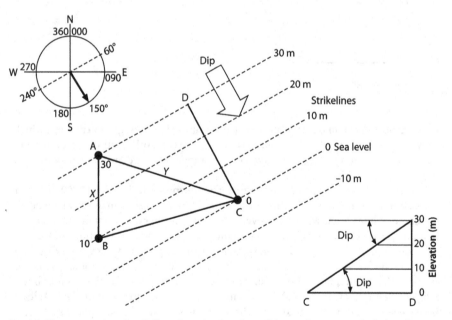

Fig. 4.2. The construction of strikelines from the elevations of three points on a geological surface

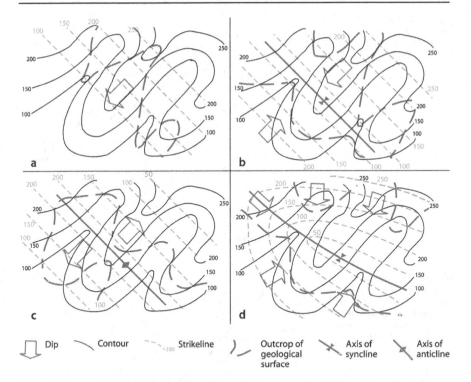

Fig. 4.3. Outcrops of geological surfaces resulting from the contact of strikelines with surface contours. The landscape is the same in each case. **a** uniformly dipping geological surface; **b** symmetrical syncline; **c** symmetrical anticline; **d** syncline plunging to the southeast

The outcrops in Fig. 4.3 are of a single geological surface but any geological layer is bounded by two surfaces. These surfaces are defined usually by changes in lithology between one layer and that underlying or overlying it. Figure 4.4 shows Fig. 4.3a with two sets of strikelines depicting the top and bottom of a dipping bed of rock. A borehole at point A (in Fig. 4.4) would encounter the top of the bed at elevations of about 180 m and the bottom at about 160 m, which, since A is at an elevation of about 225 m, are at depths of 45 m and 65 m respectively.

In the figures strikelines have been drawn between intersections of outcrops and contour lines and it has always been possible to draw at least one strikeline between two points of intersection of a particular contour and the outcrop. Unfortunately, such simple situations are seldom encountered in nature, for most outcrops are at least partially concealed under soil and vegetation. However, if dip and strike can be measured on a bedding plane within the outcrop then strikelines can be extrapolated to indicate the location of outcrops under a cover of soil or vegetation (Fig. 4.5).

In the figure the geologist has been able to map only a small part of the outcrop of a bed, the remainder being obscured by vegetation cover. Strikelines must pass through

Fig. 4.4. The outcrop of a bed is determined by the contact of top and bottom of bed strikelines with landscape contours

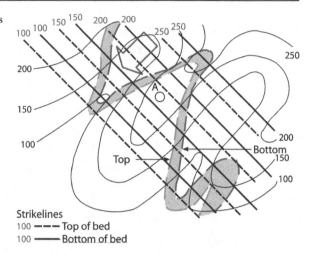

Strikelines
100 ━━━ Top of bed
100 ━━━ Bottom of bed

Fig. 4.5. Observed outcrops are few but strikelines may be drawn in the direction of strike measured on the outcrops and then interpreted to give a postulated outcrop for the bed

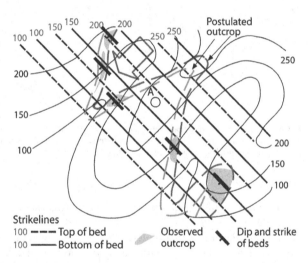

Strikelines
100 ━━━ Top of bed   Observed   Dip and strike
100 ━━━ Bottom of bed   outcrop   of beds

the intersection of the outcrop and the contours but their direction is uncertain. However, some dip and strike measurements on available exposures gives the direction of strike so that top of bed strikelines can be drawn for 150 m and 200 m and bottom of bed strikelines for 150 m. This is sufficient to draw strikelines to define the total outcrop throughout the area. The outcrop drawn is best described as 'postulated' because it is assumed that the thickness of the soil cover concealing the outcrop is negligible. If this thickness was significant, say of the order of more than three metres, then the outcrop of the bed would have to be drawn on the surface of bedrock underneath the varying thickness of soil cover.

In broad scale general geology, such as employed in the mining or oil industry, the plotting of outcrop locations in such detail is not of great importance, but it is of tremendous importance in engineering geology. Thus, if the bed outcropping in the figure was a worked coal seam containing unstable cavities and it was required to build on safe land free of hazards of potential instability then it would be obviously of vital importance to know exactly where the seam outcropped under the surface cover.

4.3.1
Interpretation of Geological Maps – Example

Figure 4.3a–d shows outcrop patterns for a single surface in circumstances of increasing structural complexity. In reality the situation may be rather more complex, for beds do not necessarily maintain uniform thickness, may not be continuous as a rock type, folding may be severe and asymmetrical and the structure may be faulted. In many complex areas, it is almost impossible to understand the structure only from the study of the outcrop pattern and the topography. Measurements of dip and strike at as many locations as possible are necessary to aid interpretation. It is also of great value and often essential to know the relative age relationships of the various layers. In sedimentary rocks, this is undertaken by palaeontological studies and because few engineering geologists have sufficient expertise to undertake such studies, they might have to be aided in the course of mapping by a palaeontologist and/or stratigrapher.

It is fortunate that most engineering geologists have to deal with relatively small areas for within a small area it may be found that the geological structure is reasonably uniform. However, greater accuracy is required for the location of boundaries than for other types of geological work. Thus in engineering geological mapping of, say, a dam site, a significant boundary may have to be surveyed (using precise survey techniques) rather than sketched on to the map.

The following theoretical example illustrates how the techniques of drawing strikelines may be used to solve problems of engineering importance. Figure 4.6 shows a simple geological map of an area in which it is proposed to build a dam; the line of the crest of the dam is shown.

The shales that would form the foundation rock are believed to be highly weathered and a 50 m deep excavation may be required to reach fresh shale. From the shape of the outcrops of the fault and the dolerite it can be deduced that the fault dips upstream and the dolerite sill down stream. The question is whether they will be encountered in the foundation excavation. To give an answer strikelines must be drawn for the fault and for the dolerite. Both top and bottom dolerite strikelines have been drawn in Fig. 4.7 although only strikelines for the top of the dolerite are necessary in this case.

The elevation of the base of the dam at the river is about 60 m a.s.l. (above sea level) so that the bottom of the excavation would be at +10 m a.s.l. Beneath this point, the strikeline for the top of the dolerite would be about +20 m a.s.l. and that for the fault at about –50 m a.s.l. Thus, it is likely that dolerite would be found below the river location in the excavation. Under the west dam abutment the top of the dolerite at about +60 m a.s.l., would not be encountered in the foundation excavation to +10 m a.s.l.; under the east abutment the dolerite lies at about –40 m a.s.l., far below the bottom of the foundation excavation. This implies that dolerite would be found in the centre and to west of centre of the excavation. Such an interpretation of the map would not be definitive for it is assumed that the dolerite dip and thickness remains uniform between the nearest outcrop and the dam site. However, it does reveal a potential prob-

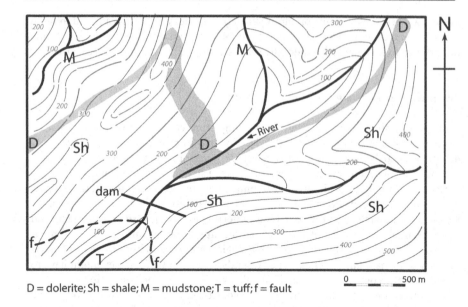

D = dolerite; Sh = shale; M = mudstone; T = tuff; f = fault

Fig. 4.6. The geological map showing the dam site

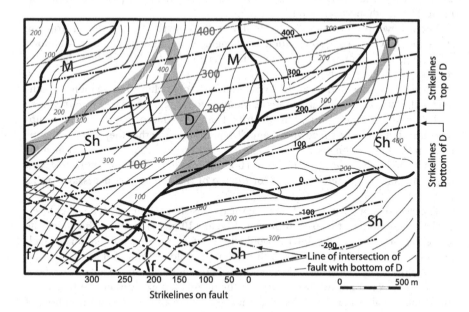

Fig. 4.7. Strikelines drawn on the fault and the top and bottom surfaces of the dolerite

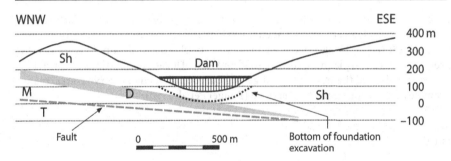

Fig. 4.8. A cross-section that could be drawn to illustrate the geology that might be encountered in the foundation excavation. Note that, in order to show sufficient detail, the scale of the cross-section is larger than that of Figs. 4.6 and 4.7

lem, which would have to be further, examined in the investigation by boreholes and other means. Another problem worthy of investigation is that of the fault. The fault and the dolerite intersect; their junction is plotted on the map (Fig. 4.7).

In the interpretation in Fig. 4.8, it has been assumed that the fault cuts through all rocks. However, it is possible that the fault existed before the intrusion of the dolerite. In practice, this would be of some importance, perhaps not so much to the dam but certainly to any diversion tunnel to be built through one or other of the valley sides. Extra geological studies would be required to establish this point. It should be noted that the answer to the problem was obtained from drawing strikelines; the cross-section was only used to illustrate the answer for the benefit of a report reader who was perhaps, unfamiliar with strikeline maps.

The discussion above has been directed mostly towards bedded rocks. Large igneous bodies, such as batholiths and laccoliths, follow no geometrical pattern that can be forecast in detail and their shape underground can mostly be interpreted only rather vaguely from a study of surface outcrops. Sedimentary deposits, such as braided stream deposits, reefs, sand channel deposits and so forth are very irregular in shape and their location and shape underground can seldom be established from a study of surface outcrops, even with the help of a reasonable number of boreholes. It is thus most important to recognise the nature of rock and soil materials to gain an idea of the way in which they were deposited or intruded. If they are so irregular that the techniques described above, either with or without the help of information from boreholes, cannot be used other techniques, generally geophysical, must be used to describe the shape of boundary surfaces underground.

4.4
Mapping at a Small Scale

Most newly graduated geologists have had some training in the rudiments of geological mapping but this is not always an adequate preparation for mapping in engineering geological practice. The following hints regarding mapping for engineering purposes on small and large scales may be of value to those beginning to undertake this fundamental geological activity. For engineering projects mapping on a scale of greater than 1:5000 (e.g. 1:25000) may be considered small scale and would be applied to problems of general planning such as, road and tunnel locations, locations of construction materials, and the description of geological conditions in a reservoir area; many examples can be found in the literature (Rengers et al. 1990).

4.4.1
Starting

The first step in such work is to assemble existing geological and contoured topographical maps, aerial photographs, and satellite imagery. An evaluation of this information will give an idea of the geology of the area, the location of outcrops and, in particular, the existence of geomorphological features, such as landslides, which may not have been recorded on the geological map. It is also of value to establish how any existing geological map was produced (whether, for example, from aerial photograph interpretation with field checks or by extensive field work) and the degree of detail shown. It is not uncommon to find older maps showing broad areas with a single formation name while an adjacent younger map divides the formation into many sub-formations as a consequence of a more recent elaboration of regional stratigraphy. The next step is to visit the area to check on the validity of the existing maps, become acquainted with the soil and rock types, consider the problems of establishing the location of observations, and examine accessibility. With this information, it is then possible to plan the ensuing work.

4.4.2
Covering the Ground

An initial problem is how to move around the area in the most effective and efficient way. This will depend upon terrain, road and path networks, climate and vegetation, but it should be remembered that the aim of mapping is not to provide the geologist with a fitness exercise but to bring his or her sensory equipment (eyes, hands and brain) into contact with the geology. Thus, vehicular transport is clearly of advantage in areas with a good road network, boats may be needed to examine riverbanks, coastlines, and so forth. The writer once found travel on horseback to be particularly useful, for the horse can carry equipment and there are few parking problems. Accessibility and transportation will divide any area into smaller zones and thence indicate a plan for working.

In each part of the area, consideration must be given to establishing the location of observations. Global positioning systems (GPS), remote sensing imagery (aerial photographs, radar images, etc.) in the field, topographical maps, and a compass allow positioning to a degree of accuracy appropriate to the scale of the final map. A good sighting compass is also required.

The map to be made must depict geology, geological structure, and geomorphological features. Geological units are established by lithology and stratigraphy. If the map is for engineering purposes then engineering geological units may replace geological units. The engineering geological units are not necessarily the same as any lithological or stratigraphical units shown on existing geological maps but may consist of rock or soil assemblages of homogeneous and uniform geotechnical characteristics. No geological mass can be truly homogeneous and uniform so a homogeneous unit here implies a mass with geotechnical features, such as strength, discontinuity spacing etc. which fall within limits specified by the investigator. The limits defining such engineering geological units could well be related to the engineering purpose for which the map is being constructed and thus might be defined by a soil or rock mass classification system (see below). Much thought must be given to the purpose, means, and methodology of establishing such units before detailed mapping begins.

The map must also show geological structure because the successful construction of many engineering works, such as tunnels, slopes, and quarries depends upon the orientation of discontinuities and location of faults. Geomorphological features, such as landslides, are probably best recognised from aerial photographs but field examination may indicate whether they are active or not. On the ground surface irregularities may result from the underlying geology or geomorphology but can also reflect the past activities of man. Such features range from Bronze Age fortifications through open pit and underground mines to World War trenches.

In the early phases of mapping, work will tend to be concentrated on the best exposures, which may be examined in some detail. Less well-exposed outcrops may then be placed in their correct unit as experience and confidence is built up. If, as is common, the best exposures are man-made, such as road cuts, they may show features that should be recorded although they do not form part of the engineering geological unit. Thus, excavated slopes may indicate the suitability of the excavation method applied and their stability, or instability, may suggest values for the strength parameters of the discontinuities they contain.

4.4.3
Hazards

Fieldwork is not without its hazards. Adequate provision must be made to overcome climatic difficulties so that in desert areas, for example, sufficient water must be available at all times. In jungle, watch must be kept for poisonous snakes and insects. In less romantic but sunny environments, it is sensible not to work when the sun is high, but to start as early as possible in the morning and thence stop when it gets too hot. Whatever the climate, suitable and perhaps conspicuous, clothing should be worn and the means of attracting attention, such as a whistle, carried. Survival equipment should be carried, remembering that two broken ankles on exposed ground in winter could be deadlier than snakebite in the jungle. However, the two greatest safety measures are always to work with an assistant (and failing that, to ensure someone knows where you are intending to work and route you will use) and to keep your mobile phone in working order.

4.5
Mapping at a Large Scale

Large-scale maps are those devoted to the depiction of relatively small surface areas in great detail and range from about 1:1 000 (for the geology of a dam site) to about

1:50 (for a rock slope or tunnel). They are prepared for an engineering geological purpose.

4.5.1
Foundation Areas and Excavations

The largest likely map is that of a proposed dam site. Here the location of engineering geological unit boundaries and discontinuities must be undertaken with accuracy so that standard survey techniques must be used to establish boundary and outcrop locations. It is probable that the dam site has already been chosen based on general geology, river hydrology and topography and that access roads and tracks for moving drilling equipment are being constructed. In such a case, if there are insufficient natural outcrops to allow the construction of an adequate map it may be made using excavation machinery or explosives. A good map is essential to the interpretation of later borehole information and outcrop examination offers three-dimensional information, which may be difficult to derive from boreholes. It is desirable that such maps are made including a grid reference system (preferably a national system) and elevations related to a stable site benchmark. In a foundation excavation, mapping may be undertaken to fulfil the requirements of a rock mass classification system, either existing or devised for the particular purpose. Location of major discontinuities must be plotted accurately.

4.5.2
Rock Slopes and Major Outcrops

In the case of the 'natural' outcrop, the first thing to do is to establish how natural it is. Half boreholes of some length and of the order of 10 cm diameter surrounded by crushed rock indicate modern blasting; shorter, usually less than a metre, boreholes of diameters of three or 4 cm and often without significant crushing indicate much older blasting using gunpowder. The former may have brought about significant disruption of the rock slope, the latter probably not. By examining the relationship of these relict blast holes to the rock blocks in the slope it may be possible to assess whether or not any recent movements have taken place in the slope. Older slopes have plants and trees growing on them. Large trees may be more than 50 years old and their roots may have penetrated deeply into the rock mass. The author was once fortunate enough to be able to trace a root passing through the rock mass for about 30 m from the only tree growing on the top of the slope. Tree roots as they grow widen rock joints and dislodge rock blocks. Tree trunks bent near their bases and thence growing vertical may indicate past slope movement. Permanent or transitory (marked by dried algal growths) seepages may issue from the rock face. Their emergence from particular joints may indicate a greater than usual aperture for that set of joints. Blasting disruption, plant growth and seepages are clearly relevant to slope stability problems but if, for example, the outcrop is being examined because it is close to a proposed tunnel alignment, any slope movement will affect the relevance of any measurements made of discontinuity orientation, aperture, persistence or infill to the unmoved ground at depth.

Having established the condition of the outcrop, the next problem to overcome is how to describe it and how to take any measurement that must be made. This is not easy to do and it is prudent to examine the whole outcrop before deciding, for example, how many rock types should be distinguished. It is no good starting at one end of the slope calling a rock simply 'sandstone' to find, as the observer progresses, there are six types of sandstone, differing by grain size, colour and strength, because then the first descriptions must be revised. The author, in his own work, has usually tried to make any examination of a significant slope last two days, the first being used to determine what should be done in the second. Few clients are entirely happy to see an engineering geologist sitting down looking at a slope and thinking; activity is required. Fortunately, there are routine tasks that must be performed, such as measuring slope dimensions, heights, etc., which, while being undertaken, allow an impression of the geology to seep into the subconscious. Scan-line surveys, undertaken often at the foot of slopes, involve recording the orientation, condition, and position of every discontinuity encountered along the line of a measuring tape at a convenient height along the base of the slope. While this work produces an impressive discontinuity database, it also allows the observer time to gain a general impression of the rock mass.

First Phase

This first phase of work would be done without the benefit of any kind of plan of the rock outcrop. It is useful, if satisfactory camera viewpoints can be found, to construct a photo mosaic of the outcrop. A photo mosaic for immediate use may be made using a digital camera and computer processing to combine the mosaic to one image. Stereo photo pairs may also be made for later analyses, for example, for discontinuity mapping. This may be annotated in the field in the next day's work. What is recorded in the field depends upon the combination of information required (to address the engineering problem) and the information available (as displayed by the rock outcrop). For many problems the basic data required will include:

- the distribution of the various rock and soil types to be seen in the outcrop and their geotechnical description,
- weathering zones,
- presence of water and seepage locations,
- information regarding the character and orientation of the discontinuities, noting the location of observations in three dimensions.

It is also important that the nature and position of major discontinuities that may define major rock blocks are recorded, together with the limits of domains of structural uniformity that may aid slope stability analysis.

Such fieldwork requires instrumentation. The following would be adequate for most outcrop descriptions:

- compass clinometers to measure dip and direction of dip on a surface,
- a sighting compass to measure directions of strike, slope etc. from a distance,
- a sighting clinometer, for slope inclinations and dips of discontinuities that cannot be reached for direct measurement,
- a thin stiff flat board to be placed on a discontinuity to allow an average dip measurement to be taken,
- a fold out 1 m long carpenter's rule, with built in level,

- several tapes, about 50 m long,
- a heavy hammer, with one large and one small strong chisels, the latter perhaps with a tungsten steel insert cutting edge,
- long nails (to drive into cracks to fix tapes to); chalk, spray paint,
- a powerful telescope or binoculars,
- a carpenter's profile *comb* (to help measure asperity and waviness angles on discontinuity surface undulations).

Besides these, but depending on outcrop size and aspect, it may be useful to have:

- a pair of high accuracy aneroid barometers, to measure heights on a rock slope (one for base station, the other for field use),
- a nautical sextant (for locating by measuring angles between markers on a surveyed base line),
- an optical rangefinder, which will measure distances within the range 10–100 m.

Perhaps the most necessary aid to outcrop examination is a field assistant, if only to carry all the clobber! Noting discontinuity data goes very slowly if it has to be done by the observer. In the author's experience, many important field observations are made with one hand tightly gripping a tree branch while balanced on a wobbly boulder half way up a cliff. In such situations, dictation is both safer and better than trying to write notes.

The comments above would relate to both natural and man-made exposures. Clearly there is an element of hazard in such work and the use of an assistant does decrease the risk. The risk is perhaps highest in recent man made exposures, particularly if enthusiastic blasting has made these. Steep rock slopes of this nature are best avoided if recent heavy rainfall has caused seepages to emerge from the outcrop. In all slopes, of whatever type, the examination should include the usually gentler slopes running back from the top of the steeper rock face. These may show tension cracks, which, if they are very close behind the rock face, would cast doubt on its short-term stability. For safety's sake, it is thus best to look at the slopes above the top and sides of an outcrop before commencing a detailed examination of the face itself. Naturally, besides recognising geological hazards, such local regulations as may exist regarding health and safety legislation should be followed.

The principal difficulty in such work is that, unless the slope can be climbed, knowledge of the upper parts is limited. Those expert in rock climbing may find that it is quite easy to get to the top but much more difficult to go to particular locations on the slope. If the survey of a rock slope shows it to be unstable, the next step may be to design stabilisation works. To do this a three-dimensional picture of the rock mass behind the slope is needed in order to determine the length, location, and number of any bolts, pins, drain holes and grout holes required for stabilisation.

Long Rock Faces

In the case of a long rock face giving problems of rock falls, one of the first tasks is to establish the relative stability of the various parts of the face. This is necessary in order to plan the stabilisation works, which might take several years to complete. The first need is a topographic survey and if finances allow this is best-undertaken using terrestrial imagery. Stereo-pairs of photographs and digital images are taken and from these a contour plan is produced. These contours are drawn relative to a vertical datum plane set in front of the rock face. The plane usually follows the general direction of the rock face; contours are then lines joining points of equal horizontal distance from and normal to the vertical datum plane. Such plans should also show locations of major fractures, patches of vegetation etc. on the rock face. If the contours are drawn at a close contour interval, say 0.2 m, the contour lines give a picture of the topography. The combination of field geological survey and study of the terrestrial topographic map and photographs should allow the definition of zones of relative instability (Price et al. 1988). Apart from the obvious purpose of seeing where work must be done the survey also helps the selection of working zones. Stabilisation work can be hazardous if approached carelessly. It is useful to locate the working zone of unstable rock between zones of better stability to which the scaffolding may be bolted and bolts set in as secure points for safety cables across selected parts of the unsafe area.

Detailed Surveys

Once zones of relative instability have been established, the areas where work is to be undertaken must be mapped in greater detail. When access scaffolding has been erected against the face, the scaffolding framework may be used as the datum plane from which measurements may be made to draw contours. Figure 4.9 shows how any point P on the rock face may be given x, y, z coordinates relative to the scaffold datum plane. If a contour plan of the face has been made it may be better to follow that. Geological survey has to be undertaken on the scaffolding including measurement of discontinuity orientation. Dip can be measured using a clinometer but because the scaffolding is magnetic, strike must be recorded by tape and chalk mark measurements on the scaffold boards to give strike angles relative to the datum plane (Fig. 4.9).

The three-dimensional picture of the rock face can be built up either mathematically, in a computer drawing program, or by the production of and measurement from cross-sections. The aim is to produce strikelines for the discontinuity planes that bound rock blocks and through which bolts or pins must pass to bind unstable rock to a secure backing mass. The vertical separation between scaffolding platforms (lifts) is usually about 2–2.25 m with a handrail at about 1 m above the platform. The convenient working zone for drilling holes into the rock face is about 1 m wide along the middle of the lift so measurements for cross-sections, contours etc. are most appropriately made at handrail level with, if topography and/or geology are complex, measurements at floor level. Cross-sections can then be constructed at handrail and floor levels to show depths of block bounding discontinuities behind the rock face.

The joints and other discontinuities that are projected back into the face and which would be considered to bound rock blocks must be selected on the basis of their appearance on surface. Thus, a fault would be considered to be a feature which would be expected to have considerable persistence. The persistence of other joints into the face would be expected to be comparable with the persistence of their trace on the face. The cross-sections should also contain some indication of the opening of the discontinuity, with particular reference to changes in the openness. Such displays are neces-

Fig. 4.9. Establishing 3-dimensional co-ordinates for points (*P*) on the rock face by measurements from the scaffolding. The front face of the scaffolding is the vertical datum plane

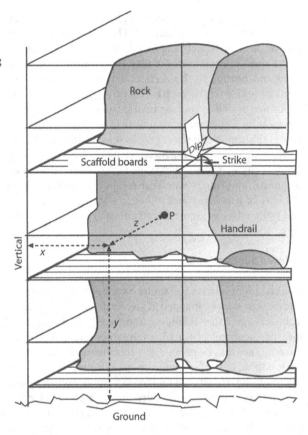

sary to discover how the movement occurred. Thus, a block that has a uniformly open backing joint may be judged to have slid forward on some basal plane. If the backing joint is more open at the top than at the bottom some rotational movement would be the more likely mode of displacement.

It is most important to try to discover how a face deformed to reach its present condition. If, for example, a block can be seen to have slid forward on a basal plane it follows that the angle of shearing resistance of the surface of the plane will have fallen below its original value because asperities on the plane have been ground down by the movement that has taken place. If backing joint openings can be seen to have resulted from movement, but no obvious basal plane of movement is present, it follows that the rock mass must be disrupted and loosened by a multiplicity of many small joint openings, so that the whole rock mass is weakened.

4.5.3
Tunnels

It has now become almost standard practice to map tunnels and chambers in engineering geological terms, as the tunnel advances. Such maps show geology, major discontinuities and may display the results of a rock mass classification system. They have a number of objectives, the most important of which are:

- to record the geology encountered, so that the site investigation data may be reinterpreted with this new data to give a better picture of the geology to be encountered in the rest of the tunnel drive;
- to check the success of any rock mass classification system used to predict support, rate of advance, cutter wear etc. against that achieved so that, if necessary, system modifications may be made.

Tunnel length, width, and height dimensions must be recorded together with roof and floor elevations. Usually little can be seen of the floor of the tunnel so that mapping is confined to sidewalls and roof. In tunnels that are lined or shotcreted, work has to be undertaken quickly between the advancing working face and the following support, giving little chance for lengthy consideration of geological details. A clear idea of the necessary data to be recorded must be established before beginning work. Orientation readings require an illuminated compass and clinometer. Data recorded need not to be restricted to a rock mass classification system and should always include the basic geology, discontinuity aperture and locations of water inflows.

Drawings presented are commonly in the form of foldout maps, prepared as shown in Fig. 4.10. The same sorts of drawings may be used vertically for shaft descriptions. These are usually large diameter holes and thus circular. The investigator is lowered

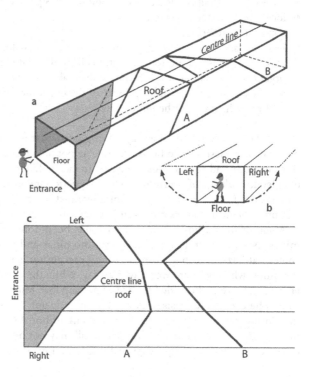

Fig. 4.10. The tunnel in (**a**) shows two lithologies, one shaded, one blank. Two major discontinuities, A and B, are found. Strata and discontinuity traces are plotted on diagrams of walls and roof. The wall planes are, as it were, 'folded out' (**b**) to produce the report drawing (**c**)

down the shaft in a steel cage to examine the sides. This precludes the use of the geological compass to measure orientation so usually four tapes are stretched from top to bottom from the cardinal points of the compass on the surface. The depths of boundaries or discontinuities crossing the tapes are noted and dips and strikes can be calculated. This type of work is subject to the normal hazards of tunnelling.

4.5.4
Mines

Engineering geologists examine mines, usually long abandoned, to determine their stability. The need to do this often stems from a recent collapse causing damage or injury. The mines are usually some variant of the room and pillar system and attention is focussed on:

- room and pillar dimensions (width, length and height) for these are necessary in order to calculate stability;
- strengths of the rock composing floor, pillar and roof and variations in strength (both compressive and tensile) throughout the mine;
- the long-term behaviour of the pillar and roof rocks i.e. creep characteristics;
- the location, frequency, persistence, and aperture of natural discontinuities in pillar and roof;
- the location, frequency, persistence, and aperture of induced discontinuities (cracks) in pillar and roof as a consequence of stress and decay;
- locations of collapses and a description of the type of collapse;
- locations of water seepages;
- zones of weathering;
- features associated with the type of rock being mined, for example, solution cavities and earth pipes in limestone, variations in coal quality in coalmines;
- organic (lichens etc.) or mineral growths (calcite etc.) whose presence on the crack surfaces might indicate the antiquity of the crack formation;
- root growths through the mine roof.

The data is presented on a map paying particular attention to roof and pillar dimensions for these should be as accurate as possible for stability calculations. Elevations of floor and roof are also needed. To calculate overburden pressure on pillars detailed knowledge of overburden characteristics and groundwater conditions is desirable and the map of surface topography must be securely linked to the mine plan. Examples of such surveys are given in Chap. 10.

Much of the author's own fieldwork has been done underground in collapsing mines, essentially undertaking surveys into pillar and roof stability. This is clearly somewhat hazardous but the main safety precaution is to inform someone (outside the mine) where you are working that day. While the usual dangers of falling rocks etc. apply, there is the added hazard of getting lost. Multiple sources of illumination should be carried. An assistant or fellow investigator is desirable perhaps not working closely together (for a roof fall could engulf both) but each knowing where the other is. Work after unusually heavy rainfall may not be a good idea. Work in abandoned coalmines may require special ventilation measures as methane may form an explosion risk. Also other poisonous gasses may form serious hazards, for example, rotting of timber support beams may have caused a deadly carbon monoxide environment. Older experienced investigators may, on a particular day, find themselves more than usually concerned over their safety in an abandoned mine (where they have worked before in reasonable safety) and have perhaps decided it was a good day to do something else. Such instincts may have yet not developed in younger investigators but they should remember that, besides gathering data, they have to survive to write the report. Caution in such fieldwork is recommended.

4.5.5
Natural Cavities

Natural cavities are most often those in *karstic* crystalline limestone, but may also be sea caves progressing by erosion in any rocks under cliff edge structures, such as lighthouses, hotels and similar coastal development, including tourist routes but in the author's experience, it is more often concerned with the stability of surface structures. Estimating their stability is usually very difficult because of their irregular shape. Clearly, a map showing elevations of roof and floor and position of caverns underground relative to surface structures must accompany any examination of such features.

Fookes and Hawkins (1988) have reviewed the problems of limestone weathering and suggested a classification. Zhu Jingmin and Xiao Zhizhong (1982) proposed a method for the evaluation of the stability of karst caverns. In essence their approach was to look at a number of multi-component geological factors and, by noting the frequency of their presence or absence assess relative stability. The core of their method has been tentatively modified by the Price (1992) to include ratings (Table 4.1). The application of such a ratings system may crudely estimate stability and the objective of mapping would be to obtain the data to apply the system throughout the caves.

4.6
Engineering Geological Maps

Between about 1965 and 1980 it was fashionable to produce maps, described as engineering geological maps, which displayed, in any area, those geological features that were considered important to the prosecution of engineering projects. The range and variety of maps produced is briefly described below; for a full description the reader is referred to the report of the IAEG Commission on Engineering Geological Mapping published by UNESCO in 1976 (International Association of Engineering Geology 1976). The character of these maps ranged from the simple presentation of factual data in terms understandable by non-geologists to the expression of opinion on specific engineering characteristics of the ground, such as its bearing capacity.

4.6.1
The Past

It is useful to review the engineering geological maps that have been published in the past. The main stream of engineering geological maps is either:

Table 4.1. Rock mass classification for the assessment of the stability of natural cavities (modified from Zhu Jingmin and Xiao Zhizhong 1982 by Price 1992)

Rock condition	Parameter and rating			
Geological structure	Not faulted or folded	Few and minor faults and folds	Significant faulting and folding, open fissures	Major faults, many folds, many open fissures
Rating	10	7	4	0
Rock properties	Thick inclined beds, strike crosses cavern axis	Medium thick inclined beds, strike crosses cavern axis	Thin beds, shallowly inclined, strike oblique to tunnel axis	Thin near horizontal beds, shale intercalations, strike parallel to cavern axis
Rating	10	7	4	0
Ground water and underground rivers	Effectively dry, no river nearby	Some water drips, no river nearby	Significant water drips, caves wet, underground water in the vicinity	Heavy water drips, underground river close by
Rating	10	7	4	0
State of the roof	No fresh collapses, continuous dripstone cover, rock structure interlocking, good arching	No fresh collapses, dripstone cover over two thirds of surface, rock mass interlocking, good arching	Some fresh collapses, surface about half covered with dripstone, interlocking rock structure but arching poor	Intermittent dripstone cover, many fresh collapses, no arching or interlocking
Rating	10	7	4	0
Rockfalls	Very few rockfalls	Some rockfalls, none presently unstable	Many rock falls, some appear unstable	Very many rockfalls, many unstable
Rating	10	7	4	0

Total rating 40–35 = stable, 35–30 = basically stable, 30–15 = less stable, <15 unstable.

i special maps prepared by government agencies which are intended to show general engineering geological conditions or to show features of particular engineering importance to the general public, or

ii maps of particular engineering sites that are arranged to show features of importance to a particular engineering work or engineering problem. Consultants, engineering geologists working as part of the site engineering staff, site investigation contractors, and so forth, generally prepare these.

The 'government agencies' include geological surveys, research institutes, universities, and so on. Their maps are generally not specific to a particular site or project. Many different types exist. They include:

i   maps that are modifications of existing geological maps and show lithology rather than stratigraphy.
ii maps that emphasise particular geological or geomorphological features (such as landslides).
iii maps that serve as a source of documentary research – these would show locations of boreholes, trial excavations, penetration soundings etc. and could be linked to an archive or data retrieval system. Use of particular symbols may indicate the type of borehole, depth to which it was sunk, number of tests performed in the borehole etc.
iv maps that give opinions about certain problems, which are of engineering significance e.g. "bad ground for foundations".
v   maps that give geotechnical data, such as depth of overburden, compressive strength of rock units etc.

In most countries there is continuing geotechnical work being undertaken that should be entered into the maps in revisions. The frequency of revision will depend upon the scale of the map and the nature of the information presented on it.

The scale of maps is very variable depending upon the purpose of the map. For large scale planning purposes scales of more than 1:50 000 are sometimes used, although many engineering geologists consider these to have no more value than to form a pretty decoration for the office of a high ranking politician. Scales between 1:5 000 and 1:20 000 are necessary for detailed engineering location planning.

4.6.2
The Present

Engineering geologists working for consulting engineers, contractors, or firms of consulting engineering geologists, generally prepare the engineering geological maps of specific sites. Their maps may:

i   depict site geology and geomorphology in such a way as to aid resolution of a particular engineering problem. Such maps may contain geotechnical data or, at the very least, the description of the features mapped will be in engineering geological terms.
ii record the site geology uncovered during construction, again using an engineering geological descriptive vocabulary rather than ordinary geological terms. Such maps are based upon data recovered during site work by an engineering geologist working for a contractor or consultant.

Any data not recovered personally by the engineering geologist should be checked for validity before being included on the map. Type (i) maps commonly accompany investigation reports; type (ii) maps may form part of the 'as built' description of the works produced at the end of contracts. Scales of these maps generally lie between 1:50 and 1:1 000.

The data collected for the construction of engineering geological maps or plans must be presented in a readable form. All maps and plans must contain the following background information if they are to be of use:

i   a legible and complete topographic background, showing roads, towns, streams, prominent buildings etc. Contour lines should be shown. The map should include marginal comment giving the date of publication of the topographic base map.

ii the grid co-ordinate network appropriate to the project or country.

iii the scale of the map given both in figures and as a linear scale. If the original map was in units of feet, miles etc. then the linear scale should be both in these units and in metres.

iv north points; these should be true north, grid north and magnetic north. The annual change of magnetic declination should be given.

v   the map should be dated giving both the date of preparation of the map and the date of publication. The dates should be accurate to the nearest month.

vi a legend showing the meaning of all the symbols shown on the map.

vii the names of those responsible for the preparation of the map.

4.6.3
Symbology

The nature of the symbols to be used on engineering geological maps merits some discussion. Attempts have been made to standardise symbols. Thus, the Working Party of the Engineering Group of the Geological Society of London produced a catalogue of recommended symbols in their report in 1972 (Geological Society Engineering Group 1972). There is obviously every advantage in using a standard symbology, for with time both map users and mapmakers will become familiar with the symbols and not have to hunt through the legend in order to discover their meaning. Unfortunately, many organisations have produced systems of standard symbology and attempts at 'standardisation' seem to have produced the confusion they sought to remove.

Moreover, any standard system of symbols may not be adequate for a particular purpose and a mapmaker will have to modify or add additional symbols to the standard list. If this is necessary, it is of vital importance that they are cartographically legible. There is always the great temptation to try to add increasingly more information to a map by making the symbology increasingly complicated. This may be very satisfying to the mapmakers who, because they are associated with the system development, have no problems in understanding their own work but can be very annoying to the map user who has to try to understand the map. It must never be forgotten that one of the main objectives of engineering geological mapping is to produce maps that are of immediate use to the engineer. One of the motivations behind the movement to produce such maps was the recognition that many geological maps were made which only geologists could use; engineering geologists must not fall into the trap of presenting maps that are only of use to engineering geologists. Maps must be 'user friendly', easily understood and easily read. To understand the character of the ground mass is difficult enough without having to cope with over-complex symbology.

4.6.4
Current Developments

The desktop computer was introduced in the early 1980s and rapidly became sufficiently powerful to allow the development of graphic systems and databases. The graphic systems allow the production of maps that reside in the system database and can be modified and updated as new information is entered. Borehole information may be retrieved from data bases linked to map co-ordinates. Thus, in theory, it should be possible to ask the map-producing agency to print out a map showing the latest information interpreted using the most up-to-date techniques. Assuming that the necessary organisation to acquire, enter and process information into the data bases can be set up and financed, a problem remains in that graphic and data base systems are continually in development as computer power increases, and data entered into an old style data base is not necessarily easily transferable to a new one.

Geographic Information Systems (GIS)

A further development is that of Computer Aided Design/Computer Aided Manufacturing (CAD/CAM) systems and the Geographic Information System (GIS). CAD/CAM systems are mostly orientated towards handling vector type of data, whereas GIS systems can handle raster data and vector data. Vector data means that the data is described as points, lines, surfaces, and volumes that define an object and the object is the entity in the CAD/CAM or GIS system. Raster data has the form of a regular grid or cell structure that together define the object, but each cell is a separate entity in the system.

GIS systems normally combine a (limited) CAD/CAM system, a database for the objects and entities with property values, a visualisation unit, and a calculation unit able to manipulate the coordinates and properties, and containing statistic routines, etc. The combination of database and visualization makes a GIS system particular handy for engineering geology. For example, it should be possible, if the separate items of data necessary for rock mass classification exist, to ask the GIS to produce a particular form of rock mass classification for a specified lithology within a given area (Fig. 4.11).

Two-, Three- and Four-Dimensional GIS

Two- and three-dimensional GIS (2D-, 3D-GIS) systems have been developed. In 2D systems, the coordinate system in the database can only store two coordinates, nor-

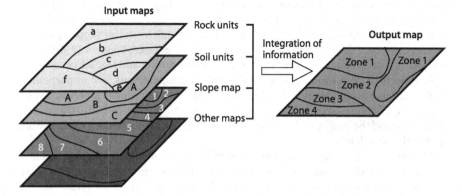

Fig. 4.11. Geographical Information Systems can combine various sources of map information to give the desired output map (after Orlic 1997)

mally x and y. Properties are then assigned to these coordinates. The 2D referenced properties can be manipulated in various ways, such as, addition and subtraction of properties, statistical calculations, etc. and be visualised. 2D systems are basically no more than a map visualisation tool with the capacity to overlay different types of zero- (point), one- (line) and two- (surface) dimensional information. The limitations of these systems in engineering geology are large as it is not possible to represent the sub-surface accurately. 2D systems can only handle a projection on a surface of the sub-surface boundaries and properties, similar to a map or cross-section made by traditional drawing methods. The advantage of the systems is that they are relatively simple and easy to handle. Sometimes the term 2.5D is used to characterise systems that allow the third coordinate (z coordinate) to be stored as a property value. A boundary model representing the sub-surface can then be made, however, the properties in the space between the boundaries are not defined in the database. In addition, it is not possible to store boundaries with multiple z values belonging to one x, y coordinate pair. In 3D systems, the coordinate system in the database consists of three coordinates, normally x, y and z. Boundary surfaces can be defined and between the boundaries properties can be defined for each point leading to the definition of volumes (so-called 'volume model'). Manipulation of the properties is similar to that for the 2D systems with the extension that also volumes can be manipulated because the statistical and calculation routines are three-dimensional. 3D-GIS systems therefore represent the sub-surface accurately. Four-dimensional systems (e.g. spatial with time) have not (yet) been made, however, some numerical calculation programs include a 3D-GIS system and can store different realisations of a numerical calculation. This apparently is a 4D system, but is at present without 4D data manipulation options, such as property addition or 4D statistical routines.

GIS in Engineering Geology

A full three-dimensional GIS system is obviously the best system to handle engineering geological data and analyses. For regional studies, this may lead to very extensive databases and consequently long calculation times, although this may become unimportant with further development of computer power. In regional studies, if property distributions are not important and the geology is relatively simple, a 2.5D system may be sufficient. A full 3D-GIS system is necessary for site-specific analysis in which either an accurate representation of the sub-surface is required and/or property distributions are required, or where the geology is more complicated. Figure 4.12 shows an example of a sub-surface property model for a tunnel project and Fig. 4.13 shows the application for a dam project. Geographical Information Systems, and in particular 3D-GIS, are able to offer considerable help to the engineering geologist however, it does not add these qualities in itself. The quality of the output is directly related to the quality of the input and the quality of the manipulation that is done with the data, e.g. "rubbish in" is still "rubbish out".

Another point that should be considered is that GIS software is complex and not always user friendly. Hence, it is often time consuming to use the programs and this extra time is certainly not always justified for all type of projects. Some professionals, probably it should be questioned whether these are professionals, cut down on time consuming operations by using simple programs, for example, 2.5D instead of the more

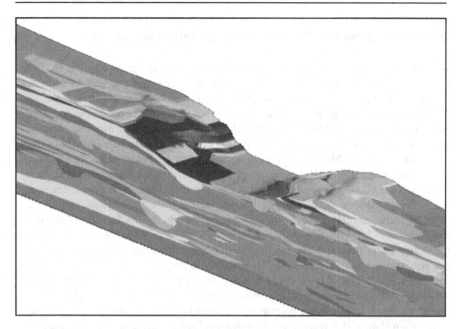

Fig. 4.12. A 3D-GIS volume model with geotechnical units for a tunnel alignment (Heinenoord Tunnel, The Netherlands; after Ozmutlu 2002)

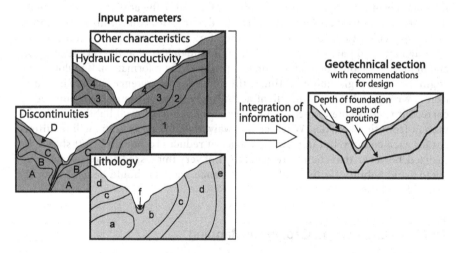

Fig. 4.13. The applicability of 3-dimensional GIS in geotechnical engineering (after Orlic 1997)

complicated 3D programs, and by using simple calculation routines when more complicated but better relations are known. Obviously, GIS used in this way can result in lower quality results than traditional methods.

4.7
Quality of Published Information and Limitation of Liability

Maps, whatever their character, contain such information as was available at the time of their compilation, which is often out of date at the time of publication and later use. Information that contains an interpretation of available data, give those opinions based on the understanding of the engineering behaviour of the ground of that time, which is not necessarily that of the time of later use. If information expresses opinion, which is virtually always the case with geology interpretations, the question arises as to who is responsible for the quality and accuracy of that opinion. The types of maps which are published by government agencies in any one country depend, therefore, upon the social, political and legal systems in that country because if engineering works are planned on the basis of the information provided by these maps, there must be some understanding as to who is responsible for the validity of the data they present. Virtually all liability is waived by most governmental agencies for the information they provide, for the quality of the information, maps, GIS models, etc., and for possible damages resulting from errors or inaccuracies. In some countries an attempt may be made to claim against a governmental agency if severe negligence can be proven, however, negligence is mostly very difficult to prove, and legal, financial, and time resources of governmental agencies are generally far larger than that of the average engineering geologist. This places the engineering geologist in an awkward position. He or she has to rely on published information as it is mostly impossible to collect and analyse the underlying data, whilst also providing the client with advice for which responsibility has to be accepted. Hence, the engineering geologist can be legally prosecuted if the advice proves to be either erroneous or inaccurate, even if this is due to erroneous or inaccurate information provided by third parties. It is therefore wise to ensure that information provided by third parties is always checked in the field and at the very least a general impression should be acquired on the quality of the information. If in doubt, the engineering geologist should estimate the possible consequences in case information is wrong and inform the client accordingly. Even if the information seems correct, it is advisable to include a waiver in the reports for all liability for information from third parties. Whether this waiver stands up in court will be different for each country, but in many countries, it reduces responsibility. It should be realized, however, that clients are generally not very impressed by consultants who try to avoid liability for their advice, and, hence, waivers should be used sparsely to avoid irritation.

4.8
An Aid to Engineering Geological Mapping

As an aid to engineering geological mapping the author developed the system discussed below. Regional scale engineering geological mapping must have purpose. The rock mass classification described below is aimed at distinguishing between those rock masses which pose no particular problems for general civil engineering to be conducted on or in them from those which will give problems. The end product is a Problem Recognition Index (PRI) (Fig. 4.14).

ITC/TU DELFT ENGINEERING GEOLOGY - FIELDFORM- ROCK MASS CLASSIFICATION MAPPING							

LOGGED BY: *DGP*   DATE: *30 March 1992*   TIME: 11:30   OUTCROP NO. *92/3/30/7*

WEATHER CONDITIONS		LOCATION	Map No. *472-1*		Outcrop size (m)		1: *10* h: *7* d: *4*
SUN: cloudy/fair/bright		Co-ordinates	Eastings *4.553.213*		Mapped on this form		1: *5* h: *7* d: *4*
RAIN: dry/drizzle/slight/heavy			Northings *323.340*		Accessibility:		Poor / fair / good

FORMATION NAME:

DESCRIPTION   (following BS5930)

Weathering	colour	grain size	structure and texture	NAME	strength
slightly	*red/brown*	*fine*	*small tabular*	*siltstone*	*moderately*

LAYER STRENGTH (LS)

    < 1.25 MPa : Crumbles in hand

    1.15 - 5 MPa : Thin slabs break easily in hand

    5 - 12.5 MPa : Thin slabs broken by heavy hand pressure

    12.5 - 50 MPa : Lumps broken by light hammer blows ✓

    50 - 100 MPa : Lumps broken by heavy hammer blows

  100 - 200 MPa : Lumps only chip with heavy hammer blows (dull ringing sound)

     > 200 MPa : Rocks ring on heavy hammer blows, sparks fly

Direction	Strength estimate	Number of sample taken for strength or durability test / Schmidt Hammer No
Perpendicular	12.5-50	*Schmidt hammer 92/3/30/75a, 5b, 5c, 5d. Slake durability: 92/3/30/7/Da, Db*
Parallel	5-12.5	

DISCONTINUITIES (DS)	B = bedding	J = joint		UNIFORMITY OF SURFACE WEATHERING	
	B1	J2	J3	PROFILE (UWP)	
Dip direction (degrees)	210	100	190	Very irregular	1
Dip (degrees)	12	85	80	Irregular	2 ✓
Spacing (m)	0.45	2.00	0.80	Uniform	3

DELETERIOUS MINERALS (tick type, location and quantity)						
	None : 1.00					
	In intact rock (%)			Fracture infill (mm)		
Type	< 5	5 - 25	>25	<2	2 - 5	>5
Easily soluble salts (halite etc.)	0.3	0.2	0	0	0	0
Gypsum, anhydrite	0.5	0.3	0.1 ✓	0.25	0.1	0.05
Swelling clays	0.5	0.3	0.1	0.25	0.1	0.05

Fig. 4.14. Field form on which basic data observed in the field is entered. The information entered is more than that required to calculate the problem recognition index and includes data for general mapping, such as discontinuity orientation (Price et al. 1996)

4.8.1
Factors in the *PRI*

The factors considered in the Problem Recognition Index are:

1. layer strength
2. layer uniformity
3. discontinuity spacing
4. uniformity of surface weathering profile
5. material sensitivity to weathering or alteration

Layer Strength Rating (LS)

Rock strength is an obvious parameter of considerable significance in most forms of engineering in rock. Ratings assigned vary from 10 for strengths less than 1.25 MPa to 100 for strengths greater than 200 MPa. However, some allowance must be given for anisotropy. In the field, layer strengths are measured in two orthogonal directions, which, in bedded rocks, would be normal and parallel to the bedding, using a Schmidt hammer or by geological hammer blows. The ratio minimum/maximum strength is calculated and the maximum rock strength rating is multiplied by this ratio to give the final strength rating for the layer; see Fig. 4.15.

Uniformity of Layer Strength (UR)

The uniformity of the geological layer is important. In an outcrop of apparently uniform lithology, strength ratings may vary or remain constant. The uniformity may be

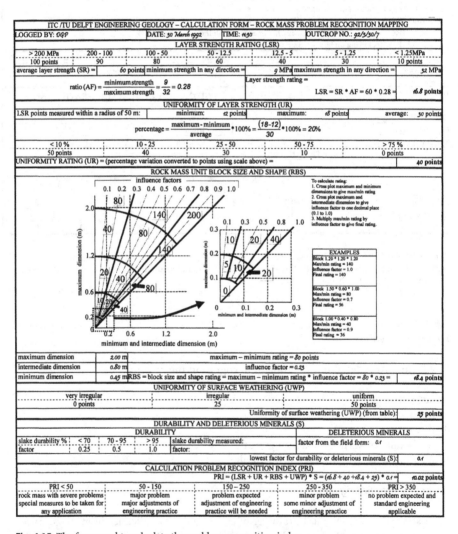

Fig. 4.15. The form used to calculate the problem recognition index

judged by making a number of strength ratings at the outcrop, at maximum within a 50 m diameter zone. The percentage variation may then be calculated on the basis of:

$$\frac{\text{Maximum rating} - \text{Minimum rating}}{\text{Average rating}}\% \tag{4.1}$$

Ratings are assigned to this parameter depending on the percentage obtained, less than 10% variation getting 50 points, more than 75% being awarded 0 points.

Discontinuity Spacing (DS)

Discontinuity spacing is important in that, for most engineering purposes, small rock blocks are a problem and large rock blocks offer fewer problems. Block shape is also important, tabular or columnar blocks causing greater problems than cubic. The figures in Fig. 4.15 offer a rating system based on block size and form, incorporating the basic idea that large cubic block blocks are best.

Uniformity of Surface Weathering Profile (UWP)

Weathering of the rock mass is important, particularly with regard to its uniformity. Experience suggests that highly irregular weathering profiles will give more problems than uniform profiles. Three simple classes are proposed, namely 'very irregular' (rating 0), fairly irregular (rating 25) and uniform (rating 50) (Fig. 4.16).

Sensitivity (S)

Some types of rock are more sensitive to weathering when newly exposed than others. This sensitivity may be established by local experience and observation or by tests, such as the slake durability test. Rocks can also contain particular minerals that may, in certain engineering processes, give problems. Minerals such as gypsum, anhydrite, smectites, and halite are examples. These problems are so important that instead of giving the subject a rating, a factor is introduced applied to the sum of all previous ratings. Factors range from 0.25 to 1 for sensitivity to weathering and from 0 to 1 for deleterious minerals. The lowest factor of the two is applied in the calculation for final rating.

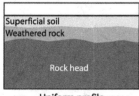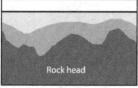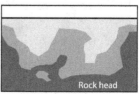

Uniform profile   Irregular profile   Very irregular profile

Fig. 4.16. Rock head surface profiles

4.8.2
Calculating the Problem Recognition Index

The end rating for the rock mass is

$$(LSR + UR + DS + UWP) \times S = PRI \qquad (4.2)$$

where *PRI* is the problem recognition index. A preliminary division has been made to give qualifying descriptions to ranges of values of the index (Fig. 4.15).

The reader will, of course, appreciate that while this description is appropriate for many engineering processes, it may be inappropriate for some. Thus 'very good' for foundation purposes could be 'very poor' in terms of difficulty of excavation. The index is intended as a guide to potential difficulty. However, a low rating could result from one of the factors, such as *UWP* being low. This may or may not be significant with regard to the particular project. The user should then look into the database to find out how the index has been calculated. Any map produced showing these indices must be accompanied by a database. Examples of the field classification form, the calculation form, and a part of a map showing *PRI* values are given in Fig. 4.14, Fig. 4.15, and Fig. 4.17.

In Fig. 4.17 *PRI* values range from 10 to 254. The lowest value relates to the engineering geological unit Tg22 (Middle Muschelkalk, which is also used as example in the forms in Figs. 4.14 and 4.15); it is composed of very weak to moderately weak clayey sandy siltstone containing large quantities of gypsum. It is highly sensitive to weathering. The highest value relates to Jurassic limestones and dolomites that are strong, and medium to very thickly bedded. Tg1 is Bunter sandstone, moderately strong and

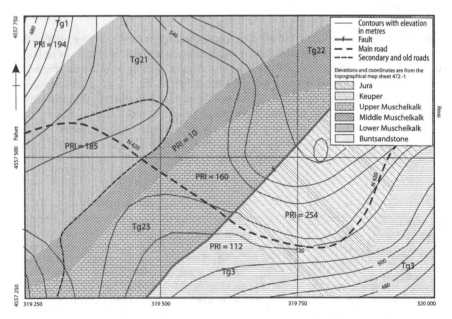

Fig. 4.17. Part of an engineering geological map of an area to the east of Falset in north-eastern Spain

very thick bedded. Tg21 and Tg23 are Lower and Upper Muschelkalk respectively, both composed of moderately strong to strong limestones and dolomites but varying in bed thickness and joint spacing. Tg3, the Keuper, contains limestones and dolomites but also weak shales. Lack of space here prevents a detailed description of each unit, but the brief descriptions given above show that the *PRI* value indicates the relative suitability of the various units for most types of engineering work.

4.9
Rock Mass Classification

A rock mass *description* gives the reader information concerning the nature and three-dimensional distribution of rock materials and discontinuities. *Characterisation* of the rock mass is achieved by undertaking in situ and laboratory testing to assign numerical values to the mass and material. A *classification* seeks to establish a rock mass quality that will determine the likely reaction of the rock mass to an engineering process to be performed on it or in it, for example, the amount of support an excavation requires to maintain stability.

4.9.1
Classification Systems

The rock mass classification systems consist of one or more relations, generally established empirically by comparing rock mass parameters to engineering parameters. This is done by:

- recognising those rock mass features which may be of importance to the successful prosecution of the engineering process,
- establishing the relative importance of these rock mass features, and
- assigning ratings to these features based on measured parameters.

The principal rock mass parameters that will be of importance for the construction of almost any engineering work in rock are:

- rock material strength,
- rock mass deformability characteristics,
- rock mass weathering,
- susceptibility to future weathering and rock material durability,
- discontinuity strength characteristics such as roughness, aperture, nature of infilling etc.,
- discontinuity persistence and spacing,
- discontinuity orientations,
- material and mass permeability,
- groundwater conditions,
- the magnitude and orientation of in situ stresses.

The first classification systems were developed for tunnelling and underground mining as an alternative for analytical and numerical stability calculations. It is very

difficult to calculate analytically the stability or support of a tunnel in a discontinuous rock mass. Many rock mass and discontinuity properties should be known in high
detail to have useful results. To justify the extra effort, the results of a numerical calculation should be better than results by other methods, This is, however, often questionable because rock mass and discontinuity properties are virtually never available
in the number and accuracy needed. Classification systems became very popular for
underground excavation (Barton et al. 1974, 1988; Bieniawski 1989; Laubscher 1990)
and have been developed and became popular for virtually every other engineering
application, such as slope stability (Hack 1998; Haines and Terbrugge 1991; Romana
1991), rippability and excavatability (Franklin et al. 1971), performance of tunnel boring machines (Laughton and Nelson 1996; Barton 2000; Poschi and Kleberger 2004),
rock mass failure criteria (Bieniawski 1989; Hack 1998; Hoek et al. 1992) and permeability.

Some systems are fully empirical and no physical or mechanical meaning can be
given to the ratings nor to the calculation methodology (for example, Bieniawski's
RMR). Other systems use relations partially based on physical and mechanical relations (for example, Hoek-Brown's failure criterion and the SSPC system). Recent developments are the incorporation of probability approaches (SSPC system, Hack 1998)
and fuzzy logic or neural networks (Cai et al. 1998; Grima et al. 2000) in classification
systems. In the following sections, some systems are briefly discussed to show the differences between them and to illustrate the most popular systems.

Deere's RQD and Franklin's Excavation Classification

Classifications may be set up on the basis of a limited number of determining parameters. Thus Deere (1968) proposed a mass classification on the single (although complex) parameter of Rock Quality Designation (RQD), as given in Table 4.2.

This classification was set up with a view to use in tunnelling and foundation engineering in which, clearly, the wider the joint spacing the greater the stability of the
tunnel and the integrity of the foundation rock mass. However, if such a classification
were applied to excavatability by digging machinery then an RQD of 90–100 would be
described as 'very poor' while that from 0–25 would be 'very good'. Other factors would
also have to be taken into account, such as whether the rock material would be likely
to break under impact of the digger bucket teeth. The two-parameter classification
proposed by Franklin et al. (1971) is general purpose (Fig. 4.18). Similarly derived
boundaries might be used to qualify ease of excavation into categories such as 'good',

Table 4.2. Rock mass quality determined by RQD

Rock mass quality	RQD (%)
Very poor	0 – 25
Poor	25 – 50
Fair	50 – 75
Good	75 – 90
Very good	90 – 100

'fair' etc. (Fig. 4.19). While the relevance of the parameters of fracture spacing and rock material strength to the process of excavation is clear there are other parameters that influence excavatability. Thus, for example, the orientation of the principal disconti-nuities relative to the direction of attack of the excavator will be important.

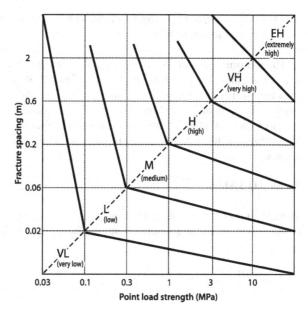

Fig. 4.18. Rock mass quality established using two param-eters (after Franklin et al. 1971)

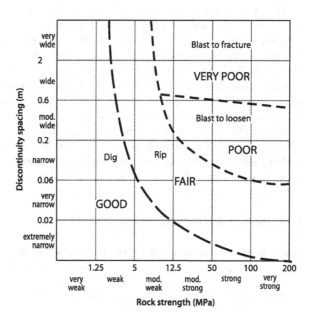

Fig. 4.19. A classification of ease and method excavation based on two parameters (after Franklin et al. 1971)

Bieniawski's RMR

Bieniawski's Rock Mass Rating (*RMR*) system (Bieniawski 1989) was developed in South Africa for underground mining. The system is based on a combination of six parameters

$$RMR = IRS + RQD + \text{Spacing} + \text{Condition} + \text{Groundwater}$$
$$+ \text{Adjustment factor} \qquad (4.3)$$

The parameters refer to Table 4.3. Each parameter is expressed in a point rating where the points are added to result in the final *RMR* ranging between 0 (very poor rock for tunnelling) and 100 (very good rock for tunnelling). The *RMR* has been related to the span and stand-up time of the excavation, and many other parameters, such as rock mass angle of internal friction and cohesion. Support of an underground excavation is determined by the *RMR* parameter and results in five different support classes.

Barton's Q-System

The Q-system of Barton et al. (Barton et al. 1974; Barton 1988) expresses the quality of the rock mass in the so-called Q-value. The Q-value is determined as follows:

$$Q = \frac{RQD}{J_n} \times \frac{J_r}{J_a} \times \frac{J_w}{SRF} \qquad (4.4)$$

The first term *RQD* (rock quality designation) divided by J_n (joint set number) is related to the size of the intact rock blocks in the rock mass. The second term J_r (joint roughness number) divided by J_a (joint alteration number) is related to the shear strength along the discontinuity planes, and the third term J_w (joint water parameter) divided by *SRF* (stress reduction factor) is related to the stress environment for the discontinuities around the tunnel opening. A multiplication of the three terms results in the 'Q' parameter, which can range between 0.00006 for an exceptionally poor rock mass and 2 666 for an exceptionally good rock mass. The numerical values of the class boundaries for the different rock mass types are subdivisions of the Q range on a logarithmic scale. The Q-value reflects the quality of the rock mass, but the support of an underground excavation is based not only on the Q-value but is also determined by the different terms in Eq. 4.4. This leads to a very extensive list of classes for support recommendations.

Laubscher's Mining Rock Mass Rating (MRMR)

Laubscher (1990) modified the *RMR* classification of Bieniawski (*RMR**). In his system, the stability and support are determined as follows:

$$RMR^* = IRS + RQD + \text{Spacing} + \text{Condition} \qquad (4.5a)$$

Table 4.3. *RMR* rock mass classification (after Bieniawski 1989)

Classification parameters

1	Strength of intact rock material (IRS) (MPa)	>250	100–250	50–100	25–50	5–25	1–5	<1
	Rating	*15*	*12*	*7*	*4*	*2*	*1*	*0*
2	RQD (%)	90–100	75–90	50–75	25–50	<25		
	Rating	*20*	*17*	*13*	*8*	*3*		
3	Spacing of discontinuities (m)	>2	0.6–2	0.2–0.6	0.06–0.2	<0.06		
	Rating	*20*	*15*	*10*	*8*	*5*		
4	Condition of discontinuities	Very rough surfaces; not continuous; no separation; unweathered wall rock	Slightly rough surfaces; separation <1 mm; slightly weathered walls	Slightly rough surfaces; separation <1 mm; highly weathered walls	Slickensided surfaces; or gouge <5 mm thick; or separation 1–5 mm; continuous	Soft gouge > 5 mm thick; or separation >5 mm; continuous		
	Rating	*30*	*25*	*20*	*10*	*0*		
5	Ground-water[a] — Inflow per 10 m tunnel length (l min^{-1})	None	<10	10–25	25–125	>125		
	Ratio[b]	0	<0.1	0.1–0.2	0.2–0.5	>0.5		
	General condition	Dry	Damp	Wet	Dripping	Flowing		
	Rating	*15*	*10*	*7*	*4*	*0*		

Rating adjustment for discontinuity orientation

Discontinuity orientation	Very favourable	Favourable	Fair	Unfavourable	Very unfavourable
Rating — Tunnels and mines	*0*	*–2*	*–5*	*–10*	*–12*
Foundations	*0*	*–2*	*–7*	*–15*	*–25*
Slopes	*0*	*–5*	*–25*	*–50*	*–60*

Rock mass classes and meaning determined from total ratings

Rating	100–81	80–61	60–41	40–21	20–0
Class no	I	II	III	IV	V
Description	Very good	Good	Fair	Poor	Very poor
Average stand-up time	20 yr for 15 m span	1 yr for 10 m span	1 wk for 5 m span	10 hr for 2.5 m span	30 min for 1 m span
Rock mass cohesion (kPa)	>400	300–400	200–300	100–200	<100
Rock mass angle of internal friction (deg)	>45	35–45	25–35	15–25	<15

[a] Groundwater should be assessed on either inflow, on ratio or on the general condition. [b] Ratio is the ratio of the water pressure in the discontinuities to the major principal stress.

$$MRMR = RMR \times \text{Adjustment factors} \qquad\qquad\qquad (4.5b)$$

The main parameters are roughly the same as for Bieniawski's RMR system (see above). The difference is the multiplication of the RMR with adjustment factors to result in the $MRMR$ value. The adjustment factors depend on future (susceptibility to) weathering, stress, orientation, method of excavation and the amount of free block faces that facilitate gravity fall. The combination of values of RMR and $MRMR$ determines the so-called 'reinforcement potential' (Table 4.4). A rock mass with a high rock mass rating before the adjustment factors are applied has a particular reinforcement potential. A high RMR rated rock mass can be reinforced by, for example, rock bolts whatever the $MRMR$ value might be after excavation. By contrast, rock bolts are not a suitable reinforcement for a rock mass with a low RMR (= has a low potential for reinforcement) even if, after excavation, the $MRMR$ is not much lower than the RMR.

The concept of adjustment factors for the rock mass before and after excavation is very attractive. This allows for compensation of local variations, which may be present in the rock mass, but might not be present at the location of the proposed excavation or vice versa. In addition, this allows for quantification of the influence of excavation and excavation induced stresses, excavation methods and the influence of past and future weathering of the rock mass to be included.

Modified Hoek-Brown Failure Criterion and Geological Strength Index (GSI)

The Modified Hoek-Brown failure criterion for jointed rock masses (Hoek et al. 1992) is formulated as follows:

$$\sigma_1' = \sigma_3' + \sigma_c \left(m_b \frac{\sigma_3'}{\sigma_c} \right)^a \qquad\qquad\qquad (4.6)$$

where σ_1' and σ_3' are the major and minor principal effective stresses at failure.

The rock mass parameter σ_c (intact rock strength) is derived from a field estimate. The parameters m_b and a are derived from a matrix describing the 'structure' and the 'surface condition' of the rock mass. The 'structure' is related to the block size and the interlocking of rock blocks while the 'surface condition' is related to weathering, persistence, and condition of discontinuities. Considerable development of this approach has been made to cater for extremely fractured rock masses where no sets of clearly defined through-going discontinuities exist; such rock is often severely tectonised; this approach has produced the Geological Strength Index (Marinos and Hoek 2000).

Hudson's Rock Engineering Systems (RES)

The Rock Engineering Systems (RES) methodology developed by Hudson (1992) and Cai et al. (1998), relates the *interaction* of parameters that have an influence on engineering in discontinuous rock masses. This includes those that have an influence on the engineering structure and those that influence other parameters. These influences are quantified and result in a rating for a parameter of the engineering structure. This

Table 4.4. Support classes following Laubscher's *MRMR* system (after Laubscher 1990)

	RMR									
	1A	1B	2A	2B	3A	3B	4A	4B	5A	5B
MRMR	⇐ Rock reinforcement							Plastic deformation ⇒		
1A										
1B										
2A										
2B	a	a								
3A	b	b	a	a						
3B	b	b	b	b	b	c				
4A	c	c	c	c	c	d	d			
4B				d	e	f	f	c-l		
5A						f/p	h-f/p	h-f/l	h-f/t	
5B							h-f/p	f/p	t	t

Rock reinforcement

a Local bolting
b Bolts at 1 m spacing
c b and straps and mesh if rock finely jointed
d b and mesh/steel-fiber reinforced shotcrete, bolts as lateral restraint
e d and straps in contact with or shotcreted in

f e and cable bolts as reinforcing and lateral restraints
g f and pinning
h Spilling
i Grouting

Rigid lining (low deformation)

j Timber
k Rigid steel sets

l Massive concrete
m k and concrete

Rigid lining

n Structurally reinforced concrete

Yielding lining, repair technique, high deformation

o Yielding steel arches

p Yielding steel arches set in concrete or shotcrete

Fill

q Fill

Spalling control

r Bolts or rope mesh lash

Rock replacement

s Rock replacement by stronger material

t Avoid development

last-named parameter can be, for instance, the stability or instability of a tunnel or slope. Parameters can describe properties of a rock mass, such as intact rock strength, discontinuity orientation, etc., and external influences on either rock mass parameters or engineering structures, such as climate, geomorphologic processes, etc. The quantification of all the interactions results in a matrix from which the required parameter, for example, the stability of a tunnel, is determined. Quantification of the interactions or influences between parameters, and between parameters and engineering structure, can have any form; for example, differential equations, binary operations (0 or 1, for example, for features that are either present or not present), classifications, or numerical calculations. How these relations are established (e.g. by engineering judgment or actually proved by testing) is of no importance. The reliability and accuracy of the final result depends on the reliability and accuracy of the relations (and obviously of the input data). The methodology resembles the working of a neural network, where the relations between input and output parameters in a neural network are normally of a simpler form. The methodology is not a classification system, but rather a methodology of thinking for engineering in or on discontinuous rock masses; the reader can be forgiven for imagining that in many respects RES "reinvents" Engineering Geology.

Slope Stability Probability Classification (SSPC)

Another approach is that adopted by Hack (Hack 1998; Hack et al. 2002) who devised a rock mass classification system to estimate the probability of the stability of existing slopes or slopes to be cut. The data required to calculate slope stability is observed on the rock mass exposed. However, this has suffered weathering and perhaps also been disturbed by the methods used to excavate it, or by slope movements, and does not represent the mass into which any new slope is to be made. Compensation factors are employed to modify the parameter values to give an imaginary 'reference rock mass' which is unweathered and undisturbed. The 'slope rock mass' is that rock mass in which a new slope is to be cut, and where, by the application of correction factors, allowance may be made for disturbance to be brought about by excavation method and weathering during the engineering lifetime of the slope. The final result is an assessment of the stability of the slope (existing or new), expressed in terms of percentage probability.

4.9.2
Discussion

Rock mass classification systems have been modified since their inception and may well be further modified with increasing experience. The failure of an intact piece of rock is very well investigated in laboratory testing and described in various so-called 'failure criteria'. The failure criteria for rock masses, however, are seldom directly described for the simple reason that a rock mass contains discontinuities and to measure the strength, including the influence, of the discontinuities requires mostly impossibly large samples. An alternative to a failure criterion can be provided by a rock mass classification. The above named Modified Hoek-Brown failure criterion (Eq. 4.6) for jointed rock masses is a "failure criterion" developed specially for this purpose; Bieniawski's RMR (Table 4.3) and SSPC systems have been related to the angle of internal friction and cohesion of a rock mass as are used in the Mohr-Coulomb failure criterion.

A potential difficulty with this thinking may be that as increasing modifications improve the systems they may also inhibit the development of new approaches and, with ever increasing complexity, become less 'user friendly'. It is evident that rock mass classification systems are established and are important in all fields of rock engineering. The use of rock mass classification to translate qualitative judgement to semi-quantitative opinion will prove a powerful weapon in the engineering geological armoury. However, while the modification and improvement of existing systems as a consequence of accumulated experience is clearly of advantage, it also means that any basic defects in the system are maintained. Increasing complexity may also imply the need for specialised training and experience to apply the system successfully. Ideally, any system is operator independent so that any two engineering geologists, without special training, should be able to apply it and arrive at more-or-less the same result. An important point to realize is that the systems are completely or partially empirical. Relations may or may not be valid for other data that is not similar to the data included in the original data set. Therefore, always check whether newer versions and modifications have been published, and whether the system is applicable to the type of rock mass or application for which it is going to be used. It is also advisable to be critical of the existing systems, and not be afraid to challenge a newer system, or develop your own system, even if older and more popular systems exist.

4.10
Further Reading

Barnes JW (2004) Basic geological mapping, 4th edn. Geological Society of London
Griffiths JS (ed) (2002) Mapping in engineering geology. Geological Society of London
Spencer EW (2000) Geologic maps; a practical guide to the preparation and interpretation of geological maps for geologists, geographers, engineers and planners, 2nd edn. Prentice Hall

Chapter 5

Recovery of Samples

5.1
Purpose and Principles

The first purpose of site investigation is to acquire the data needed to create a three-dimensional geotechnical model of the ground that will be encountered and affected by the construction of the project. This model must represent:

i   the sizes and shapes of the different bodies of materials in the ground,
ii the geotechnical properties of those materials that are relevant to the project,
iii the distribution, orientation and engineering characteristics of the structural discontinuities in the mass, and
iv the location and behaviour of groundwater.

The second purpose of site investigation is to use this information for predicting the reaction of the ground to the construction of the project.

Access to the ground must be obtained to gain information on the factors required to establish the model. This is mostly done by opening some form of excavation, by describing the materials and features encountered, and by taking samples to be tested in the laboratory. Significant data regarding boundaries, discontinuities and mass properties may also be obtained by geophysical surveys.

The excavation may be in the form of a trial pit, a shaft or a tunnel, although the first is usually employed for minor projects, such as small uncomplicated housing developments, while the last two are usually associated with in situ tests for major projects. Most commonly, access is provided by boreholes which provide both descriptive data and samples. Samples taken are commonly described as either '*disturbed*' or '*undisturbed*', the former being generally useful only for identification (or simple tests leading to identification) and the latter for formal description and testing. No sample is truly 'undisturbed' because of the process of sampling and the fact that its stress condition must have been changed by removal from its confined condition in the ground. Various classes of sample may be defined according to the method of collection, and the class required should be defined in the site investigation specification. Generally, the higher the class required, the more expensive it is to sample, and a high class should not be specified if a lower would be adequate for the purpose. Conversely, if design parameters are to be based on the results of a sophisticated testing programme, high classes of sample are necessary. No matter how sophisticated the test, the results can be no better than the quality and representativeness of the sample obtained.

Consequently, almost all site investigations at some stage involve the drilling of boreholes whose primary purpose is to gather information about the nature and distribution of geological materials and the discontinuities that ramify through the mass, and to obtain samples which can be examined and tested in the laboratory. Frequently, boreholes are also used as a means of access into the mass for in situ tests, to allow inspection of the strata on the walls of the borehole, and to install instruments for measurement of any changes which might occur in the ground conditions – e.g. piezometers, load cells, inclinometers etc.

The site investigator is necessarily faced with the problem of deciding what, where, and how to sample. Clearly, the frequency of sampling must be related to the complexity of the geology, and the type of sampling must be related to the significance of the geological conditions to the engineering project. Ideally, sampling should be continuous and provide a completely representative section of the strata penetrated by the borehole. This is possible to limited depths in soft soils by means of special long samplers pushed into the soil, for example the Delft continuous sampler, or some of the long piston samplers. In rocks, continuous samples of core may be obtained by good quality rotary drilling.

Unfortunately, many soils are either too strong to allow deep continuous sampling or too weak, or with too many boulders, for successful coring. In such circumstances, discontinuous sampling procedures must be employed, and the location of samples required must be decided before drilling the borehole, unless the site investigator intends to be present throughout the operation.

It cannot be stressed too strongly that the answer to many of these problems, and indeed to many problems of site investigation generally, is to have some advance knowledge of the geological conditions at the site. This may be obtained by *a study of existing geological information*, but in some cases may require a cheaper preliminary *reconnaissance investigation*.

It is useful to compare the volume of ground that may be affected by the engineering construction with that sampled. Thus, for example, a 50 m square building might be investigated by 5 boreholes sunk, at corners and centre, to the lowest limit of the bulb of stress (about 25% stress increase) imposed by that building. If the boreholes were continuously sampled with 100 mm diameter samples, the total volume of sample taken would amount to no more than about 0.005% of the volume of ground stressed. Major engineering decisions may be based on this very small proportion of sample although so many boreholes to such a depth might be considered a rather generous provision by many clients!

5.1.1
Drilling and Sampling

For drilling boreholes there are a number of commonly used methods, although each has many variations of equipment and technique. The most important methods are the following:

- rotary core drilling
- rotary "open hole" drilling
- percussion boring with cable tools

- power augering, of various types
- percussion "down hole hammer"
- hand augering
- probing

Site investigation frequently involves one or more of these methods, although some are more commonly used than others. There are many other techniques for drilling; for example jetting or wash boring is often used in weak alluvial sediments, but their application to site investigation is minimal. All methods have advantages and disadvantages, whether in terms of cost, information obtained or quality of samples. While it is relatively easy to sink a borehole, it requires skill and appropriate equipment to recover samples of an adequate quality for engineering tests; good core recovery requires good drillers using good equipment.

5.2
Drilling and Sampling in Rock

Very weak rocks may be drilled with a power auger using a 'fish tail' bit which overcomes the shear strength of rock during rotation. All rocks may be drilled by the percussive 'down hole hammer' which can achieve penetration rates superior to any other method. With this method the percussive bit is driven by compressed air, with the drive unit located in the hole immediately above the bit. In this case the samples of rock consist of small chips about the size of a finger nail which are carried to the surface by the exhaust air. In the previous case the samples are of similar size, but a flushing medium like water must be used to wash the cuttings to the surface. Such samples can really be used only for identification of the strata penetrated.

In most circumstances, including those described above, rock is drilled by a rotary method. Rotary rigs using a toothed roller bit overcome the compressive strength of the rock; the rock chips must be driven to the surface by a flushing medium which may be air, water or drilling "*mud*" (which can be a clay suspension or one of many polymers now available). Such a method may be employed for reconnaissance or when no samples are required for testing, as where drilling is undertaken to determine whether cavities exist beneath the site. When rock samples are required for inspection and testing, core must be obtained. For this purpose a rotary drill with an annular bit is employed.

5.2.1
Rotary Core Drilling

The principle of core drilling is essentially simple. If a tube, armed at its base with 'teeth' of a material harder than the rock to be drilled, is rotated and pushed downward, a circular annulus will be cut out of the rock and a cylindrical core will pass into the tube (Fig. 5.1a). Commonly the 'teeth' consist of industrial diamonds set into a metal matrix to form the *core bit*. This becomes heated by friction and must be cooled by a flushing medium which also removes rock cuttings and transports them to ground level. To prevent cuttings jamming between the bit and the sides of the hole, the diameter of the hole may be enlarged slightly by a diamond studded reaming shell

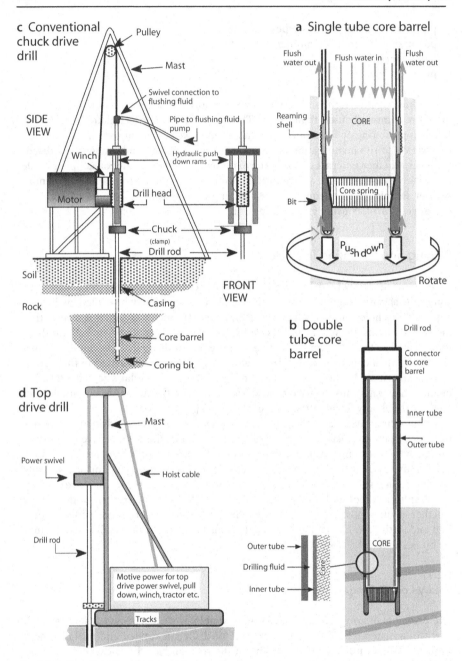

Fig. 5.1. Drills and drilling equipment for rotary coring

screwed in place above the bit. The bit contains a '*core spring*' (a split cylinder with internal teeth) which allows the core to pass upward into the core barrel but which, when the core barrel is pulled up to surface, clamps onto the core so that it breaks somewhere between core spring and the bottom of the borehole. In the single tube core barrel shown in Fig. 5.1a, the flushing fluid, commonly water, flows from above over the core and thence to the bit. This may erode the core and this, together with damage brought about by contact with the rotating barrel, reduces core recovery. For this reason the single tube core barrel is seldom used in rotary core drilling.

Most core drilling today is done with double tube core barrels (Fig. 5.1b). These contain an *inner tube* which, being mounted on bearings at the top of the *barrel,* does not rotate with the outer barrel. Flushing fluid passes between inner and outer tubes to the bit. The core enters the inner tube and is thence protected against abrasion from the rotating outer tube and erosion from flushing fluid.

5.2.2
Drills

The rotary core drilling rig (Fig. 5.1c) has to provide for the rotation of, and the vertical load on, the coring bit. Drilling rods connect the core barrel to surface and allow the passage of flushing fluid down to the core barrel. In what may now be termed the conventional chuck drive drill, the drill rod passes through a drill head to which it is clamped via a 'chuck'. Rotation of a horizontal axle driven by the motor is translated into rotation of the vertical hollow stem to which the chuck is attached by means of a gear box. Downward push on the coring bit is usually given by hydraulic rams powered by the motor. The feed stroke of the chuck drive and rams is usually of the order of 0.5 to 1.5 m, less than the length of most core barrels, which are commonly 3 m long. This means that when the chuck has reached the limit of its stroke it must be loosened from the drill rod, drawn upwards and reclamped. During this time it is possible that the sides of the borehole might partially collapse and jam against the core barrel or rods.

Top drive drills have a power swivel, commonly hydraulically operated, at the top of the drill rods and have a feed stroke equal to or greater than the length of the core barrel. Thus they lessen the chance of borehole collapse and reduce the proportion of unproductive time spent in periodically raising the rods and core-barrel to the surface. For this reason they are often preferred for deep boreholes and for investigations over water, which are vulnerable to weather and tidal changes.

Chuck drive drills have to be carried between borehole locations, although truck mounting is common. Top drive drills may be mounted on trucks or some form of tracked vehicle. Both forms of drill have to be levelled to ensure verticality of boreholes and most truck or track mounted versions incorporate levelling jacks. Both types of drill can usually drill inclined boreholes. If the weight of the drill is less than the push down force to be employed then the drill may have to be anchored down by some form of soil or rock anchors. Chuck drive drills are usually anchored down when drilling inclined boreholes.

Both forms of drill incorporate winches to pull drill rods and full core barrel to surface and these must be strong enough not only to lift the weight of these tools but also to break the rock between the core spring and the bottom of the borehole.

Most site investigation boreholes begin in soil and rock may be reached by percussion drilling. In soil the hole must be lined with casing to prevent collapse. Weak sections in rock may also have to be lined, the casing being drilled into place with the aid of a casing bit.

Core barrels come in various sizes which by convention (BS4019 1974) are designated by letters (Table 5.1) which also define the size of the associated drill rods and casing. Additional letters are used to describe the various mechanical characteristics of rods, casing and barrels. Thus, for example, casing could be described as NX or NW, the second letters relating to mechanical design of the casing. Core barrels could be described as NWF or NWM, the third letter relating to core barrel design. Other core sizes exist (or have existed). One system of core sizes relating to equipment produced by Craelius (a drilling equipment manufacturer) is given in Table 5.2.

Table 5.1. Standard core sizes (BS4019 1974)

Size	Hole diameter (mm)	Core diameter (mm)
Z	199.3	165.1
U	173.9	139.7
S	145.2	112.7
P	120.0	92.1
H	98.9	76.2
N	75.4	54.0
B	59.5	41.3
A	47.5	30.2
E	37.4	20.6
R	29.5	17.5

Table 5.2. Craelius metric sizes

Hole diameter (mm)	Core diameter (mm)	
	Thick walled bits	Thin walled bits
146	116	–
131	101	–
116	86	–
101	72	84
86	58	72
76	48	62
66	38	52
56	34	42
46	24	32
36	–	22
28	–	16

In general terms the depth to which a hole may be drilled and the size of core recovered is related to the power and weight of the rig. Manufacturers usually quote a depth rating for the machine at given hole sizes.

5.3
Drilling Tools

The ideal objective when core drilling is to achieve 100% recovery of sample. In general terms, the larger the core diameter the better the possibility of maximum recovery, so at shallow depths in the weathered and weakened rock mass coring begins using a fairly large size, perhaps S (113 mm diameter). The size may be reduced as the rock mass improves with depth, perhaps to H (76 mm diameter) and thence to N (54 mm diameter). To do this, sizes of drill rod, casing etc. have also to be reduced so that B casing fits inside N casing which fits inside H casing and so forth. There is a general 'telescopic' reduction in sizes from Z to R. Table 5.3. shows the diameters of drilling tools for N and H sizes.

5.3.1
Core Bits

Bits used for coring are cylindrical and annular. The edge is studded with hard abrasive materials, sometimes tungsten carbide, but more usually diamond. There is a large variation in the design and cost of core bits, reflecting their relative performance and durability in different rock types.

With diamond bits, the most important variations concern the shape of the abrasive face, which is called the 'kerf' or sometimes the 'crown' (and which may be anything from planar to fully rounded), the width of the kerf, the arrangement of the diamonds and the nature and composition of the matrix into which the diamonds are set. The matrix usually consists of sintered metal which may be impregnated with diamonds in a cheaper bit, or the diamonds may be hand set in a specific pattern of different sizes in more expensive types. The diamonds set into the matrix may be varied in size and quality depending on the rock type to be cut. Soft weak rocks can be cut with large diamonds of low quality while hard strong rocks require smaller high quality diamonds for efficient cutting. (Here 'soft' and 'hard' are used in the strict sense of

Table 5.3. Dimensions (in mm) of drill rods, casing, coring equipment and cores. Hole and core diameters are ± a few mm depending on wear of the equipment

Size	Drill rods O.D.	Casing I.D.	Casing O.D.
NW	66.8	76.2	88.9
HW	89.0	101.6	114.3

Core barrel	Core bit O.D.	Reaming shell O.D.	Hole diameter	Core	Size
NWF	75.3	75.7	±76	±55	N
HWF	98.8	99.2	±99	±76	H

relative abrasivity). Coring bits become worn and may reach the point where continued use is inefficient or diamonds may fall out of the matrix. They should then be withdrawn from service and returned to the manufacturer who may extract, and give credit for, those diamonds that can be re-used.

Other variations in design concern the shape and arrangement of waterway 'ports'. The face discharge design has ports which discharge the flushing fluid at the cutting edge of the bit, significantly reducing erosion of the core. Unfortunately these ports also weaken the bit which may become damaged when drilling in hard rock.

5.3.2
Core Barrels

Details of the design of barrels vary from one manufacturer to another, but they come in three types – single, double and triple tube. The requirement of the barrel is to retain the core and protect it as much as possible from damage caused by rotation of the barrel, contact with the flushing medium, raising the core to the surface and removing it from the barrel. Generally, the greater the complexity of the barrel the greater the protection, but more complex barrels are more expensive and are subject to higher maintenance costs as well as being more prone to breakdown.

The *triple tube core barrel* affords extra protection to the core and may be used where necessary, for example in coring of extremely friable rocks, such as Quaternary sequences and fault gouges, but it is not generally used in most site investigations, and may not be available for that reason. It is expensive and more subject to breakdown because of the additional complexity, but the principles of design remain the same.

Designs of core barrel vary but largely aim to reduce disturbance of the core once drilled. 'M' design core barrels have an inner tube extension which incorporates the core spring and reduces flushing fluid erosion of the core.

The inner tube may be lined with a cylindrical semi-rigid plastic liner into which the core passes. This considerably aids the successful extraction of the core from the inner tube (Binns 1998) and is particularly valuable in soft ground.

5.3.3
Flushing Media

With the exception of power auger drilling, all machine rotary drilling is carried out with the aid of a flushing fluid which may be air, water, drilling mud, polymers or a chemically based foam. The flushing fluid has the primary purpose of transporting the rock cuttings out of the borehole so that drilling may proceed in newly presented rock at the bottom of the borehole. The other important function of the flushing medium is to cool and lubricate the bit.

Water is the flushing medium most used because of its low cost and general availability, although of course this may not be the case in arid countries. It is dense enough to transport cuttings if the flow rate is reasonably high, and it is effective as a coolant and lubricant. The main disadvantages are the damage it does to the rock core by softening and erosion, and its loss in permeable strata.

Drilling mud is a mixture of water and sodium montmorillonite (bentonite), often with additives like barite (barium sulphate) to give the required properties of density and viscosity. Because of its high density, drilling mud transports cuttings at quite low velocities, and it also provides support for the sides of the borehole by counterbalancing the overburden pressure. In addition, it cakes the sides of the hole and seals any discontinuities which might otherwise reduce the fluid return. However, for obvious reasons, if the hole is to be tested for permeability, drilling mud should not be used as a flushing fluid.

Within the last decade a wide range of specialist drilling products has been developed which could, if desired, replace conventional bentonite as a flushing medium. Different types are produced (often polymers) to suit different techniques, including bio-degradable types to reduce pollution of aquifers, and others which can be rapidly degraded to solve the problem of permeability testing described above.

Core drilling may be carried out using '*stable foam*' as the flushing medium. The foam may be used to condition conventional drilling mud, or more usually, it can be used on its own by either high velocity or low velocity injection. The two main ingredients in the foam mix are a very strong wetting agent, called the foaming surfactant, and a high molecular weight polymer. Air is used to move the foam up and down the hole. Foam may be used without the polymer, but once water has been encountered or when it is necessary to support the sides or the hole, the polymer must be added. The foam works by providing the necessary cooling and lubrication, and the foam bubbles carry the rock cuttings up to the surface. It is also effective in sealing and supporting the hole.

There are many advantages in using stable foam when good quality samples in weak rocks or even unconsolidated sediments are required. The low velocity of the foam does not produce as much erosion and therefore less disturbance than other media, and the foam is degradable in a short time. Its use for site investigation work seems assured, and good results have been attained in areas where sampling is usually very difficult by other methods, for example in Hong Kong where samples must be taken of highly weathered or decomposed granites, and where sampling disturbance alters significantly the properties of the material. The main disadvantage is the relatively high cost, but if the whole purpose of the drilling is to obtain good samples, then the additional cost must obviously be worth while. In addition, much smaller quantities of water are required than for conventional mud.

5.3.4
Core and Core Barrel Sizes

Core barrels are usually 3 metres or 10 feet in length, and unless drilling is stopped for some specific reason, the length of the core barrel is the length of the core run. The size of core which it is necessary to obtain is usually governed by the purpose for which it is intended. There may be special testing requirements which demand a certain size, but in most rocks a diameter of not less than 70 mm (approximately H size) is adequate for good recovery, proper examination and testing. In massive, strong rocks, a diameter of 55 mm (N size) is usually adequate, while for very weak and/or friable rocks it may be necessary to obtain 100 mm or even 150 mm diameter (S, U, or Z size) core. It is worth noting that if large diameter core is specified from considerable depth, the lifting capacity of many drilling rigs may be exceeded, and it will be necessary to use more powerful rigs simply to enable the full core barrel to be lifted.

5.3.5
Extraction and Storage of Cores

Rock core is expensive to obtain, and it should therefore be very obvious that careful extraction from the core barrel is an absolute necessity. Careless handling and consequent core damage frequently occurs at this stage, and all suggestions to 'loosen' the core by 'tapping' the barrel with a crowbar or club hammer should be firmly rejected. Core can often be successfully extracted from the barrel using a piston extruder, provided it is done carefully. However, even this may damage the core. Alternatively split inner tubes for core barrels may be used; these open longitudinally to allow examination and measurement of the core without the need to remove it. Plastic liners for barrels may also be obtained so that the core is effectively sealed prior to removal. In all cases the barrel should be held horizontally, and when the core is extracted it should be slid on to a rigid plastic sheet corrugated to the same size as the core diameter. The core can then be placed in the channel of the core box, preferably after sealing with plastic wrapping.

   Ideas on the storage of cores vary (the writer was once presented with cores in a sack) but good practice indicates that a properly constructed core box must be available for the immediate receipt of the core. Such a box must have a hinged lid and channels of a size to fit core with plastic wrapping. The capacity of the channels should be enough to take the whole core run, usually 3 metres, plus wooden blocks which may be necessary to mark particular points in the run or to mark the top and bottom of a run if it is contained in more than one box, as is sometimes necessary. The cores should be put in the box so that they can be examined as a book is read. Boxes with core should not be so heavy that they cannot be easily lifted by two men, otherwise they will be dropped; larger diameter cores may be stored in single row boxes 1 to 1.5 m long.

   Ideally, the first examination of the core should take place on site before any transportation, but this is not always possible. In that case, the core should be transported with great care to be stored in a suitable environment protected from the weather and without drying-out or being heated or cooled, until it can be logged and described accurately

5.4
Drilling Process

The object of rotary core drilling is to obtain a continuous sample of the rock, as near as possible in the condition it was in situ before the hole was drilled. The continuity of the sample is assessed as core recovery, which is the length of core brought to the surface divided by the depth drilled in that run and expressed as a percentage. Loss of core should be attributable to the condition of the ground, and should not be due to poor drilling techniques.

5.4.1
Core Recovery

The engineering geologist may well ponder the significance of the lost core as much or even more than the properties of the core recovered. Some clues as to the nature of the material 'lost' may be gained from a study of the driller's log where gains or losses of flushing fluid, penetration rates and other information about the actual mechanics of the drilling should be recorded in detail.

There are many possible factors in drilling which can result in poor core recovery. Inappropriate drill size is one of the most common. Apart from the need to obtain core of the correct size, as outlined above, small size almost invariably makes good recovery more difficult. Holes which start in weathered rock should take large cores, reducing in size as fresher rock is drilled. Rigs which are not properly anchored may vibrate and cause core breakage and loss, which may be caused also by a number of other factors, most commonly a too high rotation speed. Another cause is that, when the barrel is lifted the core is broken off at the bottom of the hole, but if the core is already naturally broken by a discontinuity, a few pieces may fail to be retained by the catcher and remain in the hole to be over-drilled at the beginning of the next run. It is surprising how many core runs end in a natural fracture.

To achieve the best possible core recovery the coring must be undertaken with the correct blend of drilling machine, flushing media and size and type of drilling tools, core barrel and coring bit, all operated by expert drillers using rotation speeds and drilling pressures appropriate to the kind of rock mass being sampled. Such excellence used only to be achieved in the mineral exploration and extraction industry where the geology surrounding an ore deposit is well known, many boreholes have been sunk and time has been available for the development of the optimum design of drilling bits; however the investigations in the UK for sites to dispose of nuclear waste repeatedly produced core recovery of close to 100%, and good recovery is now to be expected from ground investigations.

In site investigation the geology may not be well known (otherwise there would be no need for an investigation), the number of boreholes may be relatively few and the duration of the investigation too short to allow for much experimentation or to order changes in equipment. Accordingly the best policy may be to drill cores larger than those which may seem sufficient for the assumed geology, using drills of more than adequate depth capacity. The advice of coring bit manufacturers should be sought for the selection of bits.

5.4.2
Efficiency in Drilling

Efficiency is generally judged on core recovery and the depth drilled per working shift. Once the core barrel is at the bottom of the hole the rate of penetration, bit pressure and speed of rotation chosen by the driller must be that which gives maximum core recovery in the rock type being drilled, and because the rock type is likely to change, so must the other parameters. The object is to recover the best quality samples and not just to drill a hole as quickly as possible. However, overall drilling efficiency depends not only on the speed of drilling in rock, but also the speed of hoisting the filled core barrel to the surface, removing the core, and returning the empty barrel and drill rods down the hole, with changed bit when necessary. All drilling rigs should have several core barrels available, so that a full one can be exchanged for an empty one and returned quickly while the core is carefully extracted. Tall derricks on the rig allow a longer pull and quicker disassembly of the rods.

Special rods and barrels are available to enable the barrel to be hoisted to the surface without the need to remove all the rods. The barrel is attached to a wire cable which passes through the inside of the rods which therefore have to have a larger internal diameter. Since the outside diameter is defined by the size of the hole, it follows that the walls of the rods must be thinner, and must also therefore be made of stronger materials. The barrel must also be of the same special '*wire line*' design. The whole drilling assembly is therefore more expensive, but the time saved by the facility for raising and lowering the barrel without removing the rods is very valuable. Obviously, the deeper the hole the more the time saved, and it is always a question of balancing the cost of more expensive equipment against the saving of time. Most site investigation boreholes are not deep enough to benefit from the wire line system, but it is always worth considering if the hole is about 100 m or more in depth. For marine work, where the depth of water has to be added to the depth of the drilled rock, wire line is obviously much more frequently appropriate.

5.4.3
Integral Core Sampling

Cores recovered from the borehole may have less than 100% recovery and the orientation of the cores is uncertain. In particular it may be very uncertain whether or not discontinuities are open or closed. Integral core sampling is a method of binding the core together so that the core emerges from the borehole as one solid stick. If a large diameter borehole is drilled (say H size) a smaller diameter pilot borehole may be drilled down from the centre of the larger hole through the length of the next core run. A steel bar or tube is thence grouted into the smaller hole, some of the grout flowing into any open discontinuities. The reinforced rock is thence overcored at the larger size. If the orientation of the reinforcing bar is known at surface then the orientation of the discontinuities may be established and the presence of grout in the discontinuities may indicate their openness. The process is illustrated in Fig. 5.2.

The process is expensive, for within a particular depth range two holes have to be drilled instead of one conventional hole, and the installation of the bar and the grouting involves additional work and time; geophysical borehole logging has now largely removed the need for such sampling because it can record the depth and orientation of fracture traces on borehole walls.

5.4.4
Oriented Cores

There are various devices which will impose marks on the rock at the bottom of a borehole which may be recovered in the next core run. If their orientation is known and the cores fit well together discontinuity orientation may be established by measurement relative to the marks. Much more sophisticated but relatively costly methods for assessing orientation are available within borehole geophysical logging packages but these are usually justified only in special projects either requiring deep boreholes or crucially affected by rock microstructure.

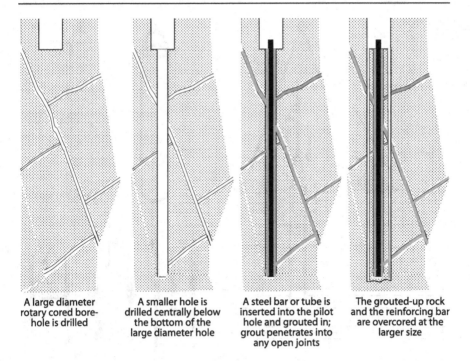

| A large diameter rotary cored bore-hole is drilled | A smaller hole is drilled centrally below the bottom of the large diameter hole | A steel bar or tube is inserted into the pilot hole and grouted in; grout penetrates into any open joints | The grouted-up rock and the reinforcing bar are overcored at the larger size |

Fig. 5.2. Integral core sampling

5.5
Drilling and Sampling in Soil

A hole may be made by repeatedly dropping a pointed pole on the ground. If the pole is hollow, soil will enter the pipe and a sample will be recovered. Thus the most common method of recovering a sample of soil is to push or hammer a tube into the ground. One of the earliest forms of drilling rig to do this was the Banka drill, which was entirely man powered and used to prospect for alluvial minerals and, in the writer's experience, for gold and diamonds. Casing was driven into the ground and samples extracted from within it using a sampling tool (the shell) raised and dropped to depths of the order of 10 m by man power. Drillers came sometimes from families devoted to this work and started at an early age; by maturity they had achieved a physique suggesting that it would be unwise to quarrel with them.

However, today, while drilling is still heavy work, it is usually undertaken by a motorised light percussion rig, properly called a cable tool percussion rig, but often still referred to by the traditional 'shell and auger' name, despite the fact that the bucket auger is now rarely used. A typical site investigation light percussion rig with tools is illustrated in Fig. 5.3a. The essential features are the collapsible quadruped A-frame with a pulley at the top, an engine to provide the power, and a winch to raise and lower

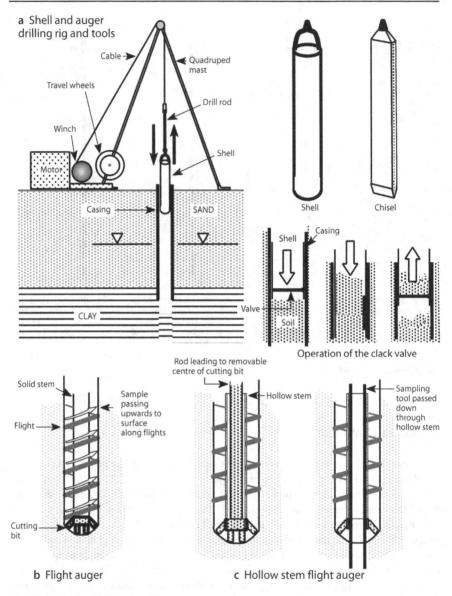

a Shell and auger drilling rig and tools

Cable
Quadruped mast
Travel wheels
Drill rod
Winch
Shell
Motor
Casing
SAND
Shell
Chisel
Shell
Casing
CLAY
Valve
Soil
Operation of the clack valve

Solid stem
Rod leading to removable centre of cutting bit
Sample passing upwards to surface along flights
Flight
Hollow stem
Sampling tool passed down through hollow stem
Cutting bit

b Flight auger   **c** Hollow stem flight auger

Fig. 5.3. Methods of drilling in soil

the wire cable to which the tools are connected. The A-frame is equipped with wheels so that the whole machine can be towed as a trailer behind a light truck or 4-wheel drive vehicle. Such rigs may be obtained in several sizes, each bigger and more powerful than the last.

5.5.1
Percussion ("Shell and Auger") Boring and Driven Tube Samples

Site investigation cable tool rigs generally bore holes between 150 and 450 mm in diameter, with about 6 sizes of tools available within this range. Since the holes are bored in soil, they are not self supporting and of course there is no drilling mud or foam to assist. Consequently, the holes must be lined with steel tube, usually in threaded lengths but the sections may be welded, particularly if for some reason the casing is to be left in place after boring has finished. The casing diameter must be slightly bigger on the inside than the boring tools, and it has at the leading edge a cutting 'shoe' to facilitate driving. A strong driving shoe is fixed to the upper end while driving by drop hammer.

There are many variations of boring tools, but the most frequently used are the '*shell*' or '*baler*' and the '*clay cutter*'. The clay cutter is a heavy steel tube with a long elliptical slot cut into the side. It is dropped into the hole, with additional weight being given by a '*sinker bar*' which is a heavy rod, immediately attached. The cutter is hauled to the surface after several strokes, and the clay removed with a metal bar through the slot. In granular materials the shell is surged in water at the bottom of the hole. The rapid movement disturbs the granular soil and puts it into suspension in the water which enters the shell on the down stroke. There is a non-return valve (*clack valve*) situated near the base of the shell, and this closes on the upstroke to retain the water and suspended soil (Fig. 5.3). The shell is drawn to the surface and emptied. It is common practice to add some water to the borehole to facilitate the progress of boring. This combination of shell and clay cutting is widely, but inaccurately, termed 'shell and auger' boring, although the auger is seldom involved.

Given time, this method can drill through almost any material, if necessary using a chisel tool. However, gravel and boulders, particularly if the boulders are not firmly held, create difficulties. The gravel cannot be suspended in water and therefore not removed by the shell. Boulders may be broken by the chisel, but there remains the problem of removing the coarse debris. Driving casing in such materials is very difficult, and sometimes, because the casing cannot be kept at or near the bottom of the hole, the bore has to be abandoned. Clearly, samples obtained by use of these tools will be greatly disturbed and may be used only for identification, or in the case of clays, for simple index tests.

The most common type of 'undisturbed' sample for laboratory testing is the driven tube (or '*open drive*') sample. These are taken whenever there are changes of strata or at regular intervals otherwise. Usually, not less than 1 m or more than 1.5 m between sample tops is the frequency of sampling. The boring tools are detached and replaced with rods equipped with a drop hammer. The sample tube in Britain is usually 450 mm long and 100 mm in diameter (Fig. 5.4). It is equipped with a hard steel cutting shoe at the front end, and a driving shoe at the other. It is then driven into the ground with the drop hammer. A clack valve allows the releasing of air and/or water as the tube is driven. When full, the assembly is returned to the surface where both shoes are removed and replaced with threaded caps after the sample has been sealed with wax to preserve as far as possible the original condition. The 100 mm diameter sampler is generally known as the U100 open drive sampler. U stands for 'undisturbed' and the

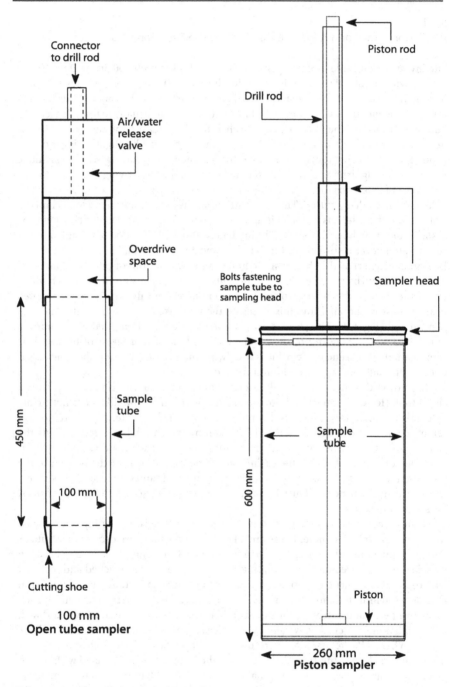

Fig. 5.4. Open tube and piston samplers

figure refers to the diameter. They were previously known as U4 tubes, when diameters were in Imperial units (4 inches). Smaller size drive tubes are also available and widely used in other countries. The type of tool employed may depend upon the testing favoured in the country of use.

This type of sampler is easy to use and cheap. The tubes are used many times, with more frequent replacements of the cutting shoe. However, they are fairly crude devices and considerable disturbance is likely at the periphery of the sample. Samples taken for triaxial testing are best taken from near the centre of the U100. Granular soils are difficult to sample with the open drive tube and tend to fall out when the tube is lifted.

There are also several types of piston drive samplers, some of which may be pushed into the ground as well as driven in boreholes. The purpose of the piston is to prevent loss of sample by providing an air tight seal above the sample in the tube, which also helps to increase the length of sample that it is possible to obtain from a given diameter. The piston sampler shown in Fig. 5.4 is of large diameter developed by the Asian Institute of Technology to sample the soft clays of the alluvial plain around Bangkok. As the tube is pushed into the ground (using the hydraulic push of a rotary core drill when seen by the writer) the piston rises. Suction reduces the chance of the sample falling from the tube when pulled to surface.

5.5.2
Continuous Soil Sampling

Continuous soil sampling is highly desirable, provided that the sample can be taken without too much disturbance. The stronger the soil, the more force is required to push the sampler, and the more the soil is disturbed. While the sampler is designed to reduce disturbance as much as possible, continuous sampling is successful only in fairly weak soils, although in some cases continuous samples have been obtained with the use of special rigs where the sampling is done in conjunction with rotary drilling. In these cases the process is more like coring than soil sampling, although the material would not be classified as rock.

Probably the best known and most widely used continuous sampling system is that developed by the Laboratorium voor Grondmechanica at Delft, The Netherlands. The Delft system takes a sample of 29 or 66 mm diameter. For the former, the 2 Mg Dutch Deep Sounding Machine is used, and for the latter the much bigger 17.5 Mg machine is required.

The sampler head has a serrated cutting shoe and as the tube is pushed into the ground the soil is fed into a woven sleeve, or stocking, and prevented from collapsing by a bentonite-barite supporting fluid of similar density to that of the ground. The stocking is surrounded by a vulcanising fluid which acts to make it impervious as the sleeve unrolls and fills. The filled sleeve is fixed in position by a tension cable attached to the top while it is supported by bentonite. In the larger sampler a plastic tube is used as an inner liner, into which the sample in the stockinette sleeve is fed. In this case, bentonite is no longer used for support but still acts as a lubricant.

With both sizes of sampler, a continuous sample about 18 m long can be taken under the right conditions. The sample is removed in 1 m lengths within the sampler extension rods or within its own plastic tube, as appropriate. Samples are placed in purpose made cases and should be described and sampled for moisture content and other index tests as soon as possible. The samples are usually split longitudinally; one half is used for examination and testing, while the other half is retained for reference. Photographs are taken after some drying out has been permitted, when soil structures are usually more obvious. The small samples are useful only for index tests, but the larger samples are suitable for a wide range of laboratory soils tests.

The Swedish Foil Sampler was developed by the Swedish Geotechnical Institute. In this sampler the soil is received by very thin high strength steel foil which unrolls to surround the sample as the apparatus is advanced by pushing. The 40 mm sampler holds a length of 12 m of foil, while the 68 mm sampler will take 30 m. This gives a maximum length to the sample which can be taken.

5.5.3
Power Augers

A drill for wood, metal, ceramics, plastic and so forth, feeds the material it cuts up its shaft and out of the hole; augers do just the same. Flight augers have a screw thread and are fed into the ground so that material from the bottom of the auger progresses upwards to the surface (Fig. 5.3b). Obviously, the sample is disturbed. If the auger is hollow, (Fig. 5.3c) a sample can be taken by some sort of sampling device pushed through the inside. However, to obtain a 100 mm diameter sample, the internal diameter must be at least 150 mm, and that in turn requires a powerful rig. Another possibility with large diameter hollow flight augers is to use the augers as casing through soil to rock, followed by rotary core drilling via the hollow stem.

5.5.4
Hand Augers

These are hand tools operated by one or two people. In cohesive soil they may penetrate to depths of about 5 m, but their limitation is man-power. There are many designs usually including some form of bucket auger blade to extract the soil which can be removed and replaced by a small sampling tube (usually 35 mm diameter, 150 mm long), which can be pushed or hammered into the ground. There are a variety of blades, or bits, to suit cohesive or non-cohesive soils. Two bits, a T-piece handle, one or two extension rods and the necessary tools can be carried by one person and used for reconnaissance work and/or mapping of near surface conditions.

5.5.5
Special Samplers

There is a wide variety of special samplers for use in very weak ground or to obtain samples with minimal disturbance. The Bishop sand sampler (Bishop 1948) uses compressed air to aid recovery of an undisturbed sample of sand, and is useful in assessing the true in situ densities of sands. Rotary core drilling with special core barrels may also be used for soil as well as rock sampling (Binns 1998), but the practical difficulties of using double or even triple tube barrels can be considerable.

5.6
Daily Drilling Records

The final borehole record (Chap. 7) is composed of information derived from the description of the samples, the testing of the samples and the daily drilling record prepared by the driller.

This last source of information is of vital importance. In percussion boring in soils, the description of the strata encountered comes from the engineer's or engineering geologist's examination of the samples; boundaries between strata are those given by the driller. In shell and auger drilling it is customary to take undisturbed tube samples at set intervals, usually something between every 1 or 1.5 m, or at '*changes in strata*' at the discretion of the driller. From the disturbed drilling section of the borehole tube or bag, samples are provided as selected by the driller. Consequently the driller has a heavy responsibility in the preparation of the final borehole record. As it happens most drillers are qualified by experience rather than training so it is of some importance that, before drilling commences, the driller is well briefed regarding the geology that may be expected and the significance of that geology to the engineering project. Thus, for example, in an alluvial plain a few centimetres of peat found in a borehole may be of little engineering significance at that spot but could be the feather edge of a deeply peat filled old river meander. Clearly such a feature should be noted in the daily log.

In rotary coring the investigator hopes for continuous samples and will complain of core loss. However, calculations of core loss depend on the driller's accuracy in measuring the depths of the tops and bottoms of core runs. If joint openings of the order of centimetres, as assessed by core recovery, are of importance, then core run depths must be of equal accuracy.

Core disturbance may be a consequence of mechanical problems with the drilling equipment. Such difficulties should be noted in the daily log.

It is obviously of great importance that the depths at which water is met or lost be noted. Since groundwater levels are disturbed by drilling, it is customary to note not only what happens during the course of boring but also to measure the depth to groundwater before the start of drilling in the new working day. This may change as the borehole progresses, depending on the depth of any casing. Thus, in Fig. 5.3a the groundwater table would have been encountered in the sand but, once the casing had been well established in the underlying clay further entry of natural water into the hole would have been reduced, perhaps stopped. Thus, to understand apparent variations in groundwater levels the daily record must include an accurate account of the position of the casing at any time during the drilling. In rotary core drilling losses of flushing water may indicate permeable strata or cavities. Samples of water are usually taken from the borehole for chemical analysis: it is necessary to know whether they are of groundwater or of water introduced to assist drilling.

5.7
Probe Drilling

It is sometimes necessary, having discovered the basic geological situation on a site, to seek to establish a boundary with greater accuracy than given by the boreholes already sunk. More boreholes by shell and auger or rotary core drilling may be unjusti-

fied financially for such limited objectives and cheaper and quicker methods of boring should be employed. A common problem of this type is determining the depth to
rockhead. In soils (usually the softer soils) wash boring or jetting (the former with
rotary action the latter without) using large quantities of water at high pressure may
serve to reach rockhead and establish its elevation.

There are many types of rotary or rotary percussive drill, used mainly for drilling
blast holes, which will progress rapidly through rock at rates perhaps as high as 1 m
per minute in weak rocks. Cuttings are brought back to surface by water or air flush
and, while small in size, are usually sufficient to determine the basic character of a
rock. Thus this type of boring may be utilised to examine particular geological situations in detail at relatively low cost.

A common use is to detect cavities in limestone or abandoned mines. Figure 5.5
gives an example of the use of probe boreholes: Borehole 1 has found an open cavity
in a coal seam and identified the strata above the seam, Borehole 2 has passed through
a pillar and identified strata below the seam, whilst Boreholes 3 and 4 indicate that
the coal seam has been displaced by a fault and further boreholes (Borehole 5 is un-
derway) are put down to locate the fault or seam outcrop at rock head. Boring would
continue until this was done.

Such investigations require the continuous presence of an engineering geologist
to log the cuttings and continuously evaluate the boreholes so far sunk to establish
the location of further boreholes. There is a significant time lag between the cuttings
being cut and their arrival at surface and because the borehole progresses rapidly there
will be some uncertainty as to the exact depth of strata boundaries. It is useful if the

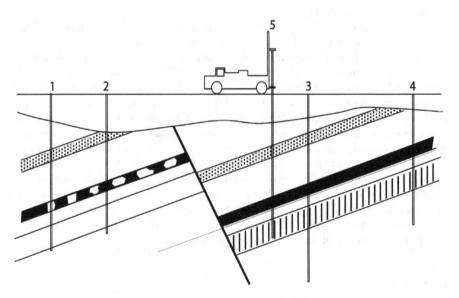

Fig. 5.5. The use of probe boreholes to determine the location of a fault boundary between worked and
unworked parts of a coal seam

site investigation includes a rotary cored borehole to establish the sequence of lithologies before such probing investigations begin.

5.7.1
Trial Excavations

One of the most useful, cheapest, and quickest ways of gaining access to formations for inspection and sampling is by trial pit or trench. In the simplest form, the excavations are hand dug with a spade. Deeper excavations may be made by machine. Relatively recent advances in the design of excavators enable trenches up to 6 m to be excavated rapidly. The use of telescopic shoring frames enables safe inspection.

The advantages of using trial trenches are many. The soil can be inspected in three dimensions in an in situ state, although some careful hand excavation of the sides of the machine excavated trench may be necessary because of disturbance. Bulk, block or driven tube samples can be taken from the sides or floor of the trench. In situ tests may be carried out.

Block samples are cut by hand from material exposed in excavation. They may be taken in both rock or cohesive soil. The procedure is often used for obtaining orientated samples, in which case both the location and orientation should be recorded before the sample is separated from the ground.

The cutting of a block sample often takes a considerable time, during which the moisture content may change. The following precautions should be taken:

- no water should be allowed to come into contact with the sample;
- the sample should be protected from the wind and the direct rays of the sun;
- immediately after the sample has been cut, it should be wrapped in plastic and bound with waterproof tape before placing carefully in a closely fitting box.

Trial pit samples are particularly useful when heterogeneous material is to be used for construction fill. Boreholes, because of their limited diameter, cannot sample cobbles or boulders to gain some idea of the proportions of coarser fractions present in a soil. This can only be done with very large samples taken from trial pits.

Shafts and tunnels are usually excavated with the primary purpose of reaching locations for in situ testing. However, any trial excavation in rock has the advantage that the investigator can record the orientation and spacing of discontinuities, which is not easily done on rock cores.

In weak rocks the large diameter augers used to make the holes for bored-and-cast-in-place piles may be employed to drill shafts for geological investigation. These may be of the order of one metre diameter and the geological conditions found in them may be described by an observer lowered into them in a steel cage, using methods explained in Chap. 4.

One of the problems in examining trial excavations is safety, for trial pits and trenches are liable to collapse. Safety regulations require that such excavations be supported but such support obscures the view of the ground mass exposed and may thus make it difficult to fulfil the objective of making the excavation. This gives the temptation to examine the pit without any support, particularly if the investigator has had no previous experience of pit collapse. Those who have had experience of this will testify to the great rapidity with which collapse may occur. Trial pits which are particularly hazardous are those which may ooze groundwater from some point in their sides.

5.8
Boring and Sampling over Water

Boring and sampling over water is always more difficult and very much more expensive than on land. Apart from the technical problems associated with water depth and movement, there is likely to be a major logistical problem involving personnel, materials and supplies, particularly if the work is undertaken far off-shore.

Boring over water would be carried out from a reasonably stable platform such as a pontoon, barge or ship. The nature of the platform necessary would be dictated by such conditions as water depth, tidal influence, distance from shore and whether or not the water was sheltered or open. Platforms may vary in sophistication from scaffold staging built out from shore to moored pontoons, jack up platforms or ships. Floating platforms and ships require proper anchoring, or even thruster devices if the water is very deep; e.g. about 80 m or more.

All proper boring and sampling programmes must be planned, but because of the relative cost of mobilisation over water, it is virtually certain the mobilisation will take place on one occasion only. Particularly careful consideration of the requirements of the investigation must therefore be given.

It is often possible to use conventional equipment for boring and sampling over water from a stable platform. The borehole may be sunk through a conductor pipe which is fixed between the platform and the water bottom. In situations where the boring is done in open water and the platform is subject to heave from wave action or gravity, a spring loaded 'heave compensator' may be incorporated between the top of the drill pipe and the pulley on the tripod or A-frame.

Because the costs are so high, wire line methods are used for sampling, to speed the process. Indeed, the need for faster drilling and sampling in off shore situations has stimulated the development of wireline equipment which is now entering its third generation. Even so, sampling in deep water is constrained by the lack of reaction available to push or drive a sampling tool, and a range of new techniques using submerged equipment has developed.

5.8.1
Pushed and Driven Wire Line Samplers

Boring and sampling by wire line are still the most commonly used methods of recovering samples over water from a stable platform. Samples of soil are taken by driving or pushing a tube through the drilling bit which has first had the central 'insert' removed by wire line, and then been raised one or two metres from the bottom of the hole. The sample tube may be driven by a sliding weight striking an anvil which is connected to the tube. The technique is very similar to that used in the Standard Penetration Test. When driven to full depth or to refusal, the tube plus anvil and weight are recovered by wire line.

A similar technique is used to push the tube, rather than drive it, in an attempt to lessen the sample disturbance. After removing the insert and raising the bit, the tube is lowered and clamped below the bit. The whole drill string is then dropped, so that the weight pushes the sample tube into the soil. Recovery is by wire line.

5.8.2
Submerged Rotary Rigs

An alternative to sampling by wire line from a stable platform is to use submerged rotary rigs operated by divers. The technique is limited to about 40 m of water because of the operating requirements of the divers. The rigs are driven by two hydraulic motors powered from a support vessel. Both rock and soil may be sampled, using wire line methods and double or triple core barrels, including the Mazier type. Recovered samples are usually of a higher quality than would be obtained by pushed or driven tubes.

5.8.3
Vibro-Corers

These devices are driven by electric, hydraulic or pneumatic power from the surface. The sample tube is vibrated into the sea bed until full depth or refusal is reached. Some convert the vibration into percussive power to assist driving the tube, and in either mode they may be used for driving piles as well as sampling. Penetration of 31 m in 1½ hours in very soft soils has been reported, with core recovery of 80% in mud and 100% in lodgement till.

5.8.4
Electro-Corers

The ocean floor electro-corer is a submerged remote controlled device for drilling and sampling in water up to 100 m in depth, although a variation of it can be used in water up to 300 m deep. The device uses the flexo-drilling technique, in which a rigid drill stem is used that incorporates a corer and an electrically driven drill. The drill stem is connected to a flexible rod which is fed from a large drum as the drill penetrates. Drilling fluid is always sea water, which is discharged on to the sea bed. The procedure is to drill to the required depth and then activate the corer, which operates on the rotary principle and can obtain 1 m length of core of between 106 and 108 mm diameter.

Similar apparatus can be used from the surface with a specially equipped ship. The length of flexible drill pipe is increased by the easier access on ship, and the depth of water in which it can operate is consequently much greater.

5.8.5
Submerged Remote Control Rotary Corers

A number of these devices are now in use, and further development can be expected. They can operate in deep water, several hundreds of metres in some cases.

One of the earliest of these devices is the Maricor, built for Wimpey by Atlas Copco, which can produce continuous core of 60 m in a water depth of 200 m. It is hydraulically powered and adds drill pipe from a magazine store as drilling proceeds. After each coring run, the core barrel is lifted by wire line to the surface. Core size is about 57 mm diameter.

The U.S. Navy Civil Engineering Laboratory sub-sea remote control coring device is designed to operate in water up to 1 800 m deep. It is hydraulically driven and has two magazines, one for drill pipe, the other for core barrels. Each contains 10 lengths of 1.5 m each, which gives a 15 m depth capacity. Each core barrel is retrieved and stored in the magazine when full.

The Stingray has been developed by McClelland Engineers in collaboration with the Norwegian Geotechnical Institute. It uses 75 mm diameter tubes for soil sampling; the tubes are pushed into the material by hydraulic jacks, but rock core may also be taken. The Stingray is designed to perform in situ cone penetrometer tests as well as obtain material samples.

5.8.6
Gravity Samplers and Stationary Piston Corers

The simple gravity sampler is essentially an open ended tube which is dropped from a limited height and penetrates the soil. Such devices are obviously somewhat limited in capability.

A development of these simple samplers is the stationary piston corer. This is dropped in free fall also, but as penetration begins, a piston which encloses the lower end remains stationary because it is connected to the main cable by a wire that becomes taught when the sampler hits the bottom. The piston creates suction, or negative pressure, in the coring tube, and as with conventional piston samplers used on land, the result is a better recovery of sample.

5.8.7
Sample Quantity and Quality

Samples of geological materials are difficult and expensive to obtain. There should be a specific reason for requesting samples, such as needing values for strength that can only come from the laboratory testing samples, in order to analyse stability of the structure to be built. The quantity and quality of samples required for that purpose should be defined *before* boring starts. No advantage can be gained if, for the tests intended, inadequate quantities or qualities are recovered. Guidelines for British site investigation practice are shown in Table 5.4. The amounts given are the amounts required to be recovered. The actual test may be performed on smaller fractions of the original, provided the smaller fraction is representative of the whole.

Table 5.4. Quantities of sample required for some soil tests (BS1377 1990)

Test	Quantities of material required (kg)		
	Clay, silt, sand	Fine to medium gravel	Coarse gravel
Moisture content, Atterberg limits, sieving	1	5	30
Compaction tests	25 – 60	25 – 60	25 – 60
Soil stabilisation tests	100	130	160

Table 5.5. Sample quality classes (DIN 4021)

Quality class	Description of soil
1	No geometric distortion. Shear strength and compressibility are unaffected
2	Geometric distortion. Density and water content unaffected
3	Density altered. Water content and particle size distribution unaffected
4	Water content and density altered. Particle size distribution unaffected
5	Particle size distribution affected by loss of fines or grains crushing

Sample 'class' is descriptive of the quality, which in turn defines the degree of disturbance of the material during the process of recovery. The German Standard DIN 4021 defines five classes (Table 5.5). Every sampling procedure produces some disturbance, and, generally, for any given material the cost of recovery rises with the increase in class. The table shows the class of sample which it is possible to obtain, according to the nature of the material and the method adopted.

5.9
Contaminated Land

The redevelopment of sites used previously for other purposes has meant that drilling and sampling often have to be completed in ground that is contaminated. There are many hazards associated with this which have to be avoided, notably:

- the contamination of personnel involved,
- the careless dispatch of apparently "clean" samples that are in fact contaminated to laboratories who test them not suspecting contaminants to be present, but later find their laboratory is contaminated,
- the unknowing communication of contaminated ground with uncontaminated ground, made possible by a drill hole that passes through both and acts as a flow path that permits water from contaminated ground to enter uncontaminated ground.

Many countries have now developed guidelines for work at sites where contamination is suspected. Failure to follow these guidelines will normally result in a personal prosecution of those responsible for ignoring them from organisations such as an Environment Agency, who themselves have a duty to protect the health of individual employees and the public at large.

5.10
Further Reading

Clayton CRI, Mathews MC, Simons NE (1995) Site investigation, 2nd edn. Blackwell Scientific Ltd., Oxford
Day RW (1998) Soil testing manual: proedures, classification data and sampling practices. McGraw Hill Publishing
Site Investigation Steering Group (1993) Guidelines for the safe investigation by drilling of landfills and contaminated land. Thomas Telford, London

Chapter 6

Field Tests and Measurements

6.1
Introduction

Because of the difficulties of obtaining good quality samples of certain materials (such as sands and soft clays) for laboratory testing some tests are considered best done in boreholes on 'samples' in situ. In materials whose coarse grain size prohibits taking samples large enough to test in the laboratory, the only way to obtain test results is by testing in situ. In rocks, whose mass properties are known to be much affected by the presence of discontinuities, reasonable assessments of geotechnical parameters can only be obtained by testing in situ. Access to the body of ground to be tested, the in situ sample, is via a borehole or trial excavation. It should be noted that the excavation that gives access to the 'sample' itself disturbs the ground and may alter its properties. Thus opening the excavation changes the stress conditions within the 'sample', sometimes opening discontinuities and loosening the mass around the excavation. Water or air used for flushing boreholes may alter the moisture content of the sample. Such in situ tests are often thought to be tests that establish 'mass' properties because the sample tested is substantially larger than that can be tested in the laboratory. In some cases this might be true but it is prudent to think of these tests as tests on a sample of the mass, which may be no more representative of the mass than tests on materials are necessarily representative of the materials.

Tests of this type fall into two broad classes, namely those conducted in boreholes, in which the common range of borehole diameters limits the volume of sample tested, and tests in excavations which are on a larger scale, the upper limits being determined by the mechanical difficulties of undertaking large scale testing.

The behaviour of ground masses may be monitored before, during and after engineering construction by in situ instrumentation. The assessment of the structure of ground masses may be substantially aided by the use of geophysical surveys.

6.2
Tests in Boreholes

The many tests conducted in boreholes are described here under four headings; those that measure the resistance to penetration through the base of a borehole, the man-made force required to deform the hole (strength and deformation tests), the amount of fluid that can be passed to and from a hole (permeability tests), and the natural forces that deform a hole (measurements of in situ stress).

6.2.1
Resistance to Penetration

Two commonly used tests are the Standard Penetration Test and the Cone Penetrometer.

Standard Penetration Test

This is a dynamic penetration test carried out using a standard procedure and standard equipment and thus may be described as empirical. It was introduced by the Raymond Pile Company in the United States to give an estimate of the degree of compaction (the relative density) of sand. The test is used for soils investigations all over the world for the apparatus required is not complex nor is there any great skill required in the execution of the test. A great mass of experience has been built up justifying its use for foundation design purposes.

The test is undertaken at the bottom of boreholes sunk in soils by methods such as shell and auger boring or rotary boring. The equipment consists of a '*split spoon sampler*' or '*split barrel*' (Fig. 6.1a) which is a sample tube split longitudinally into two halves and held together by a screwed coupling and a lower cutting shoe, all of standard dimensions. This is driven into the bottom of the borehole by an automatic trip hammer, which allows a weight of 622.7 N (140 lb) to fall on to a drive head from a height of 762 mm (30 in). The number of blows is counted to cause the sampler to penetrate to a depth of 450 mm (1 ft 6 in) into the soil. The blows for the last 300 mm penetration are described as the '*N*'-*value*. Those for the first 150 mm penetration are neglected, being considered to have been in soil disturbed by the boring process. It is common good practice for the driller to record the number of blows for each 75 mm penetration to see whether there was any change in resistance within the drive depth. Thus, on a field boring record the number of blows would be recorded as:

2 / 3 / 4 / 5 / 4 / 5

giving a *N*-value of 18. It is essential that the bottom of the borehole is carefully cleaned out, removing any loose material, before the test is undertaken.

It has long been recognised that the *N*-value reflects not only the relative density of the sand but also the overburden pressure at the depth of the test. Gibbs and Holtz (1957) showed that, for a sand of uniform relative density, *N*-values increased with increasing overburden pressure. However, at the bottom of a borehole there is no overburden pressure on the sand being tested so measured *N*-values tend to be rather low. *N*-values should thence be corrected. They are generally corrected to a relative density/*N*-value relationship proposed by Terzaghi and Peck which seems to be valid for an effective overburden pressure of the order of 150 to 250 kPa. Smith (1982) has produced graphs to help make the correction; these have been modified by the author into a single figure, Fig. 6.1b, used in the following way.

Assume that the SPT-*N*-value measured in the borehole is 10 and that the effective overburden pressure at the level of the test is 150 kPa. Use the horizontal axis and the right hand vertical axis to plot the 10/150 point. The diagonal lines indicate relative

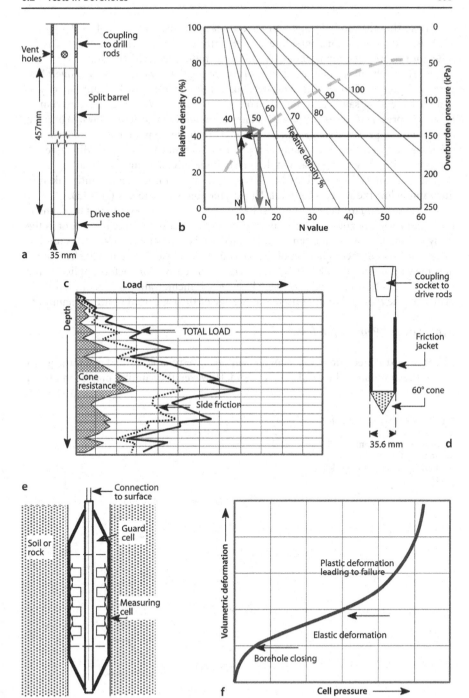

Fig. 6.1. Penetration and cell tests

density so that the plotted point will have a relative density of about 43%. Using the left hand axis draw a horizontal line at 43% until it meets the dashed curve. The N-value for this intersection is about 15. This is then the corrected N'-value.

Unfortunately this is not the only way of correcting the N-value to the N'-value. Most give an N' greater than N. N' is extensively used in calculating the bearing capacity of and settlement on sands and thus the correction for N to N' is of great importance. Figure 6.2 presents curves relating N' values to foundation width to allowable bearing pressure for approximately 25 mm (originally 1 inch) settlement. Because alternative methods of correction exist many companies issue internal company directives to ensure a uniformity of approach in designing foundations using the N'-value.

A second correction, attributed to Terzaghi, is applied if the test is undertaken in fine sand below the water table level. This correction is $N_f = 15 + \frac{1}{2}(N' - 15)$.

While the Standard Penetration Test has always been an important test for foundation engineering purposes it has, since the early 1970s, also become important in the study of liquefaction of fine sands under earthquake tremors; the N-value is used to assess the degree of compaction of the sand deposit. N-values are usually 'corrected' in some way and Soydemir (1987) has described the correction factors applied to the raw N-value in New England practice. These factors include allowing for the length of drill rods, the type of hammer used and the age of the rope used to lift the hammer.

Dynamic Cone Penetration Test

The Standard Penetration Test sampler will not give reliable results when the grain size of the granular soil exceeds that of sand. However, the sampler may be modified

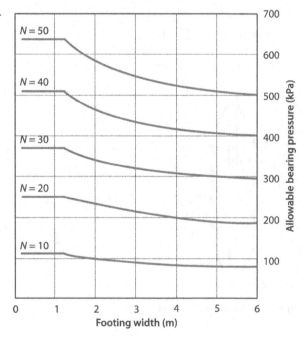

Fig. 6.2. N'-values used to determine the allowable bearing pressure to give settlement of ≤25 mm for foundations on sand (after Terzaghi and Peck 1948)

by replacing the driving shoe by a solid steel cone with a 60° apex cone; the test is then conducted in the same way as the SPT. It is generally considered that cone N-values are comparable to standard N-values on sands of similar relative density and that these cone N-values may be used in foundation design calculations. However, if the granular deposit contains grains coarser than gravel size the results can become very unreliable, for the cone may be driving a cobble or boulder down the borehole. The test is thus not suitable for very coarsely granular materials.

Other Uses of the SPT

The simple nature of the Standard Penetration Test and of its results makes it attractive as an investigation tool and attempts have been made to correlate N-values with properties of soils other than granular soils and with the properties of weak rocks, especially chalk. The latter is sensitive to disturbance and, while this makes it relatively easy to penetrate when boring, it is difficult to recover samples of the weaker chalks that are sufficiently undisturbed to be suitable for representative laboratory strength testing. However, empirical correlations have been proposed between N-values and the compressibility of chalk, and widely applied in foundation design. Relationships have been published between the strength of clays and N-values. The author considers such relationships as but very approximate and does not recommend their use.

Static Cone Penetrometer

All static cone penetrometer tests consist of pushing a pointed rod into the ground and measuring resistance to progress. The stronger the ground the greater is its the resistance to progress. A large variety of static penetrometers exist but that most widely used and with which engineers have the greatest experience was developed in the Netherlands.

The test was designed to enable a cheap and rapid assessment to be made of the length and diameter of piles necessary to carry a given load. If a pointed rod is pushed into the ground the resistance to penetration will come from resistance to the point and side friction resistance to the passage of the rods. As the point is pushed deeper into the ground the side friction component of resistance will continuously increase. To determine the resistance of each layer to cone penetration some way has to be found to distinguish side friction from point resistance. In the mechanical types of cone penetrometer the end cone is connected to the surface by a rod passing within the outer rods. The penetrometer machine hydraulically or mechanically pushes rods and cone into the ground; resistance to progress is measured by a hydraulic or electrical load cell. The apparatus may be used in various ways. One of the simplest is to push the combined outer rods and cone into the ground for a short distance, say 0.2 m. The cone alone is then advanced by pushing on the inner rod for an equal distance, again measuring the resistance. The outer tube is advanced to meet the cone and both pushed forward for another short penetration. The results are plotted on a depth/resistance to penetration graph, in which side friction and cone resistance are plotted (Fig. 6.1c).

The general measurement of side friction is useful but it is much better if both local side friction and cone resistance can be measured for any depth, for knowledge of both parameters may serve to identify soil type. This can readily be done using the electric friction (Fig. 6.1d) cone penetrometer. In this strain gauges are mounted in the cone and a short friction sleeve section (the friction jacket) behind it so that both side friction (f) and cone resistance (c) are continually recorded during penetration. The *friction ratio*,

$$R_f = \frac{f}{c} \times 100\% \tag{6.1}$$

may be plotted against cone resistance to give an indication of the nature of the soils being penetrated. In general clays have high friction ratios while sands and gravels have low friction ratios.

The test is standardised with regard to apparatus and procedure, the most important recommendations being that:

- the cone apex angle must be 60°;
- the cone cross-sectional area should be 100 mm^2;
- the friction jacket area should be 15 000 mm^2;
- the rate of penetration of the cone should be 20 mm s^{-1};
- force measurements should be accurate to within 5%.

The cone penetration test is best suited to soft soils in which tests to depths of the order of 50 m may be undertaken. Sufficient force can not be generated to penetrate through very dense granular soils or hard clays and clearly the test is unsuitable to penetrate coarsely granular deposits or clays with boulders. This implies that the test can be used only in areas for which the geology is reasonably well known; it is not a tool for primary investigation on sites of unknown geology.

The various parameters derived may be applied to foundation engineering, particularly with regard to piled foundations. This is done via empirical formulae and much depends upon experience.

Both foundation engineers working using SPT values and foundation engineers working using static cone penetrometer results have developed foundation design formulae incorporating values derived from one or the other of the two tests. In general the formulae using cone penetrometer results appear rather more elegant, particularly with regard to the calculation of settlements. Accordingly some effort has been put into attempting to correlate SPT-N-values with cone penetrometer results. Some correlations are given in Table 6.1. These correlations can only be regarded as approximate.

In the author's opinion the cone penetrometer is a test giving results which are much superior to the SPT for they are continuous rather than intermittent (thus allowing the exact location of boundaries between layers of different properties) and the foundation design procedures which can be undertaken with cone resistance values are much superior to those possible with SPT results. It would be better to regard the SPT as a first step exploration tool and the static cone penetrometer as an in situ test.

Table 6.1. Correlation between N-value and cone rsistance for clean fine to medium sands and slightly silty sands

Relative density (%)			N value	Cone resistance q_c (MPa)	ϕ (deg)
Very loose	0 –	15	< 4	< 2	< 30
Loose	15 –	35	4 – 10	2 – 4	30 – 35
Medium dense	35 –	65	10 – 30	4 – 12	35 – 40
Dense	65 –	85	30 – 50	12 – 20	40 – 45
Very dense	85 –	100	> 50	> 20	> 45

6.2.2
Strength and Deformation Tests

Two commonly used tests of this type are the vane test and the pressuremeter test.

The Vane Test

In this test a cruciform vane at the end of a solid drill rod is forced into the soil at the bottom of a borehole (generally a shell and auger hole) and then rotated until the torque on the vane causes the soil to fail in shear. The test is usually undertaken in clays that are too soft or too sensitive to allow recovery of satisfactory tube samples. The apparatus commonly used will shear clays up to about 100 kPa shear strength. The bottom of the borehole must be thoroughly cleaned out before inserting the vane to ensure that the test is in undisturbed ground. Smaller hand vanes are available for use in the sides of trial pits and vane equipment is also available for laboratory use.

The shear strength of the clay may be calculated if the torque (T) at the point of failure is measured when:

$$T = c\frac{\pi d^2 h}{2}\left(1 + \frac{d}{3h}\right) \tag{6.2}$$

where; h = length of the vane (m), d = diameter of the vane (m), c = undrained shear strength of the soil (N m^{-2}), (assuming $\varphi = 0$).

In the formula the first term relates to the shear strength of the soil on the vertical sides of the vane and the second term to the shear strength of the circular discs of soil at both ends of the vane.

While the hand vane is in quite common use as a rough guide to soil strength when examining trial pits and the laboratory vane is useful in examining samples, the field vane is now seldom used because the results do not justify the time and effort required to set up the apparatus and undertake the tests. The static cone penetrometer is of much greater value for soft soils.

Pressuremeter Tests

Pressuremeters are cylindrical cells that expand, put pressure on the sides of the borehole and cause the borehole to deform. All types of pressuremeter incorporate instrumentation to measure borehole deformation and cell pressure so that a stress/strain graph may be drawn. Study of this graph allows assessment of the deformation behaviour of the volume of ground stressed by the cell. In soils it is possible to apply sufficient pressure to shear the ground. This is seldom possible in fresh rock so that pressuremeters in rock mostly measure deformation characteristics up to a particular stress level.

Pressuremeters may operate by the all-round expansion of a rubber cylinder or by the opening of a split cylinder. Pressure is generated hydraulically. Borehole dilation is measured by diametrical electrical transducers or by volume displacement.

Menard Test

The Menard pressuremeter is the best known type of pressuremeter for use in soils. It consists of two main components; a Probe and a Pressure-Volume meter.

The 'Probe' is the cell that fits in the borehole and consists of a steel cylinder with rubber membranes stretched over it so as to form three independent cells (Fig. 6.1e). The central measuring cell contains a liquid under gas pressure while the top and bottom 'guard' cells are pressurised by gas only. Volume deformation of the borehole is measured only in the central cell adjacent to which conditions of uniform stress are considered to exist. The rubber membranes are protected by thin steel strips to prevent their being punctured by sharp stones. The cells are made to fit standard borehole sizes, such as AX, BX etc.

The action of the probe in the borehole is governed by the 'Pressure Volumeter' at the surface. This allows water under pressure to be pumped into the central cell and gas into the guard cells. The water is pressurised by carbon dioxide gas brought on site in compressed gas bottles. The pressure volumeter is connected to the cell by tubes through which water and gas pass. The system contains a known amount of water. As the borehole deforms the cell expands, taking in more water and the water level in a 'sight tube' at the surface correspondingly alters thus providing an assessment of the volumetric deformation of the borehole. Pressure is usually applied in increments and the corresponding volume changes noted. A typical pressure/volumetric deformation curve is given in (Fig. 6.1f).

The curve generally shows three sections. The first part of the curve shows quite large volumetric deformation for limited pressure and represents the recompression of the relaxed soil around the borehole. The second straight line part represents 'elastic' deformation of the soil. The third section shows increasing volumetric deformation for equal increments of pressure and represents plastic deformation leading to shear failure. This may reach the 'limit pressure' at which the soil is deemed to have failed. The pressuremeter is capable of applying pressures of the order of 100 kPa.

Generally the pressuremeter is used in boreholes that will stand without casing. If the borehole shows a slight tendency to collapse support may be given by drilling mud. In granular soils the pressuremeter may be used within special slotted casing which is strong enough to support the borehole yet sufficiently flexible to deform significantly under the loads that the pressuremeter can apply. Results must be corrected to allow for the influence of the casing.

The results obtained may be used to assess bearing capacity and settlement characteristics of the soil.

The results may be influenced by the disturbance of the ground during boring. To overcome this Menard has developed a pressuremeter which may be wash bored into place. A self-boring pressuremeter, the 'Camkometer' has also been developed. Because the device bores itself into place the soil tested can be held not to have been stress relieved so that soil load on the cell is equal to the in situ stress.

Pressuremeter testing in rock requires the generation of much larger pressures on the sides of the borehole if significant deformation is to be measured. Pressuremeters for work in rock thus tend to be more robust than those in soil, will expand less and have much more delicate instrumentation to measure very small diametrical displacements.

Dilatometer Test

The '*dilatometer*' devised by the National Laboratory for Civil Engineering in Lisbon (LNEC) is an example of the above. This consists of a steel cylinder 66 mm in diameter wrapped in a rubber jacket 4 mm thick. This gives an overall diameter of 72 mm which gives a 2 mm annulus around the cell in a 76 mm diameter borehole (NX size). Oil is pumped under pressure between the steel cylinder and the rubber membrane; deformation is measured by four differential transformers set 32 mm apart in the centre of the 540 mm long load unit which itself lies in the 755 mm long cell. Pressures up to about 15 MPa may be applied to the rock.

Other types of pressuremeter exist. Their pressure/deformation curves are similar to that shown in Fig. 6.1f, except that in a test in rock the first part of the pressure/deformation curve reflects the recovery of borehole relaxation while further load may serve to close any discontinuities which may be open within the stress field imposed by the cell. Further application of load will reflect the deformation of 'intact' rock. In the second phase of loading rock above the cell may fall into the hole and jam above the cell. Unless the cell can be closed back to its original size it may be difficult to withdraw from the borehole.

6.2.3
Permeability Tests

Values of mass permeability are required to calculate flow of water into excavations, into tunnels, under dams and from reservoirs. *Mass permeability* (k) is defined by:

$$k = \frac{q}{Ai} \tag{6.3}$$

where k is in units of velocity (m s^{-1}) and measured in the direction of flow, A is the area at 90° to the macroscopic direction of flow and i is the hydraulic gradient associated with that flow.

Pumping Tests

Assessments of permeability can be made from well pumping tests (Fig. 6.3a). Pumping tests are not undertaken lightly for they are tests which take days to perform, require the installation of submersible pumps, observations of the fall of the water levels around the well, made with the help of observation wells, and arrangements to dispose of large quantities of water! Such tests are thus lengthy and expensive. Having established a "stable" water surface, with a uniform pumping rate, the permeability of the unconfined aquifer shown in Fig. 6.3a is given by

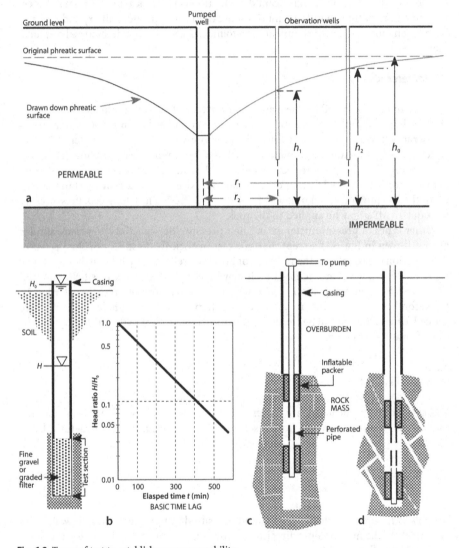

Fig. 6.3. Types of test to establish mass permeability

$$k = \frac{2.3Q}{\pi\left(h_2^2 - h_1^2\right)} \log_{10} \frac{r_2}{r_1} \tag{6.4}$$

Simpler tests can be undertaken in boreholes (Fig. 6.3b). Two kinds of test are possible; 'pumping out' tests, in which the groundwater table is lowered by pumping and the rate of restoration of the water table back to its original level is observed, and 'pumping in' tests in which water is pumped into the borehole and the water intake is measured. As an alternative to this last procedure the water head may be raised and its rate of fall back to the original level may be observed. This is the 'falling head' test.

To undertake this test the section, whose permeability is to be known, is penetrated by a borehole. Sections above are isolated by casing and the test section is filled with a coarse gravel filter material to support the sides of the hole. The location of the natural groundwater table is carefully observed during boring. The simplest routine is to raise the water table by an initial head H_0 which is then allowed to dissipate. The water falling to various levels of H are measured against time using an electric water level meter. The permeability is determined as;

$$k = \frac{A}{FT} \text{ (m s}^{-1}) \tag{6.5}$$

where A = cross-sectional area of the borehole at the water table level, F = dimensionless intake factor relating to the geometry of the test, and T = basic time lag. T is determined by plotting the ratio H/H_0 against t (time elapsed in the test to any reading of H). In the beginning of the test $t = 0$ when $H/H_0 = 1$.

In the constant head pumping in test water is added and kept at a constant level. The amount of water added will initially be large but will eventually reach a constant value, Q. The calculation of the permeability depends upon the relationship of the test section to the water table, but when the test section lies below the natural water table level the equation determining k will be of the from:

$$k = \frac{Q}{FH_0} \text{ (m s}^{-1}) \tag{6.6}$$

where Q = the rate of flow, H_0 = the constant head and F is a dimensionless factor obtained from charts.

Packer Tests

The tests described above are mostly undertaken in soils or weak rock. Permeabilities of rock are often required in order to assess leakage under dams or to estimate grout take and test pressures similar to actual expected pressures cannot be achieved by the methods described above. Accordingly in rock 'packer tests' are undertaken in which a section of a borehole in rock is sealed off by packers above and below the test section (the double packer test, Fig. 6.3c) or by one packer at some level above the bottom of the borehole 'the single packer' test). The 'packer' is a rubber seal, usually about 5 times as long as the diameter of the borehole, inflated by compressed air. Boreholes are mostly rotary core drilled.

Water is pumped into the test section under pressure until the water inflow Q is steady. Then permeability is given by the formula:

$$k = \frac{Q}{2\pi AH} \log_e \frac{A}{r} \tag{6.7}$$

for $A \geq 10r$; where A = length of the test section, r = radius of the borehole and Q = steady quantity of flow under a pressure head H; this is not necessarily that pressure recorded by surface pressure gauges for there will be pressure losses due to friction in the pipes (usually drill rods) conducting water to the test section.

Usually three pressures are applied for one measurement of k. Results are usually given in m s^{-1} but sometimes in '*Lugeon units*'. A rock mass is said to have a permeability of one Lugeon if under a pressure head of 100 m of water above groundwater level a 1 m length of NX size borehole accepts 1 litre of water a minute. It is very difficult to convert standard permeability units to Lugeon units because the Lugeon test (a type of packer test) is standardised. An equivalence of 1 Lugeon $\approx 10^{-7}$ m s^{-1} is sometimes applied.

Packer tests are not easy to undertake because of the difficulty of isolating the test section in the borehole. If the packer seal is not tight leakage will give a higher water take and false values of permeability. Similarly leaks from drill rod connections, hose connections etc. can also lead to major errors. No one can be quite certain about what is happening in the borehole; Fig. 6.3d. shows how locally open joints around test sections may allow free passage of the water from out of the test section and give results which do not reflect the general permeability of the rock mass. Accordingly many tests are needed to achieve a reasonable value of permeability and it is sometimes thought better to quote permeabilities to the nearest order of magnitude or perhaps to the nearest whole number times the order of magnitude.

Meaning of Results

It should be noted that permeability formulae depend for their application on the assumption that the strata tested are *homogeneous and isotropic*. However, it is well known that in bedded strata there may be a considerable difference between permeabilities along the layers and permeabilities across the layers. In rocks one particular joint set in a joint pattern may be open and the others closed so that maximum values of permeability will be in a particular direction related to geological structure. These factors cast doubts as to the validity of using pumping tests to obtain general values for permeability; they are essentially *well-tests* designed to supply hydraulic parameters (*Transmissivity and Storage*) suitable for predicting drawdown for a given discharge *at that well*. These restrictions should be taken into account in both planning a programme of permeability tests and also in assessing the value of the results obtained from the test programme.

Permeability measurements may serve as a guide to rock quality in a uniform rock material whose quality depends upon weathering and joint frequency.

The text given above is intended to give the general idea of permeability testing. There are clearly many possible relationships between test sections, groundwater levels, aquicludes and aquifers. Solutions to many possible situations have been formulated. Reference may be made to publications such as the 'Earth Manual' published by the U.S. Department of the Interior and British Standard 5930 (1999) for further details.

It must be emphasized that in order to get reliable assessments of permeability there must be accurate knowledge of the positions of water tables and the boundaries of aquifers and aquicludes. This means that permeability tests are best undertaken in boreholes sunk specially for these tests after earlier borings have established geology and groundwater conditions.

6.2.4
Measurements of In Situ Stress

The stresses that exist in the ground before the construction of an engineering work are referred to as the *initial* (or virgin) *in situ stresses*. These stresses are the result of the weight of overlying strata, residual stresses related to the geological history of the ground mass and the present-day tectonic stress vector.

Most in situ stress measurements are undertaken in rock and involve fixing a strain measuring device in a borehole and over-coring the device which then records strains opposite to the stresses existing in situ (Fig. 6.4). The device has mostly to be calibrated in the laboratory in a block of rock taken from the test site.

Stress may be measured two dimensionally by fastening a strain measuring device at the bottom of a borehole and then over-coring (Fig. 6.4a). Devices include photo-elastic strain gauges, photo-elastic glass plugs and electrical strain gauge rosettes. Of this last type the *Leeman 'doorstopper'* cell has been widely used and measures biaxial stress in one plane. The Leeman triaxial cell glues three strain gauge rosettes to the

Fig. 6.4. Overcoring tests to measure in situ stress

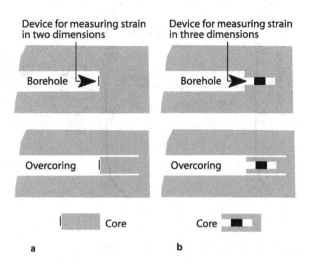

sides of a borehole (Fig. 6.4b) and the strains recorded after stress relief may be interpreted to give the triaxial state of in situ stress.

Measurements of stress in one direction may be made using a flat jack (Fig. 6.5). In this the distances between measuring points on a rock surface are recorded. A slot is excavated between the points by saw cut or by overlapping drill holes. The previously existing stresses will be relieved as the rock moves into the slot. A flat jack is cemented into the slot and pumped up until the points return to their original position. Jack pressure is then equal to in the situ stress normal to the slot.

None of the techniques described above are easy to undertake and require specialised equipment and experience. Many observations are necessary to give confidence that the in situ stress has been measured with reasonable accuracy with regard to both magnitude and direction. Particular problems are that the in situ stress field is altered by the excavation which gives access to the test site, so that the stresses recorded may not be the original in situ stresses. In fractured rock with open discontinuities the stress field measurement on the scale of the stress meter may not be representative of the overall stress pattern. If the rock has anisotropic elastic properties it may be very difficult to calibrate the stress measuring device.

6.3
Tests in Large Diameter Boreholes, Shafts and Tunnels

These tests are undertaken to determine deformation and strength characteristics of the ground mass. The 'sample' tested is in situ and contains discontinuities. If useful results are to be obtained the pressure put upon the sample must significantly exceed the working pressure to be imposed by the proposed civil engineering work. One of

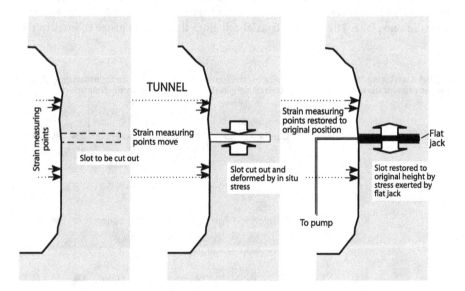

Fig. 6.5. Measuring in situ stress using a flat jack

the principal difficulties of large scale testing is to arrange for the reaction against which loading systems may push in order to subject the sample to pressure. Another difficulty is that the sample to be tested may be well below surface and has to be reached by shaft, large diameter borehole or exploratory tunnel. There may be water seepages at the test location while cramped working conditions may make it difficult to set up the test equipment satisfactorily.

A further problem is to be able to define the shape and the nature of the sample that has been tested. The results of many well-undertaken in situ tests have been spoiled for lack of a good description of the sample. The volume of ground tested should be dug out after each test and material and mass characteristics described. The most commonly used type of large scale load test is the plate bearing test.

6.3.1
Plate Bearing Test

The test consists of loading a rigid circular or, less commonly, a square plate and measuring the penetration of the plate into the ground. Plates may vary in diameter from about 0.3 m to about 2 m. In general the smaller sizes are used on soils and the larger on rocks. In rocks larger plates are preferred in the hope that the volume of rock tested will contain a representative sample of the discontinuity pattern in the rock mass.

The smaller size tests are often conducted at the bottom of large diameter boreholes, so that the plate is connected to the surface by a load carrying column. The bottom of the borehole must be very carefully cleaned out to ensure that the plate bears on to undisturbed ground. Some type of 'filler' material, generally cement mortar or plaster, is placed carefully at the bottom of the borehole to ensure the even transfer of load to the ground. Reaction may be given by a mass of kentledge (iron billets or concrete blocks) erected on a frame whose load bearing legs lie well away from the hole (Fig. 6.6a). Tests on large diameter plates may require hundreds of tonnes of kentledge to give the required test pressures. It has sometimes been found economic to jack against a cross beam between two tension piles or ground anchorages, but the beam must be long (and therefore massive) to assure that the stress field around the piles or anchorages does not influence the zone around the plate. A hydraulic jack between the column and the kentledge is pumped up to press the plate into the ground. Plate deflection is measured by dial gauges in contact with the top of a measuring rod leading from the plate up through the load carrying column.

The test is undertaken by applying increments of load and observing the subsequent plate settlements. Alternatively in soils the test to determine soil maximum bearing capacity may be conducted at a constant rate of penetration, rates being of the order of 2.5 mm min^{-1}. BS1377 (1990) suggests that the maximum bearing capacity may be considered to be the bearing pressure applied to achieve a settlement of 15% of the plate diameter.

If the test load is applied in increments then it is common practice to wait until settlement from the application of one increment is complete (or substantially complete) before applying the next increment. Cycles of loading and unloading may be undertaken in the course of the test in order to assess how much settlement is irreversible and how much 'elastic'. The magnitude of the load to be applied in each increment, the rate of application of that load, the need to undertake cyclic loading and

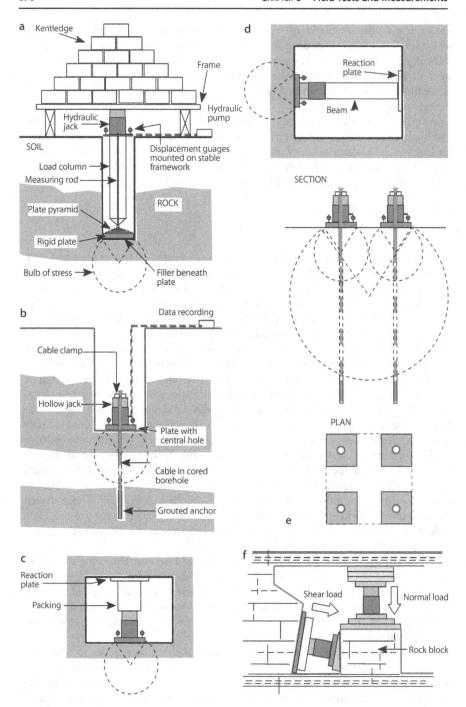

Fig. 6.6. Plate and in situ shear tests

so forth, depend upon the nature of the project for which the test is being undertaken and the nature of the materials being tested. It is very expensive to undertake plate bearing tests and each one must be carefully planned to get the data required. The tests, which require continuous supervision, may take days to perform but time spent undertaking the test is often a relatively small investment in comparison with the cost of setting the test up.

In uniformly weathered rock the objective is to find the depth of the weathered layer which has deformation characteristics adequate enough to fulfil construction criteria. This means multiple plate bearing tests which may be achieved if the reaction to plate load is given by a cable anchorage anchored at some considerable depth below the shallowest test level (Fig. 6.6b). Having executed one test the excavation may be deepened to below the level of the sample significantly affected by the previous test and the test repeated.

Tests on rock may be conducted in tunnels. In such cases the roof of the tunnel may provide the reaction to the load given by the jack (Fig. 6.6c). Tests may be also undertaken across the diameter of a vertical shaft or horizontally across tunnels (Fig. 6.6d). Preparation of the bearing surface is much easier in such conditions. If the rock can be easily cut then smaller scale tests can be undertaken in niches hacked out of a tunnel wall. These can be orientated to load in any required direction.

The volume of sample affected by the test depends on the size of the plate and as can be seen from Fig. 6.6, in which approximate bulb of stress dimensions are shown, this is small relative to the size of most foundation widths. In soil and rock masses which contain discontinuities whose presence will significantly affect their behaviour as a foundation medium this means that the test sample includes significantly fewer discontinuities than the mass to be loaded by the foundation so that the test result can be considered to be only a guide to foundation medium characteristics.

Stagg and Zienkiewicz (1968) approached the problem of providing reaction for the applied load by using a cable anchorage passing through the centre of the plate and used multiple plates to expand the volume of ground loaded. In their tests four concrete loading pads were cast on to the surface of the rock mass. The principle of this approach is shown in Fig. 6.6e. Cable anchorages were installed in 12 m deep boreholes so that at least 4.5 m of cable was free above the grouted anchor cable. Vertical loads were applied by three 100 ton hydraulic jacks on each pad; the safe working load of the cables was of the order of 275 tonnes. In order to be able to assess horizontal moduli of deformation one pair of loading pads was linked together by bars to allow them to be pulled together by two hollow jacks, while the other two pads could be pushed apart by a jack. Stagg and Zienkiewicz concluded that this reaction system gave satisfactory results provided that the top of the anchorage was 6 to 8 times the diameter of the pad below the base of the pad and that the ratio of plate diameter to hole diameter could readily be made sufficiently large to justify ignoring the possible effects of the hole on the test results; in their tests the pad diameter was eight times the borehole diameter.

The advantages of using cable anchorages as reaction are that they can be installed in any direction, so that moduli may be determined in the direction of stress applied by an engineering work.

If the free cable above the anchorage is made long enough tests may be undertaken at various levels in a weathered rock profile by deepening the access shaft around the cable.

Installation of the cable requires drilling a deep hole below the test location from which cores may be taken and tested. In situ pressuremeter tests may also be done in the borehole before anchor installation and examination of the cores should lead to be best choice of test levels.

Modulus values from the plate bearing test may be calculated from the formula:

$$S = \frac{\pi q}{4} B \frac{(1 - v^2)}{E} \qquad (6.8)$$

where E = elastic modulus; q = pressure applied to the plate; S = average settlement of the plate; B = plate diameter and v = Poisson's ratio (usually assumed to be between 0.10 (hard rock) and 0.25 (soft rock).

For tests conducted at the bottom of a borehole the formula becomes

$$S = \frac{\pi}{4} B \frac{(1 - v^2)}{E} I_d \qquad (6.9)$$

where I_d is a depth correction factor (Table 6.2).

6.3.2
In Situ Shear Tests

It may be desirable to undertake shear tests on large samples in situ if the scale of laboratory tests is too small to give reliable values of shear strength. Such tests may be undertaken on soils or soft rocks in situ but the most common application is to determine the shear strength of discontinuities in rock in connection with the imposition of inclined loads by dams. The investigation of most dams includes the excavation of trial adits and the shear test may then use roof and walls of the tunnel for reaction to jacking (Fig. 6.6f). Samples have to be cut out of the rock mass so that the discontinuity to be tested falls on the plane of shearing. Samples may be from about 0.5 to 1.5 m square and are clad with reinforced concrete above the shear plane in order to distribute normal and shear loads without damaging the sample. Only peak and residual strengths under one normal load can be determined. Cable anchorages, passing through the centre of the block, may be used as reaction to apply the normal force.

Table 6.2. Depth correction factors for plate bearing tests in boreholes (for $v = 0.25$)	I_d	Test depth / B
	1	0
	0.9	1
	0.85	2
	0.76	5
	0.74	>10

The reader will note that in Fig. 6.6f the shear force is directed slightly downwards rather than horizontally. The reason for this is that if the shearing jack is horizontal, the spread of shear stress through the sample is not uniform and failure, that is the splitting of the sample along the discontinuity, may begin first near to the jack applying shear stress, and then proceed away from it. This may cause the sample to 'ride' upwards near the shear jack, with consequent complications to the test results. The application of a slightly downwards shear load may be held to overcome this, but introduces the complication of having to resolve the shearing force into horizontal and vertical components.

6.3.3
Other Tests

The largest scale test that is commonly undertaken is the pressure tunnel test. In this a section of tunnel is sealed off and subjected to internal hydraulic pressure. Diametrically placed extensometers measure tunnel diameter extensions. The modulus of deformation may be calculated using the expression:

$$E = \frac{\psi p d (1 - v)}{\delta} \tag{6.10}$$

where E = modulus of deformation; d = internal diameter of tunnel; p = internal pressure; v = Poisson's ratio; δ = diametrical strain at centre of loaded length and ψ = coefficient depending on ratio of tunnel diameter to loaded length (ψ = 1 if loaded length > 2 × tunnel diameter).

6.4
Measurements in Boreholes and Excavations

Various types of instrument at the site investigation stage may also be used during and after construction for the measurement of certain geotechnical parameters, for example, pore water pressure. Continuous measurement from before construction, during construction and after construction, constitutes *monitoring* of the measured parameters.

Typical examples include monitoring of trial embankments to optimise design, use in projects under construction to provide information upon which to base design changes or to confirm design assumptions, and use in completed structures to monitor performance.

The fundamental requirement for any instrument is that it must be capable of measuring the required parameter, without the very fact of its physical presence causing alteration to that parameter. It must also be sufficiently robust to withstand a certain amount of damage which could easily be sustained on a construction site and still perform satisfactorily for the lifetime of the engineering structure. Much instrumentation now in common use has not been in existence long enough to know whether or not it can fulfil all of these requirements, particularly with regard to very long-term performance. With the present advances in micro-electronics and computerised systems it is probably true to say that there is nothing that cannot be measured continually and with great accuracy. If the equipment does not presently exist it could be developed, providing that there is sufficient economic incentive for the developer. Great advances in instrumentation may be expected in the coming years. For this reason this section restricts itself to the description of fundamental instrumentation and principles.

A few broad categories of instruments which have been in use for many years can be distinguished:

- *Standpipes and Piezometers:* to measure groundwater levels and pore water pressure
- *Extensometers:* to measure small changes in length
- *Inclinometers:* to measure changes in slope or angle
- *Settlement gauges:* to measure settlement under structures
- *Pressure cells:* to measure total loads on buried structures

6.4.1
Standpipes and Piezometers

Standpipes
The standpipe (Fig. 6.7a) consists of a perforated or slotted plastic pipe inserted into a borehole with suitable granular packing or filter material to prevent fine grains of soil from blocking the pipe openings. Water level is measured by a dip meter, which usually consists of twin electric cable connected to a battery and some device, visual or audible, such as a light or bell, to record the closure of the circuit when the tip touches the water.

The diameter of the pipe is not critical, but is usually in the range 20 to 50 mm. The system is simple and cheap, but there are two major disadvantages. Firstly, the standpipe can measure only a simple groundwater regime and could not measure pore pressures in different strata because the borehole and granular filter will produce hydraulic continuity between strata. Such a regime may not have existed prior to the drilling of the borehole, and the existence of strata with different pore pressures might very well be the critical geotechnical factor which could cause alteration to some fundamental design parameter. Secondly, the response time is often long, especially in low permeability soils.

The first disadvantage may be overcome by using a piezometer: standpipes are used to locate the water table. The second disadvantage is reduced by using a small diameter pipe so that the piezometer will respond to a lower volume of water flowing into the system.

Piezometers
There are many different types of piezometer; the appropriate type is selected according to the circumstances and the requirement. All types measure groundwater pressure but they differ in other respects, of which perhaps the most important is "*response time*". This is the time taken for any change in water level to be registered by the instrument. The time taken for water to flow from the soil to the piezometer is determined by the hydraulic conductivity (permeability) of the soil. Some types of piezom-

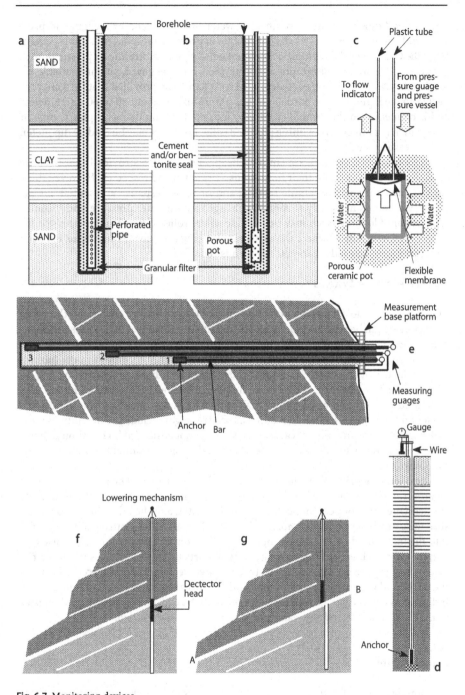

Fig. 6.7. Monitoring devices

eter respond quickly to small volumes of water flow produced by a change in pressure, while other types require a larger volume to produce a response. A simple piezometer is the Casagrande design, using either a perforated pipe or a ceramic or plastic porous pot (Fig. 6.7b). Some types of pot are protected by a steel guard, and are suitable for driving direct into the ground to depths of less than 10 m, provided that they are connected to steel piping. The porous pot has a pore size of about 50 to 60 micrometers and is connected to the ground surface by means of 10 mm diameter plastic pipe or tube, which is protected by a lining pipe or casing. Pressures may be read by dip meter, or by pressure gauge.

These open piezometers are simple, cheap, and effective, provided that they are carefully installed and that their limitations are known and respected. However, there are circumstances where remote reading and/or rapid response is essential, for example, where it is necessary to monitor change in pore water pressure under conditions of rapid loading. Such may be the case where rolled clay is being placed for the core of an earth dam, or where the earth dam is being filled for the first time. For such circumstances, a closed piezometer system is required.

Closed Piezometer Systems

The essence of a closed piezometer system is that water pressure changes in the registering device can be read at some distance from it. An example of this type is the pneumatic piezometer (Fig. 6.7c). This may sometimes be preferred in situations where, for example, freezing of water in hydraulic piezometer tubes becomes a major hazard, despite the use of anti-freeze mixtures. There are other advantages, including fast response time, low cost, no need to make height corrections, and terminal panels may be housed in protected boxes rather than expensive gauge houses. A single, portable readout unit may be used for all readings. Disadvantages include the inability to measure negative pore pressures because the system cannot be de-aired. They should therefore be used only below the water table, which makes them unsuitable for many earthfill applications.

The operating principle is that of a pneumatic pressure cell. The porous tip houses an air activated pressure cell and is connected by twin nylon tubes to apparatus at the surface. One tube (outlet) connects to a flow indicator, and the other (inlet) to a pressure measuring device and compressed gas bottle. Gas, usually nitrogen or air, is fed into the inlet and down to the tip, but cannot flow direct into the outlet because of a diaphragm which is held fast against the tube ports by the water pressure in the tip. When the gas pressure exceeds the water pressure, the diaphragm is forced away from the ports, and gas enters the outlet tube and produces measurable reaction in the flow indicator. When the gas and water pressures are balanced, no gas bubbles will register in the flow indicator and the water pressure can then be recorded.

6.4.2
Extensometers

The simplest type of "extensometer" is a tape measure, which can measure relative movements between structural members, such as parts of tunnel linings and sides of excavations.

To establish the fact that some differential movement is taking place across a joint, fracture or similar specific location, a simple glass "*tell-tale*" is adequate. Often a microscope slide is used for this purpose, and is cemented across the fracture. If movement takes place, the glass will break. It is not possible to determine the amount of movement, only to establish the fact that movement has taken place.

Extensometers measure differential movement. To measure movement of structure, for example, it is necessary to examine movement of the structure relative to the surroundings. Fixed points are established on the structure, and these are monitored by accurate surveying techniques from some distance away. The simplest type of borehole extensometer is nothing more than a metal wire or tape anchored to the bottom of a borehole with a device to measure tape extension or shortening at the surface (Fig. 6.7d). The relative movement between two points may be measured by the rod extensometer, which can be installed horizontally, vertically, for example in a borehole or at any angle (Fig. 6.7e). The use of such extensometers in subsidence problems is explained in Chap. 13. They are reliable and simple, but require correction for temperature changes.

There are many different models, but the principle depends upon fixing the rod and surrounding sleeve at one end and measuring movement at the other. The movement may be read by dial gauge, or remotely by electronic methods. Several extensometer rods and sleeves may be installed together, to give relative movements between different sections. Such instruments may be used for measuring relative movements on slopes, in tunnels, foundations, retaining walls, or in soils and rocks stressed by anchors or affected by excavation or other works. They are reliable and accurate, movements of up to 50 mm being recorded to an accuracy of perhaps 0.01 mm. The rods may be up to 100 m in length, although with such very long rods accuracy will be much less.

6.4.3
Inclinometers

Inclinometers are used to measure horizontal movements at various levels, usually within earth fills, and may be used to monitor slopes to give an indication of movement within the slope before it becomes visible at the surface. In earth fills, they are often used in conjunction with the magnet settlement gauge. Inclinometers have often been very successful in identifying movement zones and measuring direction, magnitude and rate of movement in both slopes and embankments. A very simple form of inclinometer for registering the location of zones of movement is shown in Fig. 6.7f. This is no more than a borehole with a plastic pipe grouted into it down which a simple close-fitting cylindrical weight may slide. If the strata move and bend the borehole the weight will not slide past the bend, thus detecting the upper level of the plane of movement (in Fig. 6.7f for clarity the borehole is shown as sheared).

This simple detector can be made much more complex and there are a number of different types, and within each type there are variations produced by different manufacturers. However, the basic principle is the same. A guide tube is installed in a borehole or fixed to a structure, and the inclination of the guide tube from the vertical measured at pre- determined intervals by means of a pendulum enclosed in a watertight probe which is lowered through the tube. The tilt of the pendulum is measured by electronic devices.

6.4.4
Settlement Gauges

The simplest way to measure settlement of either the ground or placed fill is to grout a reference rod into a block of concrete and measure vertical movement by conventional surveying techniques. This is expensive in terms of manpower but remote reading, multiple point systems have been developed and are now in general use. Some of these are similar in most respects to vertically installed extensometers.

6.4.5
Pressure Cells

Pressure cells are used to measure the loads beneath foundations, loads on piles, tensions in ground anchors, loads on the sides of lined and unlined tunnels, pressures within earth fill structures and so on. As such, some types may be described as load cells. There are many different types, most electronic, but liquid pressure cells record pressure on the sides of a flexible box which consists essentially of two flat steel plates, which is transferred by means of contained liquid to a measuring device. Liquid pressure forces the diaphragm against an outlet tube until gas or liquid is pumped under pressure to force the diaphragm back. When the pressure of the liquid on one side is equalled by the gas or liquid pressure on the other, liquid will just flow to the pressure gauge. In this way, the pressure inside the pressure cell can be determined. The principle is clearly similar to that of the pneumatic piezometer.

6.5
Engineering Geophysics

Site investigation techniques may be divided into two categories, destructive and non-destructive. Destructive techniques are those which, in their execution, in some way change the nature of the ground they are investigating. Thus, for example, plate bearing tests at the bottom of a shaft test ground influenced by the excavation of the shaft rather than undisturbed ground. Non-destructive techniques examine the ground without permanently altering its characteristics and are generally of either an observational or a geophysical nature.

Geophysical techniques fall into two main streams. In the first, naturally occurring phenomena, such as gravity, the earth's magnetism or telluric currents are measured with great accuracy, for values anomalous within general trends will reflect local geological conditions. In the second stream fall methods in which some form of signal is passed into the ground and changes or responses to this signal, resulting from geological conditions, are observed. Seismic and geo-electric techniques are the best known of this group.

Geological structures or anomalies are detectible because different types of ground materials have different physical properties such as density, elasticity, conductivity etc. Geophysical methods detect differences in these properties which may also coincide with other geological boundaries. The first and 'traditional' use of geophysics was to locate boundaries. These techniques are now quite well known and are continually improving. In the shallow depth geophysics utilised in the investigations for engineering projects the accuracy of location of boundaries is usually somewhat less than that desired for design purposes. However it is often rather more important to know the shape of a boundary than its exact depth. Figure 6.8 gives a simple example.

Two boreholes in Fig. 6.8a show soil over weathered rock over fresh rock. Rockhead would appear to be relatively level between the two boreholes. After a seismic refraction survey on a line joining the two boreholes together the rockhead is seen to be in the form of a buried valley. The depth of rockhead found by the geophysics is not that as found in the boreholes, probably because the velocity of the upper weathered rock is much the same as that of the overlying soil. Depths to rockhead are thus not accurate but the shape of the boundary between soil and rock is indicated.

If geophysical anomalies caused by material properties allow the definition of geological bodies then it would seem possible to establish material or mass properties by studying the geophysical characteristics of known and defined geological bodies. Thus, some 30 years ago work was begun relating the excavatability of rock to the velocity of shock waves through that rock. Attempts have been and are being made to relate ground mass deformability to shock wave parameters.

While the experience of particular engineers has lead them to enthusiasm for one of the techniques or to distrust them all, many engineers are beginning to realise that each geophysical method is but an investigation tool, which, if properly applied, will give data. The interpretation of this data may be difficult and the accuracy of the results obtained may be limited, depending on the complexity of the geological situation, the quality of the data, the skill of the interpreter and the sophistication of the interpretative techniques used. However, it must not be forgotten that the results, providing that they have been properly obtained, will always have meaning. Thus on one particular site a seismic refraction survey may give rather accurate depths to rockhead. On the next site the field data may be difficult to interpret and the results of limited

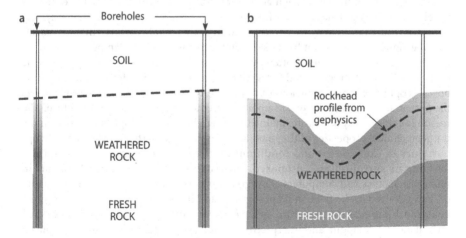

Fig. 6.8. The shape of rockhead between boreholes revealed by geophysics

value because of the complexity of the geology. This then demonstrates that additional work, perhaps using other investigation techniques, is required in order to define site geology accurately.

6.5.1
Geophysical Surveys

Geophysical surveys may be considered to take place in five phases of work. The first of these is the acquisition of data in the field, the second is the processing of these data to derive the maximum value from them and the third is the preliminary interpretation of the processed data. In the forth phase the interpretation is checked against physical observations of site conditions (from the results of boreholes, trial excavations and so forth) perhaps made after the initial geophysical work. The fifth phase is the re-interpretation, perhaps after re-processing, of initial data in the light of the physical information.

All five phases of work have to be undertaken to an equal level of quality to achieve a satisfactory end result. Basic data has to be acquired using suitable equipment and with an intensity commensurate with the required accuracy of the end result. Data must be stored in a form that allows it to be processed which generally means that the data must be in digital form and stored on discs so that it can be fed into and processed by a computer.

The most common interpretation discovers boundaries. Most non-geologists immediately assume that the boundaries are those of a geological body for all non-geologists are accustomed to looking at geological maps and sections which show the ground neatly divided into distinct, separately coloured layers. However, this tidiness is imposed by man rather than nature – all geologists have seen layers or bodies of ground which grade into each other with no clear and distinct separation. Geologists have to make such separations in order to map geological units but the boundaries used for mapping may be determined by such properties as fossil content, which have no geophysical significance. When geophysical results are compared with records of boreholes it does not necessarily follow that an apparently incorrect geophysical boundary must necessarily be corrected to fit the borehole record; the boundaries "seen" geophysically may not be those "seen by eye" in borehole samples.

While geophysical methods often produce results that seem to be of questionable value, certainly with regard to the degree of accuracy required for engineering purposes, geophysics is becoming a most important tool in engineering geology and has the potential to become the most important method investigation in the future. The reason for this is that one of the main aims of engineering geologists is to determine the geotechnical properties of the mass and geophysical methods offer the only opportunity to examine ground conditions on the scale of the mass of ground underlying or surrounding a proposed construction. The present developments in micro-electronics and computers allow great improvements to be made in the quality of data acquired in the field, in its processing, and in its interpretation. This progress holds much promise for the future. In the early days of engineering geophysics, when most data was analogue and interpretive techniques were primitive much geophysical work could be done by the engineering geologist with an appropriate background in mathematics and physics. Today most data is recorded digitally and data processing has much improved the accuracy of interpretation. Such data processing and manipulation is beyond the capacity of most engineering geologists. The problems whose resolution may be aided with the help of geophysics are becoming more complex especially in environmental geology. Accordingly, before a site investigation begins, there is an increasing need for dialogue between engineering geologist and geophysicist to determine the best techniques to be employed and to design the geophysical survey. To some degree this problem has been resolved by the development of *'engineering geophysics'* and the *'engineering geophysicist'* but the need for dialogue remains.

The section that follows cannot go deeply into geophysics but confines itself to indicating what methods are available, discussing their advantages and limitations and indicating particular uses to which they may be applied.

Not all geophysical methods are of equal utility. An analysis made by Higginbottom (1976) is given in Fig. 6.9 and shows the relative popularity of methods at that time in Great Britain.

At that time there was much emphasis on marine construction associated with the development of North Sea oil, so the most popular method was continuous seismic profiling overwater. The techniques there recorded were used mostly to determine boundary conditions. However, in the 1983 IAEG conference there was a section devoted to the use of geophysical methods to establish mass and material properties. Price (1983) analysed the papers presented to indicate relative popularity. Seismic methods remained the most popular but a new method, geo-radar, had entered the field and

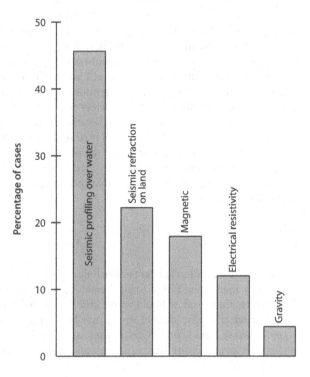

Fig. 6.9. The relative importance of greophysical methods in engineering geology (Higginbottom 1976)

the use of several methods in combination had also become popular. The Working Party Report on Engineering produced by the Engineering Group of the Geological Society Geophysics (Geological Society Engineering Group 1988) has reviewed the applicability of geophysical methods in engineering geology. The methods they list that have been used include:

- Seismic   Refraction
                 Reflection (land and marine)
                 Cross hole
- Electrical   Resistivity sounding and imaging (2 and 3D)
                 Induced polarisation (IP)
                 Electromagnetic (EP)
- Other   Ground probing radar
                 Gravity
                 Magnetic
- Borehole   A variety of logging techniques

6.5.2
Choice of Methods

In engineering geology, geophysics is used to determine boundary conditions, locate bodies of unusual properties (voids, buried pipes etc.) or to determine the engineering properties of materials and mass. The choice of method will depend much upon the particular circumstances of the work which may be not only geological but also environmental. Thus seismic refraction work using explosives to generate the seismic pulse is unlikely to be allowed in city centres and magnetometry would be unlikely to be successful in cities due to magnetic disturbances from surrounding buildings, power lines, pipes and so forth. The investigator must also not assume that only one method is appropriate for a particular purpose. Thus while seismic refraction methods are relatively useless as a means of finding underground cavities on a site of which nothing is known, they might be able to detect zones of disturbance caused by the presence of cavities in a coal seam whose location is known. Economics also influences the methods chosen; many geophysical methods have been automated where possible so as to reduce the time they take and the manpower they require, man power being the most expensive component of many methods and the basis upon which some may be chosen. Investigators must be flexible and imaginative in their application of geophysical techniques.

6.5.3
Contaminated Land

The investigation of buried containers holding hazardous waste, of buried industrial basements abandoned full of toxic chemicals, of unrecorded drains and sumps used for the illegal disposal of liquid waste, and the existence of contaminated plumes in groundwater at depth, some of which may be floating on the water table (light non-aqueous phase liquids or *L-NAPLS*) some sinking through water bearing ground to rest on low permeability layers (dense non-aqueous phase liquids or *D-NAPLS*) are part of the ground that now has to be investigated. Geophysical methods are used to help define and quantify these conditions, and their applicability is still being realised. Many techniques can be used; metal detectors for revealing buried metal containers, electrical resistivity for detecting changes in water quality, ground penetrating radar for locating trenches in which drains may be located etc. It is prudent to discuss the problem to be solved with the engineering geophysicist so that the *appropriate combination of techniques* may be applied.

6.6
Seismic Methods

If a seismic impulse is generated at surface by artificial means, for example by explosives or hammer blows, shock waves radiate out into the ground and are reflected back to the surface from any seismic boundaries in the ground mass. These boundaries may coincide with the geological boundaries between different types of materials. Seismic boundaries are distinguished by contrasts in velocity between two media and can refract waves as well as reflect them. Seismic waves are attenuated as they pass through the ground; the depths to which the ground may be investigated by seismic geophysics is related to the strength of the shock given by the seismic source.

The seismic shock generates various types of waves. Those of greatest interest are:

- the compression (longitudinal) wave, in which the particles vibrate in the direction of wave travel,
- the shear (transverse) wave, in which the particles vibrate at right angles to the direction of wave travel,
- surface waves, which travel on a boundary and are distinguished depending on the style of motion of the particle. In Raleigh waves the particles describe an elliptical path; in Love waves the particle follows a transverse motion parallel to the surface. The Raleigh wave is of some importance in engineering geophysics. The Love wave is not generated by artificial shock waves to any significant degree but is of importance to earthquake seismologists.

In engineering geophysics it is mostly the compression (P) and the shear (S) waves that are of significance. Their velocities are related to the densities and deformation parameters of the materials through which they pass. Thus,

$$V_p = \left(\frac{K + 4/3G}{\rho} \right)^{1/2}$$

(6.11)

and

$$V_s = \left(\frac{G}{\rho} \right)^{1/2}$$

(6.12)

where K = bulk modulus; G = shear modulus (modulus of rigidity = shear stress/shear strain); ρ = density; V_p = compression wave velocity, and V_s = shear wave velocity.

The ratio V_p/V_s is given by

$$V_p/V_s = \left[\frac{2(1-v)}{(1-2v)}\right]^{1/2} \tag{6.13}$$

where v = Poisson's ratio for the material.

It should be noted that because the shear modulus (G) is zero for fluids, shear waves cannot pass through a fluid medium. Shear waves are always slower than P waves (V_p/V_s is typically about 1.7) and the velocity of both P and S waves depends upon the elastic moduli of the material or mass. Thus materials with lower deformation moduli will be expected to have lower velocities. This is borne out by observation.

Velocities quoted are usually P waves unless otherwise stated. Figure 2.12 shows how velocity varies with density and porosity in limestones. Porous rocks will have fluids or gasses in the pore spaces and velocity through the saturated rock will depend upon porosity and the acoustic velocity of the pore fluid. Media have higher velocities when saturated than when dry. This is indicated by the formula (Wyllie et al. 1958):

$$\frac{1}{V} = \frac{\phi}{V_f} + \frac{1-\phi}{V_m} \tag{6.14}$$

where V = velocity of the saturated rock; V_f = velocity of the fluid; V_m = velocity of the rock matrix, and ϕ = fractional porosity.

Velocity varies also dependant on the stress within the ground mass, higher stresses giving higher velocities. The most noticeable increase in velocity is in the stress range 0 to 100 MPa.

Velocities measured in the field on various geological masses will be dependant not only on the nature of the material but also on the intensity of fracturing and degree of weathering of the mass. Table 6.3 gives ranges of velocities quoted in the literature. It is seldom clear whether these are for materials or mass but they serve to illustrate that stronger, denser materials and masses have higher velocities. What is of importance is that the lower velocity layers lie mostly near surface and that higher velocity layers lie under them.

In Fig. 6.10 part of the energy from the seismic source passes through the ground as a surface wave, part is reflected from the V_1/V_2 interface, part is refracted into the V_2 layer or, beyond the contact of the incident ray with the surface at the critical angle of incidence i_c, travels along the V_1/V_2 interface and is refracted back to surface.

Refraction will not occur from a low velocity layer overlain by a high velocity layer. This is of some importance in engineering geophysics for such low velocity layers, which are difficult to detect, may have low strength and would be thus of some importance to engineering projects.

Table 6.3. Indicative velocities for various geological materials and masses

Group	Material	V_p (km s^{-1})
Pore fluids and gases	Air	0.3
	Water	1.4–1.5
	Ice	3.4
	Petroleum	1.3–1.4
Overburden soils	Alluvium	0.5–2.1
	Clay	1.1–2.5
	Sand	0.2–2.0
	Glacial till	0.4–2.5
Igneous rocks	Granites, granodiorites	4.5–6.0
	Gabbro, dolerite, basalt	5.0–6.7
Sedimentary rocks	Evaporites (halite, gypsum etc.)	3.5–5.5
	Limestones (chalky → crystalline)	1.7 → 6.4
	Sandstones (weak and porous → quartzite)	1.4 → 6.0
	Shale and mudstone	1.4–4.0
Metamorphic rocks	Schist and gneiss[a]	3.5–7.5

[a] Velocities may be strongly anisotropic.

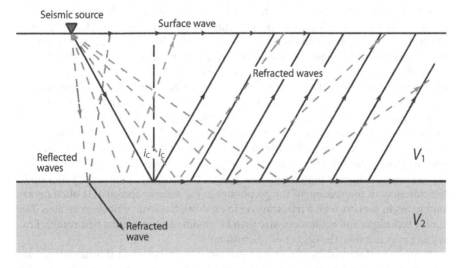

Fig. 6.10. Reflected and refracted waves from a seismic source in a two layer geology; velocity $V_2 > V_1$

6.6.1
Reflection Seismic Techniques

The seismic wave generated at the surface is reflected from boundaries underground. If the time taken for the wave to travel from surface to boundary to surface can be measured and the wave velocity is known, then the depth to that boundary can be calculated. Seismic reflection surveys always employ *'multi-channel'* equipment; in this there is one 'shot point' (where the seismic signal is generated), an array of 'geophones' which will detect the arrival of the wave form back at the surface and a multi-channel amplifier/processor unit to receive the data. Today it is common to employ more than 24 geophones in a geophone spread.

The data recovered may be displayed in a time/distance diagram. Commonly distance is on the horizontal axis and is equal to the length of the spread of geophones; the vertical axis records wave arrival in milliseconds. Arrays of geophones, of the order of some hundreds of metres long, are moved along a line of traverse which may be many kilometres long. Reflection seismic sections are composed of the many arrays along the line of traverse.

The geophone records the arrival of wave forms coming back from reflectors underground. However, not all the wave forms returning are necessarily directly from reflectors. Some of the waves may be partially refracted, some may be multiple reflections, others may be ground waves or air waves or simply noise from unidentifiable sources. Original seismic results contain all these waves; the records must be processed in order to recognise the reflections. Information is recorded digitally to allow it to be fed into a computer for processing. Frequency filtering (convolution), in which certain frequencies in the returning waves are suppressed, may help to remove noise from the record. However, this may also remove part of the reflected pulse. Inverse filtering (deconvolution) may then be applied in which other properties of the pulse are used to suppress noise and improve the character of the reflected pulse.

A reflection seismic survey programme has to be designed with regard to the objectives to be obtained. In general the important factors to be considered are the depth to which data is required and the degree of detail with which the reflector surfaces are to be examined. The depth to which the survey may be carried out and still recover useful data is determined by a number of factors of which the most important are the energy of the input signal relative to the attenuation characteristics of the ground and the level of 'noise' in a particular area which may hinder recognition of the return of reflected signals. The degree of detail (the resolution) with which reflectors may be observed depends upon wave length (pulse length) and frequency of the signal. With regard to vertical resolution the shorter the wave length the better the resolution. Seismic surveys in which higher than usual frequencies (above 100 Hz) are recorded are sometimes described as high resolution surveys. Horizontal resolution depends upon the closeness of the spacing of the geophones in the seismic spread. It is often necessary to begin surveys with a trial traverse to establish the level of noise in an area. The survey technique and equipment may then be modified to obtain the best results. Processing may improve the degree of resolution.

Noise is random and its effect may be reduced by *'stacking'*. In this the seismic array moves along the traverse so that many recordings are made above a particular point on a reflector; superimposition (or stacking) of the data from all the recordings above that point enhances the signal reflections while reducing noise.

Many seismic surveys produce profiles showing the geology along a line of vertical cross-section. Cross profiles can be used to build up a three dimensional picture of the geology. In such surveys seismic sources lie on the same line as the geophones. Other surveys may use sources on one line and arrays of geophones on other parallel lines. The data thence obtained may be used to directly interpret three-dimensional geology.

The prime customer for seismic reflection surveying has traditionally been the petroleum industry whose needs (in comparison with those of the engineering geologist) are for information on geological structure. The engineering geologist is concerned to have somewhat finer detail and so seismic reflection work on land is relatively little used for engineering geological purposes except for the investigations of deep tunnels. If these are in mountainous areas then corrections for topography pose a major problem. It is also important to note that most reflection seismic sections have *time* as the '*depth*' axis. To translate time to true depth, velocities have to be assigned to the layers and boreholes must be sunk to determine what the layers consist of *before* the velocities can be given. In engineering geology, concerned mostly with shallow (<100 m deep) geology the layers are not only naturally variable by genesis but also variably fractured and weathered, which will cause greater variations in velocities. For this reason most attention in engineering geology has been given to refraction seismic surveys in which velocity is measured and depths can thus be estimated. Geophysicists are, however, now giving some attention to shallow reflection seismic survey methods for use in engineering and hydrogeology.

The above remarks apply to reflection seismic surveying on land. There is very extensive use of continuous (reflection) seismic profiling (CSP) overwater.

6.6.2
Continuous Seismic Profiling

Continuous seismic profiling (CSP) is undertaken over water, mostly at sea. A boat tows a seismic source giving repeated seismic pulses followed by a string of hydrophones floating at shallow depth below the surface of the water. The hydrophones receive the reflected waves and after passing through a single channel amplifier/processor, their arrival is recorded on a chart. The pulse is repeated at short time intervals (of the order of 0.2 to 2 s depending on the system used) and an analogue record is built up from the repeated recordings of the returning pulses. The record has time as the vertical scale and distance as the horizontal scale. To keep the horizontal scale uniform the vessel must move at constant speed. The position of the boat is monitored by an electronic navigation system and 'fixes' may be recorded on the seismic record.

A variety of CSP techniques are available, largely distinguished by the nature of the seismic source. The seismic pulse may be generated by air or water guns, electrical spark discharge, gas explosions, rapid plate movement and so on. The sources vary in power and frequency of the pulse and are thus capable of various depths of penetration and degrees of resolution. Thus "*Pingers*" are low energy sources generated by piezoelectric transducer, operating in a frequency range of about 3 000 Hz to 12 000 Hz which give high quality resolution to depths of about 20 to 30 m. "*Boomers*" operate in a frequency range of about 500 to 10 000 Hz and may give penetration to about 50 m with good resolution. The various types of "*Sparker*" operate within frequency ranges of 50 to 5 000 Hz, are higher powered than the Boomer and may give about 500 to 600 m penetration with moderate resolution. In CSP surveys the technique is chosen relative to the purpose for which it is required. Thus for sea bed pipeline surveys a Pinger may be considered appropriate while for an inter-island tunnel a Sparker survey would yield the best information.

The results of the survey resemble a geological profile and are thus of some use in establishing the shape of reflector boundaries. Depths have to be established by correlation with boreholes. Figure 6.11 shows a profile, obtained using a sophisticated seismic source known as '*chirp*', which clearly displays layering, slumping and an infilled valley in sediments. The clarity of the profile is partly due to the absence of multiple reflections due to the absorption characteristics of the topmost sediments.

Such multiple reflections, commonly from the sea bed, may make it difficult to identify geological reflectors and interpret the records. Nevertheless the results may be of great value for particular engineering purposes, such as the location of sea bed structures, examination of the geology along the lines of undersea tunnels and so forth.

Such surveys are usually done in conjunction with other types of sea bottom profiling. Echo sounding will show the topography of the sea floor and '*side scan sonar*', a type of side-looking echo sounding, will give a picture of the sea floor on either side of the CSP track. This latter technique is of great value in surveys for dredging pur-

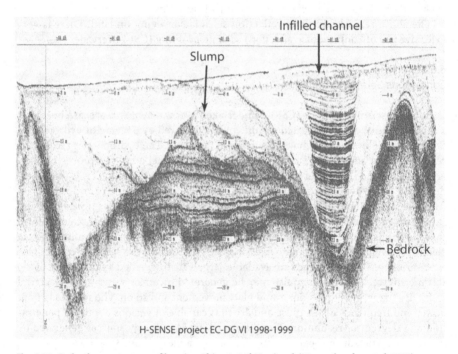

Fig. 6.11. Gothenburg estuary profile using Chirp 4–14 kHz signal (Maurenbrecher et al. 1998)

poses, for it may reveal the presence of reefs, boulders, old wrecks etc. which would not necessarily be located by echo sounding.

CSP survey methods may be limited by navigational problems. Thus, in the construction of power stations, tunnels for intake and outlet of cooling water may be driven out under the sea. Investigations by CSP methods would be ideal for such tunnels but impossible to undertake because of near shore navigational hazards.

6.6.3
Seismic Refraction Methods

Seismic surveys for engineering projects on land are mostly undertaken by the refraction technique. Twelve or more geophones are used in a traverse while the seismic pulse is provided by explosives, falling weights or hammer blows.

Results are presented in a time/distance graph (Fig. 6.12) which, in favourable conditions, will show a curve composed of straight line segments with increasing traverse distance and with sharp segment intersections. The slope of each segment represents a velocity and it is conventional to designate the velocity of the near surface layer as V_1, the next velocity as V_2 and so on. The depth (d_1) to the first horizontal refractor is given by:

$$d_1 = \frac{X_1}{2}\left(\frac{V_2 - V_1}{V_2 + V_1}\right)^{1/2} \tag{6.15}$$

An ideal time distance graph from multiple horizontal layers should show several sloping straight-line sections of the graph, each corresponding to increasing layer velocities; depths to layer interfaces and layer thicknesses may be computed by formulae of increasing complexity.

If the refractor is dipping it is possible to gather data by reverse shooting (keeping the geophones as before but putting the shot point at the other end of the traverse) to calculate the dip of the refractor. If the refractor is undulating there are interpretative

Fig. 6.12. Time/distance graph of refraction seismic survey

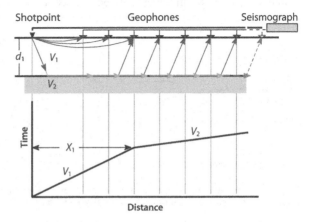

techniques which allow the calculation of the depth to the refractor under each geophone, so that, for example, an irregular rockhead surface under a soil cover may be mapped. Mostly engineering sites are examined by a grid network of traverses, plotting the levels of refractors and drawing refractor contours. However, it must be remembered that the real geological situation is three dimensional and not two dimensional. Thus, on a hillside or above an irregular rockhead the first arrival recorded by a geophone may come from a refractor not immediately beneath the geophone although it is common practice to display refraction profiles as if they were vertical cross-sections. It must also be remembered that the accuracy with which the refractor shape can be plotted depends (just as it does in surface topographical surveying) upon the density of observation points.

The depth to which the ground may be examined by the refraction method depends upon length of traverse, velocities and attenuation factors of the strata and the energy of the source. In much seismic refraction work a great part of the pulse is absorbed by near surface soils, particularly if the seismic shock is generated by falling weights or hammer blows. For engineering purposes investigations seldom require information to depths of more than 100 m and traverse lengths of the order of 400 to 500 m are suitable. Data is recorded digitally to allow for processing and computer calculation of results. The technique is particularly suitable for the investigation of 'long' engineering works involving excavation of materials, such as the construction of roads, canals, railways, etc. Thus, for example, Tan et al. (1983) have described the use of refraction surveys for the investigation of a proposed road construction in Singapore (Fig. 6.13).

Twelve, twenty-four or more geophones refraction seismic surveys are the task of a geophysicist but simple surveys using the single geophone (single channel) seismograph are now commonly undertaken by engineering geologists. As will be seen later in this chapter the reason for these surveys may be other than simply to find the depth of a refractor but many engineering geologists now use the single channel hammer seismograph as geologists use a hammer. The basic idea of this device is instead of using one shot point and twelve geophones, to use one geophone and twelve shot points the seismic pulse being given by a sledgehammer blow on a steel plate resting on the ground. The earliest model single channel seismographs simply recorded the time of the first arrival and numerous hammer blows were needed to be assured that the true first arrival had been recorded. Many an engineering geologist saw strong healthy labourers reduced to shivering blistered wrecks in the course of a day's survey in ar-

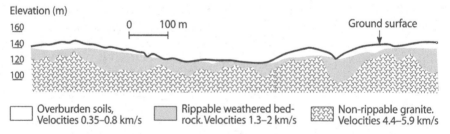

Fig. 6.13. Results of a seismic refraction survey for an express way in Singapore (Tan et al. 1983). The profile was prepared using a series of 12 channel seismic traverses

eas of seismic 'noise'. However, the invention of the enhancement seismograph, which allows successive hammer pulse records to be 'stacked', changed this and surveys now are undertaken with less effort and greater reliability. The energy that can be put into the ground by a hammer blow is, however, rather limited and single geophone surveys are employed generally when the refractor (usually rockhead under soil) lies at depths of 10 to 15 m or less below surface.

6.6.4
Down-Hole and Cross-Hole Shooting

Seismic measurements are also commonly undertaken in boreholes. Two basic techniques are used. In one a seismic impulse is generated in a borehole and passes through the ground to be picked up by geophones implanted in other boreholes (cross-hole shooting). In down-hole or up-hole shooting impulses are generated on surface or at the bottom of the borehole and are picked up by an array of geophones in the borehole. The former method can be carried out with explosives and hydrophones in a mud filled borehole or (on a smaller scale) with special pulsers and pickups which can be clamped in small diameter boreholes. Both techniques are mainly used to determine velocities (thus aiding interpretation of both reflection and refraction seismic surveys) to detect the presence of major discontinuities in rock masses or to assess quality of the rock mass. Cross-hole shooting is particularly useful in investigations for tunnels. Tunnel investigations commonly incorporate boreholes to tunnel level. Cross-hole shooting between boreholes at tunnel level will give data relating to the longitudinal homogeneity of materials along the tunnelling and is thus the only direct method available to investigate ground to be intersected by tunnels along their length.

6.6.5
Particular Applications of the Seismic Method

Seismic methods have been particularly helpful for assessing rock mass quality, excavatability and the moduli for deformability.

Evaluation of Rock Mass Quality

It is often useful to determine the "*quality*" of the rock mass with regard to the execution of a particular engineering activity involving the mass, e.g. tunnelling through it. Quality should always be assessed relative to the process. Thus a strong unweathered rock mass with widely spaced joints might be regarded to be of high quality with regard to *foundation construction* on it but of poor quality with regard to *ease of excavation* in open cut.

Seismic geophysics can be used to contribute data to aid such assessments of quality. A significant concept is the idea of the "*velocity ratio*" or "*velocity index*" (in some older articles this is referred to as the "*fracture index*"). In this the velocity measured in the field on the rock mass in situ (V_f) is contrasted with that measured in the laboratory on a fresh sample of rock material (V_l). This gives the velocity ratio (V_r):

$$V_r = \frac{V_f}{V_l} \tag{6.16}$$

which is less than 1. Low in situ velocities (and thus low velocity ratios) could be due to one or a combination of the following factors:

- weathering of the rock material (increasing weathering leads to decreasing velocity),
- close spacing of discontinuities (the closer the spacing the lower the velocity),
- openness of discontinuities (the greater the aperture the lower the velocity).

Since the influence of the factors can be assessed (although not with great accuracy) from examination of cores, outcrops etc. it is often possible to recognise the dominant cause of low velocity ratios. This can only be done on uniform rock types. The velocity ratio approach does not work well in heterogeneous rock masses.

Knill (1970) used velocity ratio to assess grout take in dam foundations. Measurements were on saturated rock samples. Velocities on laboratory specimens may be significantly higher on saturated specimens than on dry specimens and clearly tests should be undertaken on rock samples with the same moisture content as the rock in situ.

If velocities can be used to assess potential grout take then they may also be used to assess the success of the grouting operation. If openness of joints has significantly influenced V_f then filling these joints with grout should increase V_f. Clearly V_f on a grouted up mass will not reach V_l for the velocity of grout is lower than that of most rocks and part of V_f could be influenced by material weathering which would not be affected by standard grouting. An increase of dam foundation rock velocity as a consequence of grouting is recorded by Knill and Price (1972).

Field velocities are obtained by conventional refraction surveys. In rocks which display gradual changes in weathering grade the time/distance graph may have no obvious breaks to determine depths to refractors. These curves may be manipulated to allow the production of velocity contours at suitable intervals, which may indicate variations of weathering grade. This is particularly useful if the increase in quality with depth of an originally more-or-less homogeneous rock mass is being examined. Edmond and Graham (1977) have described how seismic refraction traverses, together with cross-hole and up-hole shooting provided data to help assess the quality of a granite mass in order to choose the best level to drive a cooling water tunnel.

The velocity contour idea was extended by Cosma (1983) who examined rock masses carrying ore bodies by cross-hole shooting. An automatic hammer was developed as a source and receivers were designed to receive P and S waves. A computer program was written to calculate velocities on a grid network within the area examined and then these were plotted to give velocity density sections. In these, high velocity zones were taken to indicate ore bodies while low velocity zones indicated weakened or broken rock. Such 'tomographic' techniques have clear application in engineering geology in the examination of foundation masses, potential quarries and so forth.

Evaluation of Excavatability

The velocity of seismic waves through a rock mass decreases with increasing weathering and increasing fracture intensity and openness in that rock mass. Experience has given some idea of the velocity of seismic waves through unweathered, 'normally fractured' masses composed of rocks of various types. If the velocity through a mass of a particular rock type is measured then the difference between this measured velocity and that of the 'standard' mass of a similar rock type will give an idea of the 'quality' of the mass, this quality being assessed with regard to fracture intensity and openness and to weathered condition. The ease with which rock masses may be excavated depends upon the same parameters which may be used to gauge quality, so velocity measurements on rock masses of known geology may be used to yield information on excavatability.

It is particularly important, from the viewpoint of engineering economy, to know whether a rock mass must be drilled and blasted before it can be excavated or whether it can be dug directly with standard excavation machinery. Excavation preceded by drilling and blasting is clearly much more expensive than that without such prior treatment and important contractual distinctions between engineering 'rock' and engineering 'soil' are sometimes linked to the excavation technique. Accordingly the concept of 'rippability' was introduced some decades ago. A ripper is a bulldozer equipped with a weighted steel tooth which may rip into and loosen intact rock to a depth of usually something under a metre in order to allow it to be dug or scraped out by standard machinery. Rippability relationships have been established between type of bulldozer (for clearly a more powerful and heavier machine could rip where a lighter machine could not), rock type and field velocities measured on the rock type to be excavated.

The technique has been used for many years with some success but the user has to exercise some care:

- In a dipping thinly layered mass composed of various rock types the velocity measured may be an average of high velocity unrippable rock in a greater mass of lower velocity easily rippable rock. In such a case the harder higher velocity bands would determine the rippability of the total mass.
- Anisotropy in the rock mass giving anisotropic velocities. If the direction of lowest velocity is the direction in which velocity is measured, while excavation takes place in the direction of highest velocity, then excavation will be more difficult than expected.

Adequate knowledge of site geology will give warning of the first difficulty. The best procedure is to combine rippability measurements with other techniques for evaluating rock mass characteristics (discussed in Chaps. 3 and 9) and to be sure that there is a thorough understanding of geological characteristics of the rock mass.

Seismic fan shooting would provide data on anisotropy and help resolve the second problem. In lithologically uniform rock masses anisotropy may be attributed to discontinuity frequency, persistence and openness (Hack and Price 1990). Seismic traverses in a fan array may show velocities suggesting the dominant joint direction Fig. 6.14).

Evaluation of Deformation Moduli

The relationships between seismic velocities and elastic moduli expressed by Eqs. 6.11 to 6.13 suggest that it might be possible to determine static elastic moduli of ground masses through seismic measurements. There are, however, reasons why it is necessary to distinguish between statically and dynamically derived moduli. These are that:

Fig. 6.14. The plan shows joint traces in a rock of homogeneous lithology. Joint x is persistent, frequent and sometimes open; joint y is impersistent, tight and infrequent. A fan seismic array would show a higher velocity in direction B than in direction A thus indicating the strike of the dominant joint

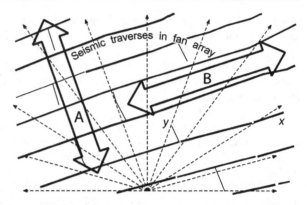

- the intensity of load imposed by the seismic wave is small and causes very much smaller strains than that imposed by a large field test (and also by the proposed construction), which are sufficient to cause large strains.
- the seismic wave used acts for a very short time on the mass while the load used to measure the static modulus is (by comparison) applied over a long period of time.

It is now well recognised in continuum mechanics that shear modulus decreases as shear strain increases and so it has been argued that there are three moduli which can be measured:

1. the dynamic modulus (E_{dyn} or E_{seis}) determined by seismic means and applicable to low stress levels;
2. the deformation modulus (E_{def}) determined by in situ testing on ground masses which do not behave elastically within the chosen stress range; and
3. the static modulus (E_{stat}) determined by in situ testing and measured on the elastic part of the stress/strain curve.

The stress range within which the modulus is calculated (perhaps as a secant modulus) should relate to the loads imposed by the proposed structure.

The object of most investigations is to determine E_{def} of the groundmass within the stress levels to be imposed by the structure; this should be a value appropriate to the mass under the whole structure, that is, at the scale of the structure. Seismic measurements offer the opportunity to measure a modulus at the correct scale; the problem is whether there is any relationship to be found between E_{def} and E_{dyn}.

Van der Schrier (1990) compared statically and dynamically determined moduli of elasticity for sedimentary rocks. E_{stat} was determined on laboratory specimens and by P and S wave train measurements in the borehole using the Schlumberger Array Sonic service. Good correlations were found but were essentially on materials, not mass. Various authors have proposed correlations between E_{def} and E_{dyn}. Stacey (1976), for example, suggested that for preliminary engineering design purposes on rock masses it was reasonable to assume that $E_{stat} = 1/4\,E_{dyn}$ and $E_{def} = 1/8\,E_{dyn}$. Aikas et al. (1983) have recommended the use of seismic measurements to measure rock mass deformation moduli under particular geological circumstances. They measured P and S wave velocities and calculated E_M (rock mass modulus) by the formula:

$$E_M = \frac{DV_s^2 \left(3\dfrac{V_P^2}{V_s} - 4\right)}{\dfrac{V_P^2}{V_s} - 1} \quad (GPa) \qquad (6.17)$$

where D = rock mass density.

Their objective was to give data for finite element models examining mine behaviour. Their conclusion was that the moduli values obtained by these methods were useful if the rock was compact, of high quality and under medium to high compression stresses.

What seems clear is that if a rock mass whose bulk characteristics approximate to those of intact rock material and which behaves elastically within the stress range to be imposed by the proposed work, then E_{dyn}, E_{stat} and E_{def} are likely to be similar. If it does not behave elastically they will differ. In most rock masses, particularly those near surface masses affected by weathering, joint opening etc. similar values for these moduli seem unlikely. It has been proposed by Bieniawski (1989) that E_{mass} (E_M) could be assessed for preliminary design by using his Rock Mass Rating (p. 127):

$$E_M = 2\,RMR - 100 \qquad (6.18)$$

It may be speculated that the modification of seismically derived mass moduli by a mass quality assessment obtained by rock mass classification offers the best chance of determining static moduli on the scale of the engineering work. However, until such research avenues are explored the most promising strategy is to compare moduli derived from a variety of sources (such as laboratory tests on intact samples, pressure-meter tests in boreholes and plate bearing tests) with seismically derived moduli and deformations measured around engineering structures. Recent advances both in the generation of seismic waves and in their recording have permitted calculation of shear modulus with depth to be made, to depths of 10 m or more depending on the soil or rock. Waves are generated by a vibrator at ground level; boreholes are not required. Values so obtained provide exceedingly good predictions of settlement based on plate loading tests (Mathews et al. 1997).

Stress Measurements

It has frequently been observed that seismic velocities change as stress on the rock changes; changes in velocities measured in chalk by cross hole measurements through the ground stressed by a plate bearing test have been reported by De Ruyter (1983). Generally the velocity increases with increasing confining stress until some other change, such as microfracturing, changes the nature of the material.

Attempts have been made to assess stresses in tunnels and mines (mostly pillars) by measuring velocities. These appear to have the greatest chance of some success if:

i   the rock is porous (porous rocks are more sensitive),
ii   the rock is dry,
iii the rock is homogeneous,
iv   the stresses to be measured are no higher than about 50 to 100 MPa, depending on
     rock type, and
v   variations in external factors (such as temperature, humidity etc.) are small.

Mostly researchers have attempted to measure changes in stress rather than an absolute value, as can be done in a mine pillar with the help of velocity measurements; when the pillar developes fractures prior to its collapse, velocities rapidly diminished.

It has also been observed that as rocks or soils reach the point of instability, as the result of landsliding or some similar agency, the development of fractures is accompanied by acoustic emissions. The onset of instability may be detected by arrays of geophones planted in and around the ground mass. Acoustic emission has been studied for at least a half century (for example by Knill 1968) and is in use in some mines as a means of monitoring stability. The advancing development of micro-electronics suggests that it will become an increasingly important monitoring technique.

Determination of other Properties by Seismic Methods

Velocity measurements on uniform rock materials give the opportunity to assess changes in material properties, such as density, porosity, strength etc., which might occur as the result of weathering or alteration. While there are very general relationships between these parameters and velocities for all rocks, velocity measurements are only useful as an alternative to other tests when dealing with a single rock type.

It is possible that seismic measurements, combined with other tests of a simple nature, may give the opportunity to determine other more complex parameters. Maris (1982) has examined the possibility of determining the relative density of sands by a combination of seismic measurements and particle size analyses as an alternative to the Standard Penetration Test. As a result of his work a very tentative relationship between P wave velocity, D_{50}/D_{10} and relative density has been proposed. It should be noted that the relationship given is for sands under low confining pressure, velocity increases quite rapidly as confining pressure increases. To arrive at relative density for a layer it would be necessary to obtain seismic data by cross hole shooting, particle size analysis data by testing samples and pressure by determining density and thence calculating pressure. Moisture content must also come into consideration but the effect of this is not yet established.

The idea is not new. General relationships between SPT-N-values and P and S wave velocities have been established by Baoshan and Chopin (1983) and Tonouchi et al. (1983) but are somewhat general.

6.7
Electrical Resistivity Methods

All materials have electrical properties. One of the most important of these is the resistance that a material offers to the passage of electrical current. It is well known that copper has low resistance (or high conductivity) while rubber has high resistance (or low conductivity). Groundwater has moderate resistance (50 to 500 ohm-metres, Ωm) but salt water is more conductive. In pure materials such as iron or aluminium the resistivity is a property of that element. In geological materials the resistivity depends upon the nature and the mixture of minerals of which the materials are composed and whether or not water is present in pore spaces. Resistivities of clay materials can range between 50 and 500 Ωm, those rich in quartz from 50 to 5 000 Ωm.

If electrical current is passed into the ground and voltage measured between two electrodes then resistivity of the ground may be measured. In the Wenner configuration of electrodes current passes between two current electrodes and voltage is measured at two voltage (or potential) electrodes. The electrodes lie on a straight line and are equally spaced. The apparent resistivity is given by the formula:

$$\rho_a = 2\pi a \frac{\Delta V}{I} \tag{6.19}$$

where ρ_a = apparent resistivity; a = electrode separation; ΔV = change in voltage between voltage electrodes; I = current between current electrodes.

Since, by Ohm's Law $V = RI$, where R = resistance, electrical resistivity is expressed in terms of ohm-units of distance. Commonly the units are ohm-cm, ohm-m and ohm-ft. The most common unit today is the ohm-m.

The depth to which the current penetrates into the ground depends upon electrode spacing; the wider the spacing the greater the depth of penetration. If the material that the current passes through is uniform to infinite depth then whatever the electrode spacing there will be no change in resistivity. However, if the ground under the electrodes consists of a layered series of materials of different resistivities then, as electrode spacing expands to give deeper current penetration, so will the measured "*apparent resistivity*" change. If the material is uniform but saturated below a certain depth (the water table level) then the apparent resistivity will also alter as the electrode separation expands. The results of a resistivity survey are often expressed as a graph plotting increasing electrode spacing against apparent resistivity.

It is possible to calculate the shape of the curve of electrode spacing against apparent resistivity that would be produced from an expanding electrode survey conducted over particular thicknesses and depths of layers of different resistivities. These curves may be compared with those found in field surveys to give a method of interpretation of survey data. Unfortunately several different models may give more-or-less the same curve so there is no unique interpretation of results. Normally additional information is required to ensure reasonable interpretation. This may be obtained from published geological profiles or from boreholes.

Increasing the *Wenner array* electrode spacing to measure the resistivity of deeper and deeper layers may be likened to the process of drilling a borehole and is sometimes referred to as '*electrical boring*'. If the electrode spacing is kept uniform (so that depth penetration is constant) but the whole array moved laterally in the direction of the array length then the process may be likened to that of digging a trench, for a change in strata will be revealed by a change in resistivity. This technique is sometimes called '*electrical trenching*'. Its primary use is to trace boundaries, faults, concealed outcrops etc. and is a useful aid to detailed mapping.

The Wenner array is probably the best known but many other arrays are also in use. The *Schlumberger array* has current electrodes symmetrically placed around the potential electrodes but at greater spacings. Different arrays are described in the literature, each of which is of value for particular geological circumstances.

If the basic geology of an area is reasonably well known then the presence of groundwater may be detected for this will modify resistivity. Since there is a difference between the resistivity of fresh and salt water it is possible to judge the quality of water located.

If a series of constant electrode separation traverses are done on a grid network then resistivity values may be contoured and anomalous areas revealed. Higginbottom (1976) has described the results of a survey conducted to investigate the possibility of locating fissures in rock.

6.8
Magnetic Methods

The earth's magnetic field is generated by electrical currents circulating in its outer core. Values of total intensity of the earth's magnetic field vary from about 74 000 gammas (1 gamma = 1/100 000 oersted) at the poles to 30 000 gammas at the equator. The major variations in field strength and polarity direction relate to major continental scale and deep seated geological factors. Minor and local variations depend upon the nature of near-surface geological materials. Iron bearing geological bodies are magnetic and the materials of which they are composed have high "*magnetic susceptibility*". The intensity of the earth's magnetic field tends to be higher over bodies of minerals of high magnetic susceptibility. Magnetic susceptibilities of common minerals and rocks are given in the Table 6.4.

Magnetic "anomalies" are values of the magnetic field intensity above (positive) or below (negative) the general regional field intensity of the area.

The size and shape of magnetic anomaly will be due not only to the contrast in susceptibility between a magnetic body and the surrounding material but also due to the orientation of the magnetic body relative to the local direction of polarity of the earth's field. Linear features generally produce poorly defined anomalies if they are orientated north-south.

The strength of the earth's magnetic field may be measured using magnetometers. Old style magnetometers measured either the vertical or horizontal components of the earth's field by measuring the deflection of bar magnets or dip needles. Measurements using such instruments involved levelling the instrument and took some time to complete. Instruments developed since World War II, the flux-gate and proton magnetometers, are electronic, require no levelling and readings may be taken very rapidly. Probably the *proton magnetometer* is the most widely used magnetometer for ground engineering; readings may be taken at intervals of about 1 second. The detector head is separate from the recording instrument and may be towed behind an aeroplane or even underwater behind a boat.

Surveys on land most often involve taking proton magnetometer readings at intersections of a survey grid. No complex calculation is required to convert readings to total intensity values. To prepare a magnetic anomaly map intensity values are subtracted from the regional field intensity to allow anomalies to be plotted. The greatest

Table 6.4. Magnetic susceptibilities of rocks and minerals
(emu = electromagnetic unit for
mass susceptibility, $m^3 kg^{-1}$,
based on Hunt et al. 1995)

Mineral or rock	Average susceptibility (emu[a] $\times 10^{-8}$)		
Quartz	−0.5	–	0.6
Calcite	−0.3	–	1.4
Coal	2		
Clay	10	–	20
Chalcopyrite	0.55	–	10
Limonite	65	–	75
Hematite	10	–	750
Magnetite	20 000	–	110 000
Limestone	0.1	–	1 200
Sandstone	0	–	930
Shale	3	–	886
Granite	0	–	1 900
Dolerite	35	–	5 600
Basalt	10	–	6 000
Andesite	6 500	–	1 300

[a] emu = electro magnetic unit.

part of the work lies in setting out the survey grid. The value of the earth's field varies diurnally and this should be measured by returning to one spot at regular intervals during the days work. The final plot may have to be corrected for this diurnal variation depending on its magnitude in comparison with the anomalies.

It is possible to link the detector head to an electronic location finding system and link both to a micro-computer which will then prepare the magnetic anomaly map as the survey proceeds.

Magnetic anomalies may reveal the presence of bodies of high magnetic susceptibility surrounded by non-magnetic rocks under a cover of soil. Dykes of basic igneous rock, which should produce anomalies, may be of engineering significance – they could be associated with changes of rockhead level, form hard barriers to excavation or present permeable zones in tunnelling. Buried landforms cut by erosion into magnetic rocks may also be detected by magnetometry.

Metal objects buried under soil may be detected by magnetometry and wrecks, anchors, unexploded bombs etc. may be found on the sea bed.

Magnetometer surveys are particularly useful in areas where abandoned mine workings are a problem. The mine workings may have many long abandoned shafts leading to them and these, if poorly backfilled, may present a localised subsidence problem. These shafts may be no more than three or four metres in diameter and thus present a target almost impossible to find by boring. However, there is sometimes sufficient magnetic contrast between the shaft lining and/or infilling and the surrounding rock or soil to produce a detectable magnetic anomaly. A magnetometer survey for an old mine shaft requires a survey on a grid network, with observations at 1 to 3 m spacings. Any anomalies found must be examined by excavations to find out what the anomalies represent; Fig. 6.15. These could also reflect the presence of old foundations, walls etc. associated with the old mine buildings. Excavations for old mine shafts should be undertaken with some caution for the infilling might collapse under the investigator.

The effectiveness of magnetometer surveys may be hampered by local magnetic disturbances which could result from buried or overhead live electric cables, buried pipes, wire fences and so forth. There is almost no chance of conducting an effective survey in an urban area.

6.8.1
Electromagnetic Techniques

The description "electromagnetic methods" encompasses a wide variety of geophysical techniques and systems. They frequently involve observing the reaction of ground masses to electromagnetic impulses of varying frequencies. Two methods are of frequently used in engineering geology. These are *induction methods* and *ground penetrating radar.*

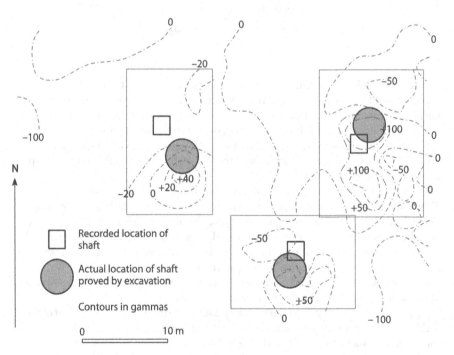

Fig. 6.15. Part of a proton magnetometer survey for the location of old infilled mine shafts and showing the association of positive anomalies with the shafts (Higginbottom 1976)

Induction Methods

In these a cyclic magnetic field is generated by passing alternating electrical current through a wire loop or through a wire grounded at both ends. The frequency of the magnetic field may be between 100 Hz and about 5 000 Hz. If any conductive material lies within the generated magnetic field electrical currents will be generated in it and these will, in their turn, generate a magnetic field. The total magnetic field then existing may be considered to consist of the primary field (generated by the source current) and the induced secondary field (generated by the conducting body). Measurement of the characteristics of the total magnetic field and interpretation of these in comparison with the primary field may lead to the detection of buried conductors.

In the field the equipment may consist of a transmitting coil (transmitting the primary field) and a receiver coil (measuring the total field). Long wires may take the place of coils. The method maybe applied on the ground or from aircraft.

Electromagnetic methods, of which there are many, are commonly described as EM methods.

Interpretation of the results often relies upon comparison of measured values with those calculated or observed to be caused by various shapes of conducting bodies.

Ground Penetrating Radar

The basic principle of the transmission and reflection of electromagnetic waves in order to establish the location of aeroplanes and aid ship navigation is well known. Radar impulses have also some capacity to penetrate into the ground and to be reflected from less "permeable" layers in a way similar to that of seismic shock waves. Because the radar signal is electrically generated it can be repeated continuously, rather like the repeated seismic pulse used in continuous seismic profiling. If both sender and receiver can be moved along the surface of the ground then the returning reflected signals can be continuously recorded to give a record similar to that received from CSP techniques (Fig. 6.11).

The radar pulse is reflected from a boundary, so that

$$V = \frac{2D}{T} \tag{6.20}$$

where V = effective propagation velocity through the material overlying the reflector; D = depth to reflecting surface, and T = elapsed time between transmitted and received pulse.

Also,

$$V = \frac{C}{E_r} \tag{6.21}$$

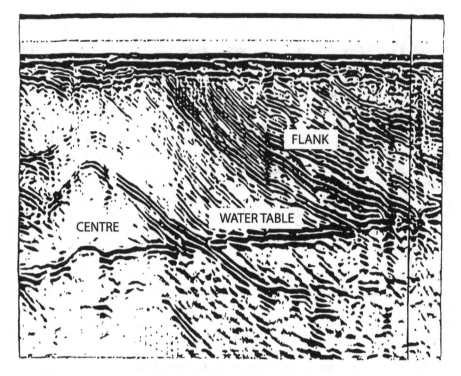

Fig. 6.16. A "radargram" of an esker clearly showing its centre, flanking deposits and the watertable (Bjelm et al. 1983)

where C = propagation velocity in free space and E_r = effective relative dielectric constant of the material. A typical radar record is shown in Fig. 6.16. The conductivity of the soil, the dielectric constant between the reflector and the surface and the shape of the reflector are factors that affect the performance of ground radar.

The technique is still developing but shows great promise. Penetrations of up to 40 m have been recorded. The method will probably supplant seismic reflection and refraction for shallow boundary detection once its development is complete. Darracott and Lake (1981) provide a good introduction to the method.

6.9
Gravity Methods

The intensity of the earth's gravitational field may be measured using gravimeters. These are, in effect, extremely sensitive balances in which the very small movements associated with variations in gravity between successive points of observation are magnified to readability by mechanical or optical devices. They are sensitive to variations of about 1 part in a hundred million of the earths field (about 0.01 milligal, where 1 milligal = g/1 000). The local value of gravity is affected by the density of the underlying geological material and by other factors, such as latitude, elevation, local topography and tidal effects. In order to determine variations in underlying geology by gravity measurements the influence of these other factors must be assessed by use of various corrections. The almost unquantifiable effects of rapid near surface density variations tend, in practice, to limit the application of gravity techniques in engineering geology to the detection of major geological structures (faults, etc.). However, in certain conditions of uniform lithology it may be possible to detect features of engineering significance, such as underground cavities.

Surveys for such purposes are sometimes described as "microgravity surveys" implying that the most accurate instruments are used to make observations on a close network (<10 m station spacing). A particular application is finding karst cavities in limestone. In theory quite small cavities at depth will give anomalies detectable by modern gravimeters (Greenfield 1979). The problem is that the corrections to the raw data required to isolate the anomalies are often much greater than the anomalies themselves and are difficult to estimate with the required accuracy. Thus, for example, while gravity methods could in theory be used to detect karst on limestone dam sites this is seldom done because the dam site topography imposes too many corrections to the raw data. The method may, however, be useful on limestone coastal platforms which could be the sites for airports or similar constructions which must be built on level ground. Bichara and Lakshmanan (1983) have estimated embankment bulk densities by this method.

6.10
Further Reading

Bell FG, Cripps JC, Culshaw, MG, Coffey JR (eds) (1990) Field testing in engineering geology. Geological Society of London (Engineering Geology Special Publication 6)
Craig C (ed) (1996) Advances in site investigation practice. Thomas Telford, London
McCann DM, Eddleston M, Fenning PJ, Reeves GM (eds) (1997) Modern geophysics in engineering geology. Geological Society of London (Engineering Group Special Publication 12)
McDowell PW, Barker RD, Butcher AP, Culshaw MG, Jackson PD, McCann DM, Skipp BO, Mathews SL, Arthur JCR (eds) (2002) Geophysics in engineering investigations. Joint publication of the Geological Society of London (Engineering Group Special Publication 12), and CIRIA, London

Chapter 7

Organisation, Design and Reporting of Site Investigations

Practical experience has shown that for a site investigation to be successful it must be well planned and undertaken in an orderly manner, using appropriate and well maintained field and laboratory equipment, operated by experienced and skilled personnel.

Expertise in the mechanics of investigation (boring, testing etc.) is a routine requirement in all investigations. The most difficult problem is how and where, and when, to use the various 'tools' available to the site investigator. A philosophy of site investigation has built up over recent years which proposes the idea of the developing investigation advancing in stages to a satisfactory conclusion, each stage being built on a sound foundation of knowledge established by the previous stage.

7.1
Stages of Investigation

The stages of investigation described below are but the expression of a principle. The stages do not need to be separate; they may merge into each other and additional stages may be inserted.

7.1.1
Project Conception Stage

After the decision to initiate a project has been taken, a desk study is undertaken of all available geotechnical, geological and topographical data. The proposed site and its environs should be examined by an experienced engineering geologist. The objective of this stage is to try to identify potential problems that may arise from site geotechnical conditions in relation to the proposed engineering work. Here it may be noted that the term *Site Investigation* is taken by many to represent the investigation of the site per-se, including its previous use, ownership, access etc. An investigation of the ground at a site is thus a *Ground Investigation* and but part of a Site Investigation. The engineering geologist is usually concerned with both these aspects of the work. The geotechnical, geological and topographical data should include:

- all available topographic maps,
- all available geological and hydrogeological maps, memoirs and published articles in the scientific journals,
- aerial photographs at all scales,

- records of natural hazards such as earthquakes, hurricanes, avalanches etc.,
- site investigation and construction reports for adjacent engineering projects; published articles on the geotechnical properties of the geological units to be found on the site,
- hydrogeological and hydrological data,
- records of any past, present and future human activities which have, are or could influence the geological environment.

The recommendation that the site be examined by an experienced engineering geologist may perhaps be considered to verge upon self-advertisement in a book written by an experienced engineering geologist. However, there is no doubt that the significance and even the presence of important geological features on the site can most quickly be recognised by those who are skilled in this work. A few days work by an experienced engineering geologist may save much time and trouble; i.e. money.

7.1.2
Preliminary Investigation Stage

The evaluation of a project at its conception stage may reveal significant gaps in basic knowledge of the site, so that no recognition of likely problems is possible. In such a case some preliminary investigation may be required to establish that basic knowledge. This would be undertaken using relatively simple and inexpensive techniques, such as existing records (maps, photographs, etc.), geological and engineering geological mapping, geophysics and perhaps some boreholes. The boreholes could be undertaken partly as an experiment to determine the best method for the boring, sampling and in situ testing to be undertaken in the main stage of investigation. At the end of this stage there should be sufficient knowledge of the site to allow design of the main ground investigation.

The first two stages of investigation are sometimes described as the *reconnaissance investigation* or *feasibility investigation*. If a number of sites are being investigated prior to choosing one for development the feasibility investigation may give sufficient information to allow the choice to be made.

7.1.3
Main Investigation Stage

In the main investigation stage the work done should recover the information required to design the engineering project. This information is obtained by whatever means are appropriate to the ground conditions and the nature of the engineering work. It is possible that some of the investigation work may be difficult and expensive to undertake because of problems of access to the locations of boreholes or in situ tests. Often these problems are easier to overcome during the construction of the project when earthmoving equipment is readily available and there is a great temptation to postpone necessary investigation until construction begins. This temptation should be resisted for it is possible that postponed items of the main investigation could reveal ground conditions which would invalidate project design. The client always pays for a ground investigation; the cheapest way is to commission one and the most expensive way is to "save" on one, but pay for it later with the delay and redesign that can follow a problem which was not foreseen. *No project should be designed on the assumption that the ground conditions will prove to be satisfactory.*

7.1.4
Construction Investigation Stage

One of the unfortunate facts of site investigation is that the prognoses made in the investigation reports resulting from the main investigation are seldom absolutely and totally correct. The construction of the project quite often reveals discrepancies between the ground conditions forecast and the ground conditions encountered. However, if the investigation is well done the client will have been warned that some variations in particular aspects of the ground (e.g. depth to bedrock) must be allowed for and in this way these variations need not cause significant project re-design and can be accommodated in the contract. The ground conditions encountered must be monitored, recorded and assessed. If no satisfactory assessment can be made on the basis of the information recorded then additional investigations must be undertaken to obtain further data and thence resolve any anomalies.

A theoretical example of what might happen is given in Fig. 7.1 which shows a geological cross-section (Fig. 7.1a), composed from the observations on the samples from two boreholes and an outcrop, and indicating uniformly dipping sandstones and shales overlain by soil. Dips were measured on cores and outcrop and appeared to be more-or-less uniform. However, when the soil over rockhead was removed a fault was observed. There was no way this could be forecast from the data available. It has probably only a small displacement but may well be found in the diversion tunnel and the question is whether or not it will have any significant effect on tunnel construction. Clearly further investigations should be initiated to gain more data on fault location, displacement and openness to assess its significance, but the prudent engineering geologist would have ensured that the client had a reserve for just this sort of "unknown". *The secret is to be aware of what you do not know* – in this case the engineering geologists should be aware that nothing was known about the location of faults.

7.1.5
Post-Construction Investigation Stage

The behaviour of the completed engineering work will have been computed on the basis of data acquired in the earlier stages of investigation. Certain features of behaviour, such as settlement, may take many years to become complete after construction of the project. If observed behaviour is not the same as anticipated behaviour this may indicate that the properties of the ground are affected by some unforeseen and previously undetected factor. Further investigations may be required to resolve this anomaly.

Monitoring of behaviour of the engineering project and comparison with predicted performance is of vital importance for all engineering works although contract arrangements often seem to hinder such monitoring. It must be remembered, however, that major engineering disasters mostly take place some time after completion of the work. The importance of post construction monitoring is thus clear. Anomalies of

Fig. 7.1. The contrast between geology forecast (**a**) and rockhead geology found after the removal of soil cover (**b**) suggests that the diversion tunnel might encounter some difficulties due to the presence of the previously unsuspected fault

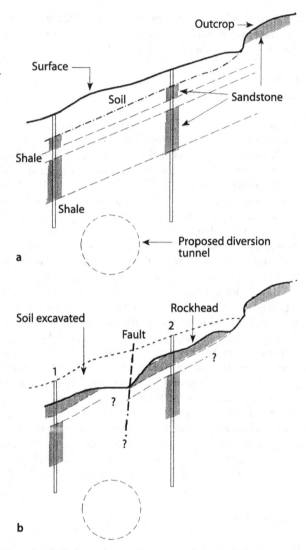

project behaviour must be observed, their cause established and remedial measures undertaken before severe damage or perhaps failure can occur.

7.2
Design of Site Investigations

While the concept of the development of site investigations in stages provides a philosophy of investigation procedure it does not answer the question "...*but what do I actually do?*" which is so often asked by engineers when they are confronted with the need to design an investigation for a particular project.

The answer to this question may be found by reviewing the purpose of investigations, *which is to determine the behaviour of the ground in response to the construction of the engineering work*. This 'behaviour' is calculated using the procedures, formulae etc. established in the various branches of engineering and applied earth sciences. Before the calculations can be made data dealing with the properties of material and mass, location of layer boundaries, and discontinuities and so forth, must be available so that it can be introduced into the calculation procedures; only then is it possible to obtain a numerical assessment of ground behaviour. What sort of data and how much data is needed depends upon the geological conditions and materials that are considered to be present, and the form of analyses that is to be used. The analysis will attempt to predict stability and deformation of the ground during construction in response to the unloading and subsequent re-loading it will experience. Most analyses require representative geological inputs. It is for this reason that engineering geologists should be acquainted with the rudiments of soil and rock mechanics and better still, a working knowledge of these subjects. Similarly, geotechnical engineers should be aware of the constraints that can be imposed by geological materials, structures and conditions upon the theories they intend to use for calculating their predictions. To all this must be added the subject of ground water as it influences the effective stress in the ground and can promote volumetric changes in some rocks and soils. In this regard fine grained sedimentary rocks and soluble sedimentary rocks should be considered with great care. Use should always be made of *Case Histories* – a much neglected and often forgotten source of invaluable knowledge. The size of this model will depend on the volume of ground which will influence or be influenced by the engineering work.

As the number of different materials present and the regularity and way with which they are distributed depends upon the geology of the site so the intensity of investigation (the number of boreholes, samples, geophysical traverses etc.) will in part depend upon the complexity of site geology. Complex geological situations will generally require a greater intensity of investigations than simple geological situations.

7.2.1
Building the Geotechnical Model

The process of building up the geotechnical model begins with a first appraisal of the geology based on available existing information, perhaps followed by geological mapping of the site on a large scale (say 1:1 000 or 1:500). This information is interpreted by standard geological procedures to establish a *prototype geological model*. A study of this model should indicate certain locations on the site where boreholes may most appropriately be sunk and perhaps geophysical work undertaken to verify or modify the geological model. The aim of this work is to determine layer boundaries and to establish an idea of the nature of these layers. If boreholes are used samples must be taken. If soils are being investigated then probably the first type of boring would be the percussion borehole, recovering tube 'undisturbed' samples and disturbed bag samples. Normally in such work the driller is instructed to take samples at regular intervals, say at every 1.5 m for tube samples, or at changes in strata. This rather low quality and quantity sampling is just to get an idea of the nature of the materials. In rock the initial boreholes are rotary core drilled, often at a somewhat larger diameter than may later prove to have been necessary in order to assure good recovery. This is done because subsequent drilling in rock may be undertaken in order to determine strata boundaries and open-hole methods could be used if there is good initial knowledge of the litho-stratigraphy.

The materials recognised in the samples belong to geological bodies which were penetrated by the boreholes. The shape of these bodies determines the distribution of materials under the site. Shape can be assessed by correlation between boreholes but this assessment can be greatly aided by the use of geophysical techniques (such as seismic refraction or reflection, ground penetrating radar or electrical resistivity) to determine boundaries. The 'model' distribution of materials built up from these data must then be tested for geological credibility. In other words whether the way in which the groundmass appears to be built up from the evidence available seems likely in the light of the known geological history and structure of the region.

This first model is then examined further by boring, geophysics or other methods, *to verify the concept* and *to obtain the samples* that may be necessary to deal with the particular problems posed by the combination of ground conditions and engineering work. Thus for some problems special samplers may be necessary to obtain high quality samples for drained triaxial tests or consolidation tests. The quality of sample is related to the degree of accuracy and reliability of test results required.

Samples may be taken from boreholes but also 'samples' may be tested in situ – access to these sample locations may be gained by boreholes, trial pits, trenches or shafts.

The degree of precision required in the assessment of ground conditions relates to the likely magnitude of ground reaction as a consequence of the construction of the engineering work. If the reaction is likely to be such that the success of the engineering project is in doubt then the quantity and quality of sampling, testing and interpretation must be high. Unfavourable ground behaviour could come about as the result of a simple structure being built in difficult geological conditions or a very complex engineering work being executed in simple geological conditions. It follows then that each investigation, whose design relates to both geology and the nature of the engineering project, is unique. *Thus there is no standard form of site investigation for a particular type of work. Each investigation is a completely new venture.*

7.2.2
Guidelines for Design of an SI

Although each investigation should be considered as individual many share common objectives and the actions needed to achieve these provide guidelines which may help the design of an investigation. These are:

- Determine the nature of the engineering work as far as is possible. It is necessary to know size, loads imposed, depths of excavations etc. desired by the client's proposals before planning the investigation.
- Determine the geology, geomorphology and hydrogeology of the area of the site and its environs.
- Determine the nature of the geological environment and the likely influence of any human activities on the site.

- Establish the size and location of the ground mass that could influence or be influenced by the engineering work.
- List the data which must be acquired to allow the necessary calculations that relate to the proposed engineering work to be made.
- Consider the best means of acquiring the data.
- Design a preliminary stage of investigation.

A good principle of design is to consider that each investigation must be economically efficient in the sense that the cost of the investigation must be money well spent. The investigator must be able to justify each and every item in the site investigation in terms of the value of that item in building up the geotechnical model. The investigator must be able to show good and sufficient reason for undertaking each part of the investigation.

While the idea of being able to justify each item of the investigation imbues the investigation with a sense of purpose and discipline (and will also prove popular with finance administrators) there is a danger that the investigator might become fixed on a wrong path and try to prove pre-conceived ideas rather than find the true situation. There is some justification for the 'random' investigation, undertaken almost at the completion of the geotechnical model, which tests the validity of this model. The more complex the geology and the more important the project, the greater is the need for this *insurance*.

7.3
Progressive Evaluation of Site Investigation Data

The data that is recovered from a site investigation must be evaluated in the course of the investigation and not left as an exercise to be undertaken once the fieldwork is over, so that, if necessary, the location of boreholes, trial pits etc., may be altered. It is easy to forget that *the principle purpose of a ground investigation is to understand the ground*, and this is best acquired whilst the investigation is in progress. The investigation data will refer mostly to two types of knowledge, first the location of engineering geological unit boundaries (layers, bands, discontinuities, etc.) and second the geotechnical properties of mass and materials.

The first sort of investigation data is most readily displayed on cross-sections and maps. Data received from boreholes or geophysics may be shown on borehole records or on profiles. Study of these records, together with the results of any testing on the samples retrieved, should allow the investigator to recognise boundaries between engineering geological units (layers of material with similar geotechnical properties) and to determine their level (relative to the datum plane of the site map). These levels may then be plotted on the map and *strikelines* (contour lines on geological surfaces) may be drawn for the boundaries. The results of the investigation in terms of the distribution of materials all over the site may then be seen.

The distribution of these strikelines, together with any other published information about the geology of the area, may give the investigator a theory about the nature of the geological structure being investigated and subsequent investigation works may be planned with the intention of proving or changing the first theory. If this idea is followed it ensures that at all times the investigation has a plan and a purpose.

Similarly, once testing has been undertaken to establish the geotechnical properties of the various materials the results of this testing must be compiled to allow assessment of the engineering geological units. This compilation is best done in tabular form. After the first compilation the data may be recompiled into units with similar geotechnical properties – property boundaries must be established by the investigator with regard to the nature of the engineering work.

7.4
Investigation Progress and Engineering Design

Engineering design of a project begins immediately after project conception and often before any significant site investigation can take place. The first design is thus based on assumed, generally favourable, ground conditions. If the assumptions are proved false then the project must be partially or wholly re-designed.

While project re-design is almost inevitable in the early phases of the development of a project, re-design due to faulty knowledge of ground conditions should be reduced once site investigations have commenced by the continuous interchange of information between investigator and project designer. *This liaison is essential to prevent wasted effort in design and in investigation.* Table 7.1 shows where, in the progress of an investigation, such liaison is appropriate.

Table 7.1. Stages of investigation and the progress of engineering design

Site investigation stage	Investigation activities	Information exchange	Design and construction progress
Project conception	Ground conditions required	←	Basic design concepts (start here)
	Basic knowledge of ground conditions		
	Recognition of major problems	→	Confirmation or amendment of basic design concept
Preliminary	Preliminary field investigations	←	
	Design of main investigation	→	Preliminary detailed design
Main	Information recovered during investigation	⇄	Modifications to detailed design
	Main investigation report	→	Final design of project
Construction			**Construction**
	Recording ground conditions as found	→	Modifications to design as necessary
	Response of ground	⇄	Modifications to design and possible need for ground treatment and/or support
			Completion of construction
Post construction	Monitoring behaviour	→	Operation
			Maintenance of works (end here)

7.5
Tender Visits

Engineering projects (including site investigations) are generally put out to tender. The first man to visit the site may then be an estimating engineer, on site to collect the contract documents and view the site with other contractors. Such a visitor should return to his office with information that will be of help to engineering geologists and geotechnical engineers for their part of the project. It is important that this information should be adequate, particularly if the site is in some far distant land which is difficult and expensive to visit. It is always useful to bring back:

i   topographical maps of the area in which the site is situated;
ii geological maps and reports;
iii aerial photographs;
iv details of transport infrastructure;
v   addresses of government agencies;
vi photographs of the site.

In some countries any of the above items could be considered a subject of national security and attempts to obtain one or more of them can result in the innocent enquirer being "detained"; be prudent!. Sometimes samples of rock or soil may be brought back, particularly if the project includes the use of natural materials for construction purposes. Any samples should be of a size and volume to be representative of the material from which they were collected (Table 7.2).

7.6
Supervision of Investigating Works

Having won the contract for a ground investigation and designed both its content and organisation, it is necessary to ensure the work is conducted properly. This requires it to be supervised by an engineering geologist. Much can happen during these investigations that needs an informed decision to be made at the time. For example, a borehole being drilled according to required standards, may collapse. Should the drilling crew (a) abandon the hole; (b) grout the hole in the region of the collapse and try to re-drill; (c) investigate the cause of the collapse or (d) just move to another point a few metres away and start again? Or again, a trial pit may be 1.5 m below ground level when one of its faces collapses and the pit half fills with water. Should the pit (a) be continued to its designated depth of 2 m; (b) be abandoned; (c) back filled immediately; (d) relocated? An engineering geologist is best able to sensibly answer these questions so as to move the investigation forward in a positive way. Unsupervised work provides, at best, an inadequate record of the ground and at worst a wrong record, from which will be made incorrect predictions of the ground and its reaction to the engineering works proposed – the precise situation the investigation was commissioned to avoid! There is no substitute for careful supervision and *a site investigation that has been supervised will always provide more trustworthy data than one that has not.*

Table 7.2. Minimum sample size required for useful testing

Purpose of sample	Material	Weight or volume of bulk sample	Diameter and length of tube or core sample (mm)	
			Diameter	Length
Chemical composition	Clays and silts	0.5–1.0 kg	38	75
	Sands	0.5–1.0 kg	38	75
	Gravels	3.0 kg	90	200
	Rocks	0.5 kg	38	75
	Groundwater	2.5 l		
Structural characteristics incl. grain size, porosity etc.	Clays and silts	0.5–1.0 kg	90	90
	Sands	1.0–2.5 kg	90	200
	Gravels	4.5–45 kg	90	200
	Rocks (coarse grained)	0.3 m^3	90	90
	Rocks (fine grained)	0.15 m^3	75	75
Strength characteristics incl. elastic moduli, shear strength, consolidation etc.	Clays and silts	$(0.3 \text{ m})^3$	38	75
	Sands	$(0.3 \text{ m})^3$	38	75
	Gravels	$(0.5 \text{ m})^3$	0.2	0.3 m
	Rocks (weathered)	2 of $(0.3 \text{ m})^3$	90	200
	Rocks (unweathered)	1 of $(0.3 \text{ m})^3$	75	150
Hydraulic characteristics incl. permeability, etc.	Clays and silts	$(0.15 \text{ m})^3$	38	75
	Sands	$(0.2 \text{ m})^3$	38	75
	Gravels	$(0.5–1.0 \text{ m})^3$	0.2	0.4 m
	Rocks (coarse grained)	$(0.3 \text{ m})^3$	90	200
	Rocks (fine grained)	$(0.15 \text{ m})^3$	75	150
Comprehensive examination	Clays and silts	20–45 kg	90	200
	Sands	20–45 kg	90	200
	Gravels	45–90 kg	0.2	0.4 m
	Rocks	2 of $(0.3 \text{ m})^3$	90	200
	Groundwater	4.5–10.0 l		

7.7
The Engineering Geological Situation

The end product of material and mass properties, of mass fabric, of geological environment and so forth is the description of the engineering geological situation. This presents to the engineer, architect, mine engineer, or whoever is concerned with the proposed project, the data needed in order to determine the reaction of the ground and thence correctly design the engineering work so the ground reaction will not damage the work. This presentation of data and opinion is no light and easy task but it is absolutely essential that it be done satisfactorily. Some clients want only the facts, a *Factual Report*. Factual data must be accurately presented in a matter such that it cannot be misunderstood. Other clients request both the facts and an interpretation of them, an *Interpretative Report*; here opinion must be presented with clarity and objectivity. The main problems in doing this are:

- that accurate presentation of data is an arduous and difficult task;
- that the presentation must be made in terms that are to be readily understood by the reader who may not have the depth of technical knowledge possessed by the writer.

However, despite the difficulties of the task of presentation it must be undertaken. It must not be forgotten that the reports in which the presentation is made (commonly reports on site investigations) may form parts of contractual documents. Claims and lawsuits may be the result of inaccurate or poor quality reporting. Worse still, *misunderstood reports could lead to fatalities* and engineering disasters if the data or opinions they appear to present are acted upon. It must also not be forgotten that the report document, together with any maps, plans, profiles etc. that it contains, represents the cost of the investigations. It is common for projects to be constructed some years after the completion of investigations, when all samples have disappeared, all trenches filled in and so forth. All that remains is the report on whose contents design is then based.

7.8
Investigation Reports

The results of a site investigation are generally expressed in the form of a report. This contains a written text to which are added tables or graphs giving test results, and maps and diagrams illustrating the location of site investigation works, site geology and perhaps the recommendations of the investigator. All reports should be of a good literary standard. Diagrams and illustrations should follow the proposals set out in the section on engineering geological mapping. Borehole records should be prepared to standards similar to those shown in Fig. 7.2. Each illustration or table may, in use, become separated from the main body of the report. They should thereafter be prepared as independent documents containing sufficient data to locate the information they contain to the site in time and space.

7.9
The Form of the Report

Every organisation has its own way of preparing a report, which may be varied depending upon the project. The following basic scheme of report writing may prove useful for those who have little previous experience of report preparation.

Title page. This should show the name of the project, the name of the organisation or individual preparing the report, the report number (as used in the filing system of the writers), the date of the publication of the report.

Report 'contents' page. This should show the titles of the main sections of the report and the page on which they may be found and also list all figures, tables, borehole records and so forth, that may be included.

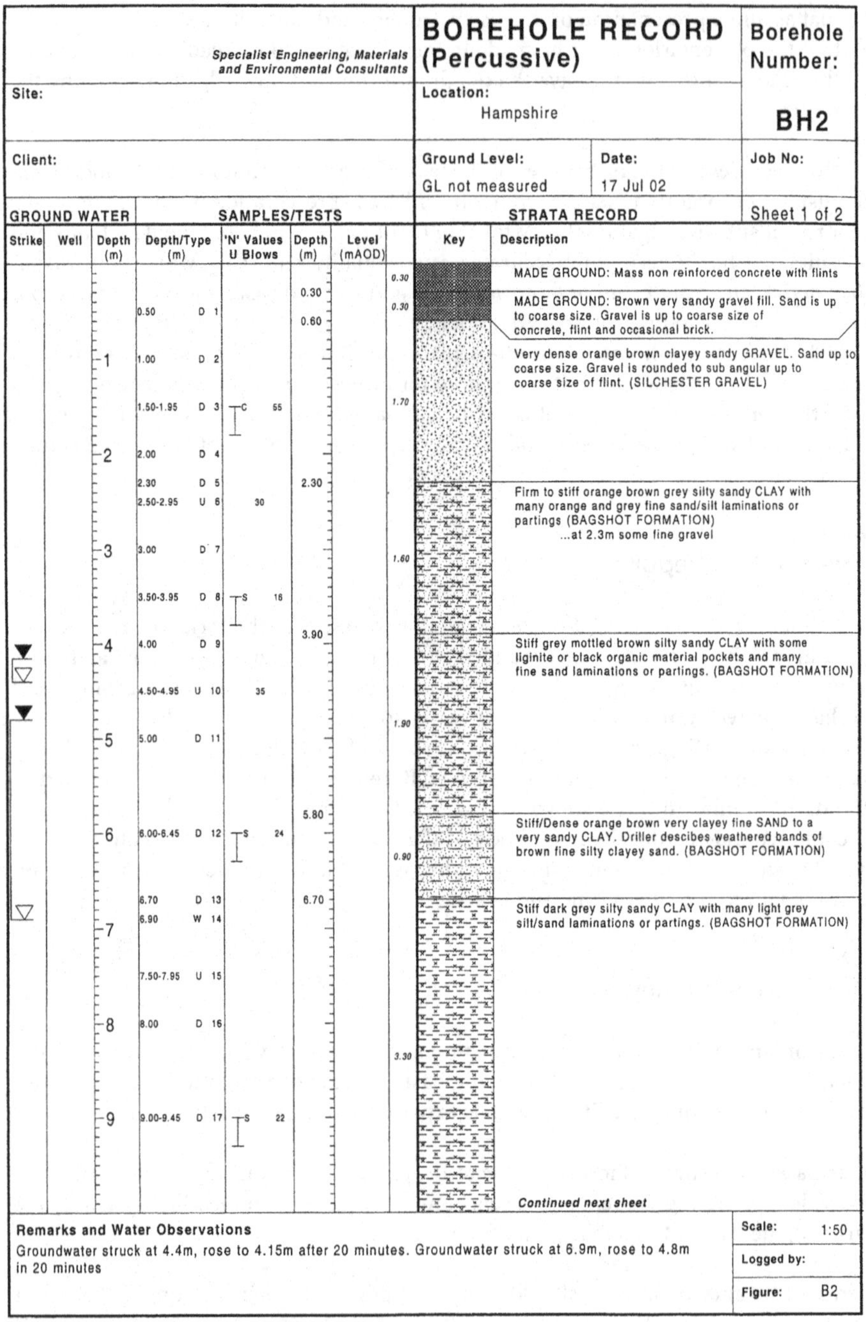

								BOREHOLE RECORD (Percussive)		Borehole Number:

Specialist Engineering, Materials and Environmental Consultants

BOREHOLE RECORD (Percussive)

Borehole Number: BH2

Site:	Location: Hampshire	

Client:	Ground Level: GL not measured	Date: 17 Jul 02	Job No:

Sheet 1 of 2

GROUND WATER			SAMPLES/TESTS				STRATA RECORD	
Strike	Well	Depth (m)	Depth/Type (m)	'N' Values U Blows	Depth (m)	Level (mAOD)	Key	Description
					0.30			MADE GROUND: Mass non reinforced concrete with flints
			0.50 D 1		0.30 0.60	0.30		MADE GROUND: Brown very sandy gravel fill. Sand is up to coarse size. Gravel is up to coarse size of concrete, flint and occasional brick.
		1	1.00 D 2					Very dense orange brown clayey sandy GRAVEL. Sand up to coarse size. Gravel is rounded to sub angular up to coarse size of flint. (SILCHESTER GRAVEL)
			1.50-1.95 D 3 C	55		1.70		
		2	2.00 D 4					
			2.30 D 5		2.30			Firm to stiff orange brown grey silty sandy CLAY with many orange and grey fine sand/silt laminations or partings (BAGSHOT FORMATION) ...at 2.3m some fine gravel
			2.50-2.95 U 6	30				
		3	3.00 D 7			1.60		
			3.50-3.95 D 8 S	16				
		4	4.00 D 9		3.90			Stiff grey mottled brown silty sandy CLAY with some lignite or black organic material pockets and many fine sand laminations or partings. (BAGSHOT FORMATION)
▼			4.50-4.95 U 10	35				
▽		5	5.00 D 11			1.90		
▼								
		6	6.00-6.45 D 12 S	24	5.80			Stiff/Dense orange brown very clayey fine SAND to a very sandy CLAY. Driller descibes weathered bands of brown fine silty clayey sand. (BAGSHOT FORMATION)
						0.90		
▽		7	6.70 D 13		6.70			Stiff dark grey silty sandy CLAY with many light grey silt/sand laminations or partings. (BAGSHOT FORMATION)
			6.90 W 14					
			7.50-7.95 U 15					
		8	8.00 D 16					
						3.30		
		9	9.00-9.45 D 17 S	22				
								Continued next sheet

Remarks and Water Observations
Groundwater struck at 4.4m, rose to 4.15m after 20 minutes. Groundwater struck at 6.9m, rose to 4.8m in 20 minutes

Scale:	1:50
Logged by:	
Figure:	B2

Fig. 7.2. Example of a borehole log through soil

Introduction. A brief description *of the purpose of the site investigation,* which should in itself explain the design of the investigation, the name of the client and of those who designed the investigation, the dates of the beginning and the end of field work and laboratory testing and any special instructions given by the client.

Description of the project. A brief description of the purpose and design of the project.

Topography and geology of the site. A description of the topography of the site at the time of the investigation, including comments on land use and access to the site. The basic geology of the site should be described making a clear distinction between information available before the investigation and that gained during the investigation. Comments regarding possible natural hazards should be made here as well as a review of site climate, if these are of significance.

Site investigation. The site investigation work done is described in this section, which may be divided into sub-sections such as "*Borehole*", "*In situ testing*", "*Laboratory testing*" as appropriate. This section must be factual and correct, for it may be used as a basis for billing the work done in the contract.

Results of site investigation. This section describes the geotechnical model built up from the data discovered in the course of the investigation. It should be divided into engineering geological units and a description of each unit (its location and properties) given.

Influence of ground conditions on project construction. This gives the opinions of the investigator on the likely behaviour of the ground as a response to the construction of the project and also the response of the construction to the behaviour of the ground. Advice on foundation design and construction, excavation techniques and so forth are presented. The section may be divided in relation to the various parts of the project.

Discussion. A discussion may be included in the report. This may consider the reliability of the investigation data, recommendations for further investigations and so forth.

7.9.1
Other Aspects of Report Preparation

Each of the sections described above should contain references to figures, tables etc. which are generally presented after the text. Tables and figures are best presented on separate pages, they should not be included in the text. Each page of the report should be numbered. This may usefully be done describing each page '*X*' of '*Y*' pages, '*X*' being the number of the page and '*Y*' the total number of pages in the report text. The contents page should be Page 1. Any standards, norms, textbooks, scientific articles etc. referred to in the report should be given a reference number in the text which relates to a list of references at the end of the report. References should be given in a uniform way, following one of the systems recommended by the scientific journals.

Appendices. Appendices may be used to give additional information essential to a proper understanding of the report. Such information could be, for example, an explanation of a method of geophysical interpretation or the execution of a laboratory test. Appendices should not contain information generated by the investigation.

7.9.2
Borehole Records

Boreholes are one of the primary tools of site investigation and the way in which borehole data is presented must be carefully considered. A good and uniform standard must be achieved. An example of a good standard of borehole record is given in Fig. 7.2. The method of presentation is to show information in columns and blocks. The blocks at the top of the record deal with the methods of drilling and the reference number and location of the borehole. Those at the bottom give the names of contractor, client, a key to some of the symbols used on the record and observations (remarks) about unusual circumstances connected with the sinking of the borehole. The data derived from the borehole is held within columns between the top and bottom blocks.

To the left, on the binding margin of the record, there are columns showing location of samples, progress of the borehole, water levels, core recoveries and/or results of penetration tests and a column for core size and/or sample type. To the right there is a column for the description of the strata found in the borehole. A centimetre scale is drawn on two of the vertical lines to allow the borehole to be drawn accurately to scale. At a scale of 1:100 the record can present up to 20 m depth of borehole. The symbolic log is placed on the right hand side next to a column in which the levels of strata boundaries (relative to the locally used horizontal datum plane) are given. If all the borehole records presented are kept to the same scale the user may overlap a series of borehole records and compare symbolic logs along a line of section.

Young engineering geologists charged with the task of preparing such records often ask why the various blocks and columns of information are necessary. There is a reason for each information block.

Starting at the top left of the record the blocks entitled "*Drilling method*", "*Core barrel and bit design*" and "*Machine*" tell the skilled borehole record reader whether the drilling equipment used was adequate to drill the strata recorded. If not then loss of core or sample might be attributable to poor drilling rather than poor ground conditions. That the drilling equipment was not suited to the strata to be drilled is not necessarily the fault of the driller or drilling contractor for the geological conditions found might have been totally unexpected. Finding the unexpected by the process of investigation is, after all, the purpose of investigation. Such information is particularly valuable if the borehole reports relate to a preliminary stage of the investigation – appropriate equipment can then be chosen for the main stage. The strata descriptions that are written in the "*Description of Strata*" column must be written with regard to a recognised system for the description of soil and rock materials, preferably that system used in the country in which the investigation was undertaken. It is useful to mention the system used in the text of the main report.

However, while descriptive standards help to decide whether, for example, a sandstone is thinly bedded or thickly laminated, they do not help the sample logger to decide what to call the materials encountered or to decide where to establish boundaries between strata of different types. This is perhaps rather more difficult a problem with rocks than with soils. There are three common problems:

1. *Establishing boundaries in strata that are uniform in nature.* Here the temptation is to say that the rock is uniform while it might be possible to describe differences of a rather more subtle nature than is common on rock description, such as, for example, between "sandy mudstone", "slightly sandy mudstone" and "very sandy mudstone". Whether such subtlety of description is necessary or not will depend upon the nature of the project. For rock dredging it could be of great importance while it could be of no significance for determining the depth of mudstone overburden over limestone to be quarried. The need for subtle distinction between rather similar strata will depend upon the nature of the project. *The real problem comes when the logger must log without much idea of what the project is about.*
2. *In some rock and soil successions there are very frequent but distinct strata differences to be seen.* One which is particularly common is the rapid alternations of shale and fine sandstones that occur, for example, in parts of the Carboniferous. If alternate bands are, say, 2 cm thick then *'very thinly bedded alternating SHALES and fine SANDSTONES'* is appropriate. If the layers are more than a metre or so thick then they could be described separately. The problem arises when the layers are, say, 20 or 30 cm thick. Again as with the first difficulty, *what is done depends upon the nature of the project for which the borehole record is being prepared.*
3. *Constraining strata descriptions* within the type space limitations of the borehole record form. The description should not be too verbose; keep to the point and ensure the essentials are recorded.

An example of a good log in rock is given in Fig. 7.3. Ground levels must be recorded for it sometimes happens that in the initial stages of construction general excavation changes site levels. If it is by chance assumed that the borehole started from the changed site level then calculations with regard to the depths of layers would be incorrect.

Orientation of the borehole must be known in the case of inclined boreholes. This is mostly given in terms of direction of the hole (relative to magnetic north) and dip of the hole from the horizontal. It is very valuable to put a co-ordinate reference, relative to the national co-ordinate reference system, on every borehole record. Most reports contain site plans showing where the borehole was drilled relative to features in the landscape which existed at the time of the investigation. However, between investigation and construction these may have disappeared, perhaps as a result of site preparation. Boreholes can always be relocated on site if the grid reference is known. This should be given to the nearest metre.

Every borehole should be numbered. The numbering system should be kept as simple as possible. If boreholes are longer than one record sheet can hold then the use of the *"sheet X of Y"* tells the reader this. Second sheets should, in principle, hold the same amount of block data as the original sheet. Every site has a name and this must be given on the record. One difficulty is that the site name given at the time of investigation is not always that of the construction site. At the bottom of the record there are blocks for the name of the contractor (if a site investigation contractor is used) and for the client. Mostly a contractor or consulting engineer prepares the record and the record would have a printed logo with name and address.

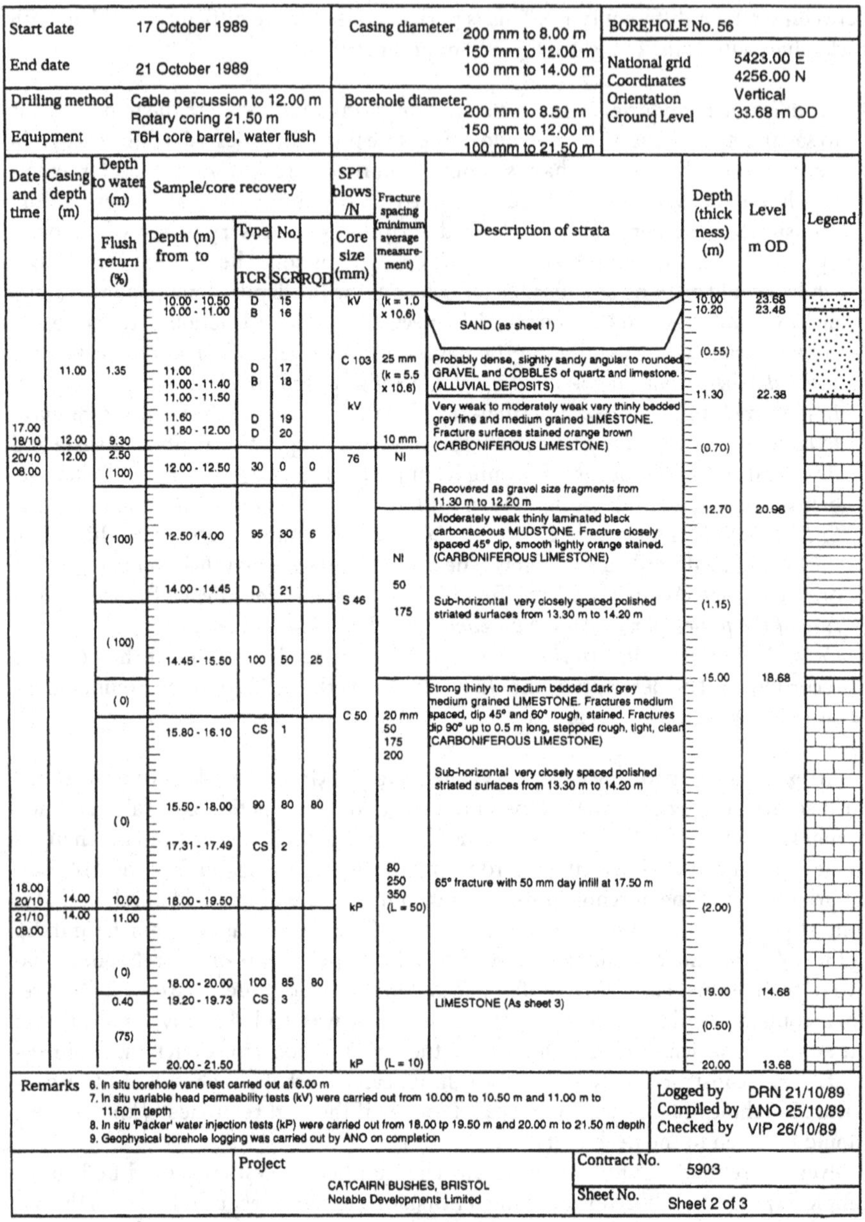

Start date	17 October 1989			Casing diameter	200 mm to 8.00 m	BOREHOLE No. 56	
End date	21 October 1989				150 mm to 12.00 m 100 mm to 14.00 m	National grid Coordinates	5423.00 E 4256.00 N
Drilling method	Cable percussion to 12.00 m Rotary coring 21.50 m			Borehole diameter	200 mm to 8.50 m	Orientation Ground Level	Vertical 33.68 m OD
Equipment	T6H core barrel, water flush				150 mm to 12.00 m 100 mm to 21.50 m		

Fig. 7.3 borehole log:

Date and time	Casing depth (m)	Depth to water (m)	Sample/core recovery				SPT blows /N	Fracture spacing (minimum average measurement)	Description of strata	Depth (thickness) (m)	Level m OD	Legend
			Flush return (%)	Depth (m) from to	Type	No.	Core size					
					TCR	SCR	RQD	(mm)				
				10.00 - 10.50 10.00 - 11.00	D B	15 16	kV	(k = 1.0 x 10.6)	SAND (as sheet 1)	10.00 10.20	23.68 23.48	
	11.00	1.35		11.00 11.00 - 11.40 11.00 - 11.50	D B	17 18	C 103 kV	25 mm (k = 5.5 x 10.6)	Probably dense, slightly sandy angular to rounded GRAVEL and COBBLES of quartz and limestone. (ALLUVIAL DEPOSITS)	(0.55) 11.30	22.38	
17.00 18/10	12.00	9.30		11.60 11.80 - 12.00	D D	19 20		10 mm	Very weak to moderately weak very thinly bedded grey fine and medium grained LIMESTONE. Fracture surfaces stained orange brown (CARBONIFEROUS LIMESTONE)	(0.70)		
20/10 08.00	12.00	2.50 (100)		12.00 - 12.50	30	0	0	76	NI	Recovered as gravel size fragments from 11.30 m to 12.20 m	12.70	20.98	
		(100)		12.50 14.00	95	30	6		NI	Moderately weak thinly laminated black carbonaceous MUDSTONE. Fracture closely spaced 45° dip, smooth lightly orange stained. (CARBONIFEROUS LIMESTONE)			
				14.00 - 14.45	D	21	S 46	50 175	Sub-horizontal very closely spaced polished striated surfaces from 13.30 m to 14.20 m	(1.15)		
		(100)		14.45 - 15.50	100	50	25				15.00	18.68	
		(0)		15.80 - 16.10	CS	1	C 50	20 mm 50 175 200	Strong thinly to medium bedded dark grey medium grained LIMESTONE. Fractures medium spaced, dip 45° and 60° rough, stained. Fractures dip 90° up to 0.5 m long, stepped rough, tight, clean (CARBONIFEROUS LIMESTONE)			
		(0)		15.50 - 18.00	90	80	80			Sub-horizontal very closely spaced polished striated surfaces from 13.30 m to 14.20 m			
				17.31 - 17.49	CS	2		80 250 350		65° fracture with 50 mm clay infill at 17.50 m			
18.00 20/10	14.00	10.00		18.00 - 19.50			kP	(L = 50)		(2.00)		
21/10 08.00	14.00	11.00										
		(0)		18.00 - 20.00	100	85	80				19.00	14.68	
		0.40		19.20 - 19.73	CS	3			LIMESTONE (As sheet 3)	(0.50)		
		(75)		20.00 - 21.50			kP	(L = 10)		20.00	13.68	

Remarks	6. In situ borehole vane test carried out at 6.00 m 7. In situ variable head permeability tests (kV) were carried out from 10.00 m to 10.50 m and 11.00 m to 11.50 m depth 8. In situ 'Packer' water injection tests (kP) were carried out from 18.00 m to 19.50 m and 20.00 m to 21.50 m depth 9. Geophysical borehole logging was carried out by ANO on completion	Logged by DRN 21/10/89 Compiled by ANO 25/10/89 Checked by VIP 26/10/89
	Project CATCAIRN BUSHES, BRISTOL Notable Developments Limited	Contract No. 5903 Sheet No. Sheet 2 of 3

Fig. 7.3. Example of a borehole log through rock (from British Standard BS5930 (1999) and in which other good examples can be found)

The form displayed in the figure has also a very small block for the name of the person responsible for the preparation of the record. It is a matter of some earnest debate to decide whether or not this should appear on a record. Mostly the company employing the logger is responsible under law for the accuracy of the record and it could be argued that omission of the logger's name ensures that he or she would suffer no personal difficulties in the event of dispute. On the other hand it may also be that the user of the record might want to contact the logger for reasons other than legal dispute, perhaps, to ask for an opinion based on personal knowledge which was not requested at the time.

7.10
Further Reading

British Standard 5930 (1999) Code of practice for site investigations. British Standards Institute, London
Clayton CRI, Mathews, MC, Simons NE (1995) Site investigation. Blackwell Scientific Ltd., Oxford

Part II
Ground Behaviour

Chapter 8

Ground Response to Engineering and Natural Processes

Three verbal equations were set out in Chap. 1 which describe the procedure whereby the engineering behaviour of the ground may be deduced. To achieve this end it is, therefore, necessary to determine *the material properties, the mass fabric and the appropriate environmental conditions*, although the same level of detail is not necessary for all projects. For example, there are many 'engineering properties' and only some may relate specifically to particular types of work. Mass fabric can, for example, be established to various degrees of detail depending on the nature of the project. By way of illustration, if it is proposed to construct a lightly trafficked road through a flat landscape underlain by unweathered gneiss then it will be unnecessary to know much about the orientation (i.e. mass fabric) and strength (i.e. engineering property) of the discontinuities in that gneiss. However, if it is proposed to drive a tunnel through the same rock mass then both the discontinuity strength and orientation are essential parameters for that tunnel design.

Each engineering project may be regarded as being composed of a series of 'engineering processes', each producing various possible ground reactions. It is important, therefore, to establish the processes which will be undertaken in any project, in order to qualitatively assess the ground reaction to each of those processes, as well as determine the quantitative data which must be acquired by investigation in order to analyse the engineering behaviour of the ground mass.

8.1
Engineering Processes

There are many different types of engineering work but, during construction and/or after completion, they impose one or more 'engineering processes' on the ground which are as follows:

- loading the ground;
- withdrawing support from the ground;
- changing fluid and/or gas pressures in the ground.

Thus, the weight of a dam loads the ground, and the excavation of a tunnel withdraws support from the ground, while pumping water from a well will change fluid pressures in the ground within the zone of the cone of depression of water levels. Each process produces a typical reaction as is summarised in Fig. 8.1. This chapter intro-

	LOADING THE GROUND		WITHDRAWAL OF SUPPORT		CHANGE IN FLUID AND GAS PRESSURES	
	STATIC	DYNAMIC	SURFACE EXCAVATIONS	UNDERGROUND EXCAVATIONS	DECREASE	INCREASE
ENGINEERING PROCESSES the resulting effects may cease shortly after completion of the engineering process	Dams, bridges, buildings of all types	Machinery vibrations, traffic, explosions from quarrying	CLOSED new continuous slopes, pits. Foundation excavations, shafts / OPEN new discontinuous slopes, Road, railway cuttings, canals	LINEAR length >> diameter. Tunnels, chambers, shafts / AREAL extending over large area underground. Mines	Pumping from wells, quarries, mines, tunnels, pits	Pumping waste fluids into wells, impounding reservoirs
POSSIBLE REACTION	Settlement or failure		Instability of excavation slopes heave of excavation floor	Collapse of underground openings, surface subsidence — and associated seismicity	Surface subsidence collapse of infilled karst	Induced seismicity slope instability, aquifer pollution
NATURAL PROCESSES operate continually and may act on completed engineering works	Deposition of sediments, snow, ice	Earthquakes, sea waves, tides	Decay of material and discontinuity strength by weathering. Undercutting by river, sea and wind erosion	Development of underground cavities by dissolution of calcite, halite, gypsum and other soluble rocks	Drought, river capture, erosion of a natural dam so lowering water levels	Impoundment of natural reservoirs by landslides. Floods, changes in river courses
	STATIC	DYNAMIC	SURFACE EXCAVATIONS	UNDERGROUND EXCAVATIONS	DECREASE	INCREASE
	LOADING THE GROUND		WITHDRAWAL OF SUPPORT		CHANGE IN FLUID AND GAS PRESSURES	

Fig. 8.1. Possible reactions to engineering and natural processes

duces this subject, and succeeding chapters deal with the basic theory of, and describe reactions that may be expected to, these three engineering processes.

In the case of static loading, the reaction of the ground may include:

- elastic deformation of the ground and thus 'immediate' settlement of the load-imposing structure;
- long-term consolidation of the ground and 'consolidation settlement' of the structure;
- failure, in shear, of the ground if the foundation pressure exceeds the bearing capacity of the ground.

Similar results will occur if the loading process is dynamic arising, for example, from manmade vibrations caused by blasting, machinery or traffic, or by earthquakes.

If an excavation is made into the ground, the ground around the excavation loses the support of the material which has been removed. Surface excavation may be classified into 'closed' or 'open' forms, the former being holes completely surrounded by cut slopes (such as foundation excavations, open pit mines and some quarries) and the latter being linear cuts through hills with one or possibly two cut slopes (such as road, railway or canal excavations).

The process of withdrawal of support may result in:

- deformation of the excavated slope;
- upwards heave of the floor of the excavation;
- rotational or sliding failure of the slope.

Support may be withdrawn underground by the excavation of tunnels, chambers or mines. The results may include:

- deformation of the rock surrounding the excavation as this will converge or collapse into the opening;
- subsidence of the surface above the opening as the result of the ground moving in to fill the void.

Fluid (and/or gas) pressures may be decreased by extraction (pumping out) or increased by addition (pumping in) of fluid. If pore fluid pressures are reduced by pumping out then the effective stress will increase, and the materials will consolidate, resulting in surface subsidence. If fluids are pumped into the ground which is already under stress then the rise in pore pressure could disturb the balance of stresses and cause failures, possibly even inducing local seismicity.

In most engineering works more than one of these processes could be active. Thus, for example, excavating a tunnel portal will withdraw support from a slope, driving the tunnel underground will withdraw support from the ground surrounding the tunnel and, since most tunnels act as drains, will also lower the groundwater table in the area around the tunnel. In the construction of a dam, the foundation excavation will either cut new slopes or steepen existing slopes, and unload the ground; building the dam will re-load the ground, and impounding the reservoir will both load the ground and raise fluid pressures.

In each of these processes there will be a response. Knowledge of the appropriate material properties, mass fabric and environmental controls are required to determine the magnitude and nature of that response.

8.2
Natural Processes

The engineering processes are paralleled by natural processes occurring within the environment, and through geological time. Thus, in nature, static load results from the progressive deposition of sediments, or the superposition of snow and ice. Dynamic loading results from earthquakes. Support may be withdrawn at the land surface by erosion associated with rivers, the sea or wind and, also (perhaps more importantly) by the reduction of the strength of materials in a slope as the result of weathering. Support is withdrawn underground by the solution of limestone, gypsum and salt. Fluids may be extracted by groundwater changes resulting from river capture, the exposure and drainage of artesian aquifers, and drought. Fluids may be added by floods, changes in river courses, and by the impoundment of lakes through landsliding. The results of the natural processes are, in the general sense, similar to those resulting from engineering processes although they can tend to be less spectacular, except in the case of earthquakes, volcanic eruptions and large landslides.

Natural processes operate continuously, whereas the reaction to the engineering process is generally finished quite soon after the construction of the engineering work. Thereafter, *natural processes operate on the complete work for its total anticipated operational life.* Thus a road cut in rock, which is stable on completion, can be influenced by progressive weathering of the material and discontinuities. Surface erosion by wind and rainfall runoff, may bring that slope to failure. How long this takes, whether 10 years or 10 000 years, will depend upon the resistance of the rock mass to weathering, groundwater penetration, erosion and climate.

What the reaction of the ground will be to both engineering and natural processes can be calculated provided the correct data is obtained. In planning an investigation in order to obtain that data it is useful to think through the whole programme of construction of a work to see what processes may act and what reactions may possibly result. A theoretical example of how such a construction project may be 'thought through' is presented in the following section.

8.3
Recognising Problems

The theoretical project is that of a tall office building in an urban area; the design of the building incorporates a basement for storage and car parking (Fig. 8.2).

The pre-existing buildings on both sides of the proposed office block are of varying ages. Buildings 1 and 3 are new, Building 4 is modern but older, and Building 2 is very old, structurally suspect and of importance historically. The desk study carried out in the project conception stage of the site investigation revealed that the bedrock in the general area is granite, characterised by some weathering and with numerous faults and shear zones. The overburden is composed of normally consolidated alluvium; groundwater is shallow. Some records regarding the investigations for the adjacent modern buildings and the foundations of those buildings are preserved in the city archives.

Nothing is known about the geological conditions or foundations underneath Building 2. Records of old boreholes exist for Building 4, and both investigation records and as-built foundation drawings exist for Buildings 1 and 3. From these data a profile showing the geology can be drawn (Fig. 8.3). The groundwater level is high in the medium dense sands and soft to firm clays of the alluvium. The granite appears to be weathered and has erosion hollows coinciding with fault or shear zones. These hollows appear to be partly filled with coarse granite gravel and sand.

It is most probable that granite underlies the proposed block, and it would also seem likely that rockhead is somewhat deeper than the bottom of the proposed basement.

Fig. 8.2. The proposed development

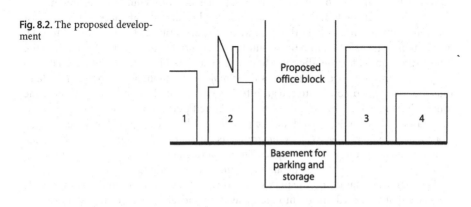

Fig. 8.3. Results of documentary research

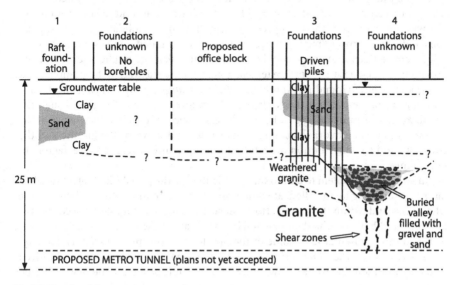

When the work goes forward it must inevitably start with an excavation for the basement, thus withdrawing support from the surrounding ground.

For many foundation excavations the problem is one of the stability of the excavated slope. This may be cut to an inclination at which the slope is stable. However, to calculate what the appropriate slope angle might be requires a knowledge of the geotechnical properties of the materials within the slope. In this particular case, the slope must be near-vertical because of the presence of the adjacent buildings, and the soil assumed to be present will inevitably be too weak to stand at such an angle. The slope must, therefore, be supported in some way. In order to design the support system it is necessary to know not only the geotechnical properties of the ground but also the foundation loads imposed by the nearby buildings, and the type, depth, and condition of the foundations upon which they are based. Depending on the nature of the alluvium, heave in the bottom of the foundation excavation could be a problem.

Assuming that the foundation excavation can be carried out successfully the cast concrete foundation will load the ground. As the structure is built so the additional weight will be transferred into the ground. It is essential to determine the depth to the top of the granite, for the alluvium below the proposed basement will be unable to support the building because failure and/or major settlement are probable. That load must be transferred down into the granite. Depending on the depth of the granite the foundations may have to extend below the planned bottom of the basement to reach the granite. This would exacerbate the problem of lateral support for the excavated slope will become higher. Also, as the granite is weathered, the bearing capacity of the near-rockhead granite may be suspect. There could, nevertheless, be a bonus in that a deeper excavation might provide an additional basement level.

The high water table and the pervious nature of some of the granular alluvial soils indicates that water will drain into the excavation and, as a result, changes in groundwater pressures will have a strong influence on the problems of excavated slope stability. The water will have to be pumped out from the excavation to permit construction to proceed. As a result the water table will be lowered around the excavation and this will in turn reduce the fluid pressures, resulting in consolidation of the alluvial soils under the surrounding buildings which are not founded on the granite. As a result severe settlement damage could take place. Such damage might be overcome by providing a watertight support system surrounding the excavation, by pumping water back into the surrounding ground ('*artificial recharge*') or by carrying out shallow grouting to compensate for the settlement. Therefore, the flow of water through the alluvium and the underlying granite into the excavation is a likely consequence of the excavation. Possible hydraulic connections within and between the granite and the alluvium, and the distribution of water within the granite are, therefore, factors of importance.

An additional problem to add to these difficulties is the possibility that a tunnelled metro system may be installed, at some unspecified time in the future, beneath the whole area, and within the granite. How will such a development affect the completed building, and how will the then-existing buildings affect the metro?

Having recognised at least some of the problems in this project (*the reader is offered, and left, with the opportunity to discover some more*) consideration may be given to the form of site investigation which would be appropriate for the new building.

Factors which would have to be taken into account in the design of the investigation would be:

- the geotechnical properties and distribution of the various types of alluvium, including such factors as permeability, shear strength and consolidation parameters;
- the depth to the granite and the shape of the bedrock surface;
- the nature and depth of weathering of the granite, its strength and mass permeability;
- groundwater levels in the various strata;
- the nature, depth and condition of the foundations of the adjacent buildings, and the distribution and character of the strata in which the foundations were placed;
- future plans with regard to the metro.

Critical factors will be the depth and shape of the rockhead surface, a basic idea of the properties of the alluvium and knowledge of the foundations of the adjacent buildings. If the granite is too deep, the alluvium too weak and deformable, and the adjacent buildings of uncertain nature and construction quality, then it may be considered prudent to modify the design and omit the basement. However, new structures have been built in conditions such as those described, so a review of the published literature of such construction in other cities (i.e. *Case Histories*) could be rewarding. Accordingly, it would be a good idea to start the first stage of intrusive investigation with one high quality borehole, obtaining samples for inspection and testing of *all* the materials present, and installing a multi-tip piezometer system to measure pore pressures with depth *and time*. It will also be necessary to locate the shape of the granite rockhead. In rural conditions, or where there was no background interference, seismic refraction or ground-penetrating radar could be appropriate methods but, in a built-up area, the use of a number of probe boreholes would be more suitable.

Depending on the results of this preliminary investigation, the feasibility of the whole project may be reviewed. If construction of the intended structure, including the basement, seem possible, then the main investigation, to study the ground conditions of the site in the surrounding area and in depth, should be carried out.

8.4
Groundwater

The perceptive reader will have realised that *many of the natural and engineering processed mentioned so far, together with their responses, will involve a change in fluid pressure* (Fig. 8.1). The fluid most normally involved is groundwater. Hydrogeology is a huge subject on its own but one which engineering geologists must appreciate in practice for the following reasons: groundwater will influence all engineering work below the water table, groundwater will invariably seep and flow into an excavation below the water table unless stringent measures are undertaken to exclude it, and the predictions made to forecast such flows are invariably wrong, despite the fearsome nature of the mathematics employed in their calculation.

Therefore, in order to consider adequately the consequences of any engineering process in terms of the ground's reaction to them, *it is necessary to consider the role of groundwater in those processes and reactions.* Here, four elements of the vocabulary used in hydrogeology need to be explained:

1. *watertable:* the level of standing water in the ground;
2. *aquifer:* ground through which water flows at rates of significance to engineering works (aquifers are described as *unconfined* if their upper surface is ground level and *confined* if they are covered – or "confined" – by less permeable material);
3. *aquitard:* ground through which water does not flow at such a rate;
4. *hydrogeological boundary:* a surface (bedding, fault, etc.) that separates ground of different hydrogeological character.

8.4.1
Common Problems With Groundwater in Engineering

Engineering below the water table needs to cope with the pressure of water at depth and the flows of water that can be encountered, especially into and around or beneath the engineering structure itself.

Water Pressure

Most of the common problems are associated with water pressure, some of which are illustrated in Fig. 8.4. Figure 8.4a shows what might be considered the starting point for all slope stability calculations, the rock block on a sloping surface. When dry, stability depends on friction between surface and block and block weight, but if the backing joint is filled with water, water pressure thrusts the block down slope and, if water flows under the block, uplift pressures result, which reduce the normal load on the surface and hence the frictional resistance the weight of the block can generate.

In soil slopes or very large slopes in rock, where rotational failure is possible (Fig. 8.4b) water pressures generated by groundwater reduce potential stability. When a reservoir is impounded behind a dam water pressure acting on the back of the dam (Fig. 8.4c) will tend to push the dam downstream. This could lead to failure if the rock foundation contains unfavourably orientated weak layers or discontinuities. If water flows under the dam (Fig. 8.4d), this will give rise to uplift pressures, further reducing dam stability. Example (*d*) may be considered as nothing more than an elaboration of (*a*).

Figure 8.4e shows an excavation being made in soil over a layer of low permeability rock, an aquitard, which confines ground containing water under pressure. Before the excavation was made the soil pressure above was sufficient to balance the pressure of water at depth. However, as the excavation is deepened confining pressure is reduced and the point may be reached when it is less than the water pressure trapped beneath it. This unbalanced pressure can rupture the layer that confines it and very rapid, often catastrophic, water inflows occur.

Excavations may be made to depths below water table level (for structures such as dry docks, underground car parks, underpass roads and the like) by lowering the water table level or excluding water from the excavation. The finished structure will be watertight and, once completed, the water table will return to normal levels imposing uplift forces on the structure (Fig. 8.4f). These may have to be resisted by anchorages or tension piles.

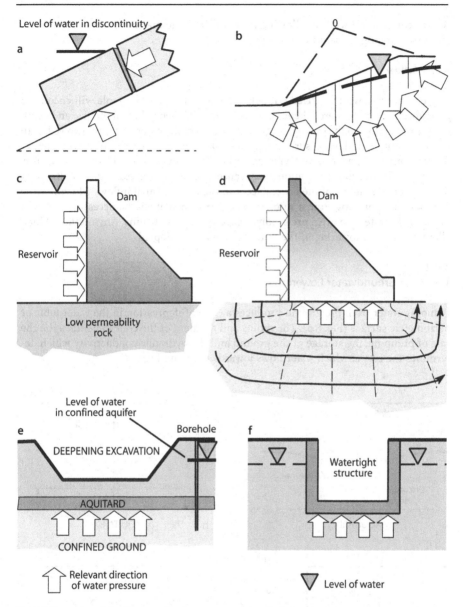

Fig. 8.4. Examples of significance of water pressures on engineering structures

In each example except (*b*) the water pressure is acting against a more-or-less impermeable surface but in (*b*) the water pressures are intergranular pore pressures. Pore pressures may be increased in many different ways to the point where *they cause* failure. Thus, for example, saturated fine sands may liquefy because of excessive pore pressures generated by earthquake tremors. In the construction of embankments excessive compaction without allowing time for drainage may raise pore pressures to levels threatening embankment stability.

Water Inflow

Any engineering works carried out below the level of the water table will encounter groundwater inflow towards them. The rate of flow will depend upon the size and depth of the excavation and the hydrogeological properties of the rocks and soils involved, both in the excavation and in the surrounding area. Dealing with the water inflow, or alternatively, adopting some means of excluding it wholly or in part, may involve major expense.

Inflow of water into excavations may cause flooding of the excavation, collapse of the sides and "boiling" of the floor. Except in the case of small inflows which may be dealt with by pumping from a sump or collector trench, it will be necessary either to drain or de-water the area, preferably before excavation begins. Any method which allows temporary lowering of the water table may be adopted.

8.4.2
Theory of Groundwater Lowering

Pumping water from the ground produces a cone of depression in the water table or piezometric surface (Fig. 8.5). The radius and gradient of the cone depends upon the rate of pumping, the storage of the ground and the hydraulic conductivity which determines the rate at which water flows towards the pumping well.

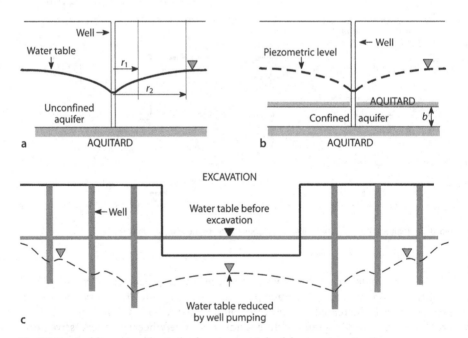

Fig. 8.5. Cones of depression of water level or piezometric level due to water extraction

The amount of discharge Q which would be necessary to lower the water table through a given depth to achieve a steady state of flow can be estimated from:

$$Q = \frac{K\pi(h_2^2 - h_1^2)}{\log_e\left(\frac{r_2}{r_1}\right)} \tag{8.1}$$

Equation 8.1 is for an unconfined aquifer, Fig. 8.5a, where h_2 is the elevation of the water table above an impermeable horizon at r_2 measured in an observation hole, and h_1 is the height of water above the same impermeable horizon at r_1 and K is the hydraulic conductivity.

For a confined aquifer, Fig. 8.5b, the equation is:

$$Q = \frac{2\pi K b (h_2 - h_1)}{\log_e\left(\frac{r_2}{r_1}\right)} \tag{8.2}$$

where b is the saturated thickness of the aquifer.

Well Points

The simplest method of de-watering is by the use of a well point system (Fig. 8.5c), which consists of a number of perforated pipes placed vertically in the ground, usually by "jetting" and connected together by a "ring main" hose which in turn leads to a pump. The depth to which the pipes are placed and the spacing between them depends upon hydrogeological conditions. The pipes must be close enough to allow intersection of individual cones of depression produced in the water table by pumping. Generally, the maximum depression of the water table which can be achieved in practice is about 5 m.

Bored Wells

These may be used when a depression of more than 5 m is required, or when the water has to be pumped from depth. Since these require installation of submersible pumps in properly drilled and screened wells, they are very much more expensive than well points.

Horizontal Drains

Drainage tubes may be installed on benches if this becomes necessary after excavation work has started. The drains are drilled to intersect the water table and have a slight inclination towards the excavation. Water emerging from them must still be pumped out of the excavation. They are useful particularly for the drainage of slopes, rather than for excavations. If the slopes are in rock then attention must be given to the direction of the drainage holes to make sure that they intercept those joints through which water may be flowing.

8.4.3
Excluding Water

Excavations in "dry" conditions may be made by excluding water from the excavation. Sheet piles may be used to create an impermeable barrier to water inflow. The piles may be bored or driven, usually concrete in the case of bored piles, usually steel if driven (Fig. 8.6). Preferably, the piles should be placed to an impermeable stratum below the depth of the excavation. Otherwise the hydraulic head between the water table outside the excavation and the floor of the excavation could cause sufficient flow of water up through the floor to cause *quicksand* conditions in soils.

If the piles cannot be placed to an impermeable layer, they should be placed deep enough so that the flow path of water underneath them is lengthened to reduce the flow into the bottom of the excavation. Sheet piles have the advantage that they can be used in any type of soil except boulder beds. Steel piles may be recovered for re-use; if a contiguous pile concrete wall is used, it may become part of whatever structure is placed in the excavation.

Grouting

Grouting may be used in porous and permeable media to create an impermeable grout curtain. The site is surrounded by closely spaced drill holes, into which is pumped grout. Usually a mixture of cement with bentonite is used since this has the necessary properties of mobility when fluid and impermeability when set. Grout curtains are

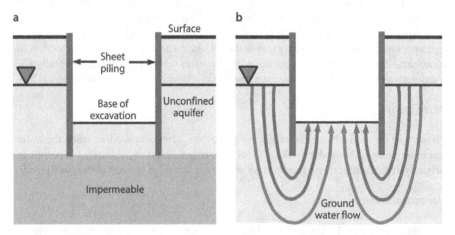

Fig. 8.6. "Dry" excavation; **a** sheet piles driven to an impermeable stratum allow an excavation to be made without great water inflow; **b** sheet piles are driven to a depth within the pervious aquifer. Water flow around them and into the excavation may be sufficient to cause quicksand conditions in the excavation

never impermeable after the first grouting. Variations in the groundmass and errors with grouting procedure result in poorly grouted zones through which leakage will occur. Remedial grouting, sometimes three or more times, is mostly necessary to create a fully impermeable curtain.

Diaphragm walls

Diaphragm walls may be constructed to exclude groundwater if these can be used as part of the permanent works. They are constructed of structural concrete and may be formed in a trench filled with bentonite slurry. In some cases the slurry trench will be enough on its own to exclude groundwater.

Freezing

Freezing of the groundwater is an expensive but extremely effective technique for excluding water. Usually, brine at $-30\,^{\circ}C$ is pumped through a ring of double pipes installed in boreholes. The returned brine, at a temperature of perhaps $-27\,^{\circ}C$ is refrigerated and re-circulated. Several weeks may be necessary before an impermeable ice wall is formed and the freeze has to be maintained until the excavation is lined or the foundation works completed. It is important to realise that *water has to be present* (it is the water that is being frozen) at the time and location of the freezing pipes before excavation starts. Otherwise only a temperature lowering is obtained and when water levels rise, an influx of (warmer) water may not be turned into ice and reach the excavation.

8.4.4
Inflow into Tunnels

With regard to groundwater, planning and construction of a tunnel must take account of the following:

- amounts and locations of water inflow into the tunnel during construction;
- effect on surface structures of the possible withdrawal of support by water table lowering during tunnel construction;
- affect of the interaction between water in the finished tunnel, if it is to be an aqueduct, and the prevailing external hydrogeological regime; and
- effect of hydrostatic pressure on the tunnel lining.

Prediction of amounts and locations of inflows are of extreme importance, not only for cost but from the safety point of view. *Many disasters suffered during tunnel construction have been the direct consequence of the unexpected interception of large water inflows.*

Drainage systems can be designed only when inflows can be predicted. Tunnels driven upgrade will serve as their own primary drainage system, but the hazard to workers of sudden inrushes of water will remain. These may be brought about by the tunnel passing through a hydrogeological boundary, of which the most dangerous may be faults. Tunnels driven down-grade or from access shafts are more difficult to drain and the hazard is increased. Inflows at the tunnel face of 1 000 litres per second have been recorded in several instances (Goodman et al. 1965).

Figure 8.7a shows a tunnel acting as a drain in a homogeneous isotropic medium. Such a simple case may occur in shallow tunnels driven in unconfined aquifers in non-lithified sediments. The rate of water inflow Q, that might be expected, per unit length of tunnel is given by:

$$Q = \frac{2\pi K h_0}{2.3\log_{10}\left(\frac{2h_0}{r}\right)}$$
(8.3)

where K = hydraulic conductivity, h_0 = depth from water table to tunnel and r = radius of tunnel. Alternatively, an elementary flow net can be constructed, from which Q can be calculated.

This simple case is unlikely in a rock mass, even if the improbable assumption of mass isotropy and homogeneity is continued, because after some time t, the water table will be drawn down, the original conditions will no longer exist, and a transient flow system would develop quickly (Fig. 8.7b). In this case, the rate of inflow $Q(t)$ per unit length of tunnel at any time t after the water level had been reduced to the level of the tunnel can be approximated by:

$$Q(t) = \sqrt{\frac{8}{3}CKh_0^3 St}$$
(8.4)

where S is the specific yield, and C is an arbitrary constant, found by experiment, to have a value of approximately 0.75.

Thus, it is clear that the *present state of the art is capable only of giving order of magnitude values to predictions of inflow*. A pre-requisite to increased accuracy in inflow prediction is obviously a greatly increased knowledge of the geological conditions. Unfortunately, this is extremely difficult to achieve, particularly for a tunnel

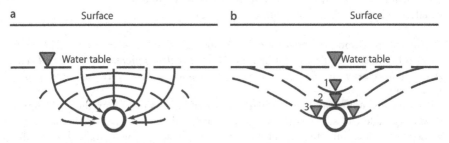

Fig. 8.7. Water influx into a tunnel

which is to be driven at depth through an area which has a complicated and rapidly changing geological structure.

Some improvement can be made by increasing the accuracy of geological knowledge and carrying out calculations similar to the above for each "geological unit" of the tunnel. For example, different rock types may provide a primary "geological unit" division; ideally hydrogeological boundaries should be used to define these units as they bound ground of similar hydraulic characters. For each primary division, subdivision could be made on the basis of fracture density and orientation, taking particular account of areas where fracture density would be expected to increase due to faulting and/or anticlinal folding.

Example

Figure 8.8. gives a theoretical example to illustrate lines of thought that may be adopted to consider such problems. The geological section has been developed from surface geological mapping and shows faulted but uniformly dipping sedimentary rocks intruded by a granite batholith and basalt dyke (the dyke is shown rather thicker than is likely in nature).

The first problem is the nature of the fault. Faults may be thin impermeable hydrogeological barriers or could possibly be wide with permeable crush zones. In the first case the fault would be a limited hazard but in the second a major danger, allowing the sandstone aquifer to discharge into the tunnel under 250 m hydraulic head. Clearly the nature and permeability of the fault would deserve further attention during the course of detailed site investigation, remembering that the condition of the fault would not necessarily be uniform but vary depending on the nature of the strata through which it passes.

Because of its chilled margin adjacent to metamorphosed rock the outer surfaces of the basalt could act as a barrier to water flow from the karstic limestone, while well developed internal jointing could act as a conduit to channel water into the tunnel. Alternatively the basalt could be a partial barrier sufficient to allow impoundment of water in the limestone but allowing passage of water into the dyke to give high hydraulic head in the basalt. Groundwater conditions in both the limestone and basalt would deserve investigation. Little inflow might be anticipated in mudstone but rather more in the granite through widely spaced, perhaps part open, joints.

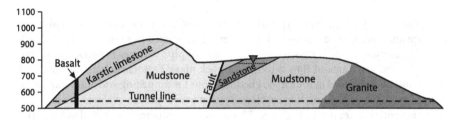

Fig. 8.8. Theoretical example of a tunnel to be driven through sedimentary strata with igneous intrusions

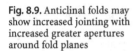 **Fig. 8.9.** Anticlinal folds may show increased jointing with increased greater apertures around fold planes

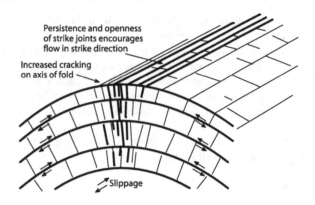

Such geological studies in practice would be three dimensional involving the interpretation of a wide zone of mapping around the tunnel line, and be followed up by physical investigation designed to better define the problems revealed by geological studies. Joint aperture at depth is difficult to estimate but in folded structures zones of potential difficulty may be recognised. In stronger rocks there may be a greater frequency of joints, perhaps with wider apertures, around fold axes (Fig. 8.9). Any tunnel passing through a fold axis might expect greater water inflow in such locations.

8.4.5
Water Outflow

Some excavations, such as canals, small reservoirs and landfill sites, are intended to keep water or fluids in and in such cases this is usually achieved by constructing some sort of impervious, water retaining, lining. Reservoirs behind dams generally rely upon geological conditions to achieve this and, before constructing a dam, studies are undertaken to assess where and how much leakage may take place from the reservoir. This will depend upon the nature of the strata, whether the river to be dammed is influent or effluent, hydrogeological boundaries and variations in water table level within the strata surrounding the reservoir.

8.4.6
Links to Rock Mass Classification

In fractured masses the quantity and rate of flow will to some degree depend upon fracture aperture and roughness. The former, being related to mass porosity, will influence quantity. It is well known that roughness of pipe surfaces reduces flow so the latter will influence rate of flow through the rock mass. Factors such as aperture and joint roughness are used in rock mass classifications for tunnelling and other purposes. Gates (1997) has utilised rock mass classification procedures to evaluate the groundwater potential in fractured bedrock. The derived parameter is the Hydro-Potential value (HP_{value}), which is given by:

$$HP_{value} = \left(\frac{RQD}{J_n}\right) \times \left(\frac{J_r}{J_k J_{af}}\right) \times J_w \qquad (8.5)$$

where J_n = joint number, J_r = joint roughness, J_k = joint hydraulic conductivity, J_{af} = joint aperture factor and J_w = joint water factor.

Clearly the factors quoted owe much to Barton et al. (1974) but ratings have been selected or modified by Gates as appropriate to the purpose. Data to complete the equation was derived from the examination of outcrops close to a series of wells. The rock masses examined were mostly igneous or volcanic.

8.5
Conclusion

The following chapters consider the processes described in Fig. 8.1 and their responses in ground that would normally contain groundwater. When a site investigation is either being designed or conducted, the engineering geologist *has to foresee* those processes and reactions that may be associated with a project, so that the information required for such reactions to be quantified is obtained.

8.6
Further Reading

Eddleston M, Walthall S, Cripps JC, Culshaw MG (eds) (1995) Engineering geology of construction. Geological Society of London (Engineering Geology Special Publication 10)
Fetter CW (1994) Applied hydrogeology, 3rd edn. Prentice Hall, Toronto
Jardine FM (ed) (2003) The response of buildings to excavation-induced movements. Construction Industry Research and Information Association (CIRIA), London

Chapter 9

Withdrawal of Support by Surface Excavations

9.1
Slope Movements

When support to the ground is withdrawn by surface excavation, a slope results. The excavation can be brought about either by natural agencies (such as river erosion, coastal erosion) or by man-made process (such as the excavations for railways roads and canals). If the slope formed is too steep to be supported by the strength of the ground in the sides of excavation then the slope will fail as a landslide.

Engineers find it convenient to divide slope failures into two sorts; *first time* (i.e. involving ground that has never failed before) and *reactivated* (i.e. remobilising a slide that had stopped moving). In the field it is often the case that these may appear as three types of landslide:

1. the natural mass movements of a slope as a consequence of erosion and weathering processes, (these could be either first time or reactivated);
2. the reactivation of old landslides on natural slopes as a result of human activities (such as cutting into a slope, building on a slope or changing the groundwater movement pattern within a slope) or possibly natural erosion by downcutting of rivers; and
3. the failure of cut slopes in otherwise unfailed ground, as a consequence of imperfect design or construction.

The possibility of landsliding is an ever present danger in any construction in hilly country. Much loss of material and life is caused each year by landsliding.

The growth of many cities has been inhibited by slope stability problems as they grew from a small riverside settlement to spread up the valley sides. An example of this is the city of Bath in Southern England, which was founded in Roman times on the alluvial flats of the River Avon (Kellaway and Taylor 1968). The area about the city consists of Jurassic rocks in which sandstones and limestones (of the Oolite Series) overlie shales and clays (of the Lias). The higher rocks form escarpments which have sometimes slid as the result of failure of the underlying shales and clays, perhaps following their softening by weathering. Many old landslides are present. The city of Bath has grown beyond the first area of settlement and has spread over the landslipped and potentially unstable areas, which give great problems in building construction.

9.2
Simple Slope Failure

Surface excavations by natural or engineering processes leave a slope whose stability is related to the slope angle, the strength of the materials and discontinuities in the slope and the groundwater situation. In Fig. 9.1 the simplest example of slope failure is considered, that of a block of rock resting, or perhaps sliding down, on an inclined discontinuity. The analysis of this situation is the basis of calculations for all types of slope stability problems. The cohesion (c), the angle of shearing resistance (φ) between the basal plane on which sliding movement may take place and the overlying block, the angle of slope (β) which is also, in this case, the dip of the basal discontinuity, the area of the base of the block (A) and the weight of the block (W) are the factors contributing to stability in dry conditions.

At limiting equilibrium the forces promoting movement equal the forces resisting movement

$$W\sin\beta = cA + W\cos\beta\tan\varphi \tag{9.1}$$

If F_s = the factor of safety of the slope, then

$$F_s = \frac{\text{Forces resisting movement}}{\text{Forces promoting movement}} \tag{9.2}$$

or

$$F_s = \frac{cA + (W\cos\beta)\tan\phi}{W\sin\beta} \tag{9.3}$$

If open joints behind the block become filled with water which thence flows under the base of the block, the stability of the block is reduced by the uplift force (U) on the base of the block and forward thrust (V) from water in the joint.

At limiting equilibrium

$$(W\sin\beta) + V = cA + (W\cos\beta - U)\tan\phi \tag{9.4}$$

and thus

$$F_s = \frac{cA + (W\cos\beta - U)\tan\phi}{(W\sin\beta) + V} \tag{9.5}$$

The forces promoting movement are greater in the wet condition while the forces resisting movement are less. Rainfall percolating through a slope and filling cracks

Fig. 9.1. The stability of a rock block on a slope

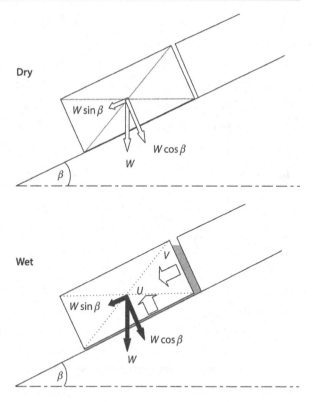

may give water pressures of significance to stability. These water pressures may be transient and apply only when rainfall is occurring and the build up of water pressure may be dependent upon the intensity of rainfall, relative to the drainage capacity of the rockface. Figure 9.2 shows four possible conditions of a slope. In (*a*) the slope is dry and in (*b*), the "*free draining*" slope, the permeability is sufficiently large to prevent significant water pressures building up as the result of transient infiltration from rainfall of limited intensity. In (*c*) the rainfall has increased to fill the discontinuities but free drainage occurs from the toe of the block. Here the water pressure, the distribution of which is represented by the shape of the shaded trapezoids, may build up sufficiently to cause failure. In (*d*) the toe seepage is obstructed, perhaps by ice or soil infill, to allow the water pressure to build up still further.

Thus the possible failure of this type of slope as a consequence of water pressures depends on a combination of circumstances. For example, a slope which has been stable for many years could fail due to exceptionally heavy rainfall or due to ice developing on the face in an unusually severe winter.

Sliding blocks do exist in nature but most slopes are more irregular in shape. However, the failure process by sliding on a discontinuity is the same as that of the sliding block, except that the shape of the 'block' is more complex. This is the most common mode of failure in rocks where masses *translate* down slope on well defined surfaces,

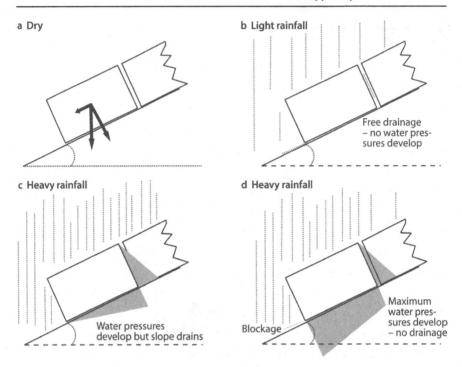

Fig. 9.2. The development of water pressures in a rock slope

although it can occur in soils, for example, in older clays that contain shear zones from tectonic movements or previous landslides.

Many slopes have been observed to fail along approximately circular slip surfaces so that the moving mass *rotates*. These *rotational slips* occur mostly in soils but may occur in weak rocks or in very large rock slopes. The curved slip surface of a rotational failure cuts through intact material and is curved as a consequence of the distribution of stress in the slope. A classical experiment was carried out in 1940 to illustrate the failure of a vertical cut (Harroun 1940). In this experiment (Fig. 9.3) gelatine was poured and allowed to set between two vertical glass plates in a box and held as a vertical cut in horizontal 'ground' by a temporary support. Gelatine was used because of its low strength, its property of losing strength with increasing temperature and because of its photoelastic properties. By using these last properties it was possible to observe the development of stress in, and the deformation of, the model.

To begin the experiment the temporary support was removed and the model immediately slumped from position 1 (with the support) to position 2. The vertical wall bulged outwards at its base while the ground surface level fell. Photoelastic analysis indicated the development of a curved surface of shear failure, leading to the toe of the cut. The temperature of the model was then raised so that the gelatine became weaker. The model slumped to position 3, in which tension cracks on the ground surface appeared. Material began to flow out of the toe of the cut. A further increase in temperature produced stage 4, in which sliding occurred on the curved failure surface

Fig. 9.3. An experiment on a gelatine model to simulate the stages of a slope (from Harroun 1940). The measurements of tensile strains come from almost contemporary work by Terzaghi

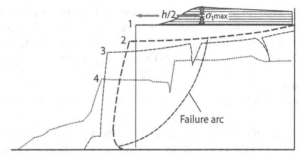

and the toe of the cut flowed forward. The idea of the experiment was to simulate the effect of increasing the effective height of the cut by decreasing the strength of the material. However, the decrease in strength of the model material also simulates loss of strength due to weathering, the opening of fissures, general disruption of structure etc. which many materials experience once slope movement (as in stage 2) begins.

Later experiments on similar materials measured tensile strain (and thence stress) on the surface (σ_t in the figure). Maximum stress occurred at a distance of about half the height of the cut from its edge; the point of maximum tensile stress is the point of development of the tension crack. The slope deformations shown in these experiments on the model have been observed by the author on slopes in rock and soil.

9.3
The Stability of Slopes in Soil

9.3.1
Dry Slopes in Granular Soils

Materials with no cohesion, such as sands and gravels, are frequently known as granular materials. A slope in such materials inclined at angle β to the horizontal may be considered to be composed of a series of sliding blocks, whose stability is described by Eq. 9.3. Because the granular material has no cohesion,

$$F_s = \frac{(W\cos\beta)\tan\phi}{(W\sin\beta)}$$

(9.6)

where ϕ = the angle of shearing resistance of the soil.
At equilibrium, $F_s = 1$ and $\beta = \phi$.

9.3.2
Seepage Parallel to Slope

Equation 9.6 shows that the stability of the slope depends upon β and ϕ. The weight has no influence either in the dry or the totally submerged condition. However, if there is seepage in the slope, giving a water table below the surface, then water pressures will reduce its stability. Such a case might arise if a pit is dug through granular materials to below the water table level. If the water is then pumped out rapidly seepage pressures in the slope could bring it to failure. Coastal slopes in granular materials have to be designed to allow for seepage pressures developed by tidal sea level changes.

9.3.3
Rotational Failure

Most slopes in cohesive soils fail totally or partially in rotation. The failure surface cuts through intact soil and has the approximate shape of the arc of a circle. Failure, or the potential for failure, may be analysed in various ways.

One of these ways has been to divide up the slope into a series of sliding blocks (Fig. 9.4), the base of each being a small part of the arc which, for analysis, is taken to be straight line. The weight of the block depends upon its height and the density of the material in it. Resistance to movement depends upon the shear strength of the base of the block and the support (Y_u) of the downslope block; the forces promoting movement are the downslope component of the block's weight and the down slope push (Y_d) of the upslope block. For the whole slope to move in rotation each block would have develop forces (X_u and X_d) along its vertical sides in order to move relative to adjacent blocks.

If the influence of X and Y forces is ignored the factor of safety of the slope, F_s, is

$$F_s = \frac{\sum(cl + W\cos\beta\tan\phi)}{\sum W\sin\beta} \qquad (9.7)$$

Of course, as in the sliding block analysis, if water emerges from the slope water pressures will act on the blocks, reducing factors of safety. Thus for example, rapid drawdown of water in an excavation will give a curved water table within the slope, introducing water pressures. Alternatively a new slope, showing some seepage, may be stable in the dry season but a rise in water table level in the wet season may bring about failure.

Fig. 9.4. Rotational failure analysed by the method of slices

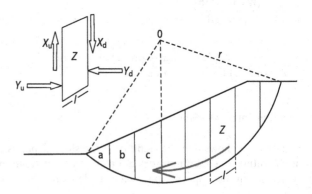

Should all the soil in the slope be uniform saturated clay under undrained conditions ($\phi = 0$), moment equilibrium is considered in the analysis along a failure surface assumed to be a circular arc. A trial failure surface (centre 0, radius r and length L) is shown in the figure (Fig. 9.5). Potential instability is due to the total weight of the soil mass (W) above the failure surface. For such a slope, as for all other slopes

$$F_s = \frac{\text{Forces resisting movement}}{\text{Forces promoting movement}} \qquad (9.8)$$

but, because the failure surface is the arc of a circle, forces may be expressed as moments about the centre of the circle. Thus:

$$F_s = \frac{c_u L r}{W d} \qquad (9.9)$$

where c_u is the undrained shear strength of the clay. This total stress analysis may be held to pertain to the case of a slope in saturated clay soil immediately after construction.

This method of analysis is of limited value in that it can deal only with uniform soils and groundwater conditions. Consequently the analysis based on block movement (the Swedish or Fellenius method of slices) is the most widely used analysis technique. In block movement analysis, as in rock masses, the plane on which movement may take place is known from discontinuity studies, the unit weight of rock materials is usually more or less uniform, and water movement governed by discontinuities. However, in soils, unit weights, shear strength parameters, permeabilities and degrees of saturation may vary significantly from soil layer to soil layer. Thus, no single analysis of an assumed rotational failure surface in a slope is sufficient to determine the lowest factor of safety. *It is necessary to analyse the slope for a number of trial failure arcs about different circle centres in order to determine the minimum factor of safety.*

Fig. 9.5. Rotational failure of a slope in clay

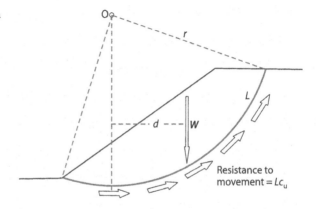

Resistance to movement $= L c_u$

When the analysis methods described above were devised a stability analysis using slide rules, logarithms or mechanical calculators was a massive undertaking requiring many days work, during which one had the opportunity to *think* about the slope and its analyses – if only to relieve the tedium of repeated calculations. Today computer programmes complete the calculations in seconds and the analysis requires little time or effort; the time so saved seems rarely to be given to reviewing the slope.

It is essential to put the correct data into the calculation. Thus, in the layered soil depicted in Fig. 9.6a the shear strength parameters and unit weights of each layer must be determined by investigation to allow the stability analysis of the slope to be made. Groundwater levels are of particular importance. Layers with significantly different properties may control the location of the slip circle which could emerge on the slope or beyond the foot of the slope (Fig. 9.6a). It may be noted from Fig. 9.4 that slices a, b and c, to the left of the failure arc act to resist the downward movement of the main soil mass. Should attempts be made to cut away at the bottom of the slope, this resistance will be diminished; failure could result. Total removal of the slope toe is not nec-

Fig. 9.6. Some geological conditions which will influence soil slope stability

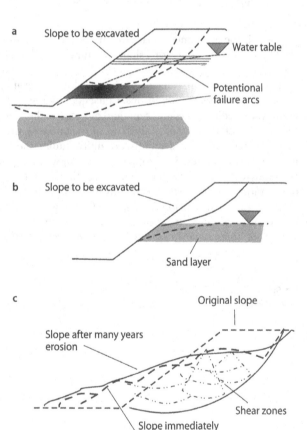

a   Slope to be excavated   Water table

Potentional failure arcs

b   Slope to be excavated

Sand layer

c   Original slope

Slope after many years erosion

Shear zones

Slope immediately after failure

essary to give problems. If a continuous trench is dug at the foot of the slope, if only as a temporary measure to, say, insert a drain, this may isolate part of the mass resisting movement from that promoting movement and bring about failure. This can be significant in highway slope design.

A slope might fail from other causes. The ground in Fig. 9.6b contains a water bearing sand layer. When the slope is excavated water will discharge from this layer bringing sand with it and undercutting the upper part of the slope. Any change in permeability between layers which gives rise to seepage on the slope could bring about this form of failure.

Perhaps the greatest hazard is that of the old landslide. Figure 9.6c shows a slope which failed and, from its hummocky shape was clearly a landslide. With time and erosion the slope becomes more gentle, but contains the shear surfaces under and within the mass which moved. The stability of any new slope excavated in this ground will depend not only on the strength of the slope materials but also on that of the shear surfaces included in it.

Not all soil slides are rotational. In many areas soil, either deposited or residual, overlies rock at shallow depth. If rockhead slopes then, perhaps in wet periods, translational soil slips over rockhead may occur. Should these extend upwards into thicker soils then undercutting may bring about rotational failures in the soils above.

9.3.4
The Stability of Tips and Spoil Heaps

Tips and spoil heaps are man-made and often carelessly constructed. They may occur in industrial areas and the manner and rapidity with which any failure may take place is of considerable importance. The problem of the stability of tips and spoil heaps was brought to public attention with dramatic force in 1966 by the disaster at Aberfan, when a slide involving only some 140 000 m^3 of colliery rubbish resulted in the loss of 144 lives.

Not only to the public, but also to the most professional engineers and geologists – even to those concerned with mining – it came as a problem to which they had given little, if any, serious attention. It would appear that in most respects the problems encountered with tips and spoil heaps are similar in many ways to engineering constructions, such as road and railway embankments and water retaining structures. However, unlike fills for engineering purposes, most mining waste was placed, until recently, without compaction. Slurry and tailings in settling lagoons are likewise in a similar loose state of packing. Often the material in tips and spoil heaps may contain much moisture. There is thus a risk that any slip may develop into a rapidly moving moisture laden *flowslide or mudflow*. Ordinary rotational landslides involving the relatively slow movement of soil masses may cause damage to services and properties, but seldom involve loss of life. The greatest danger lies in the subsequent development of the moving mass into a swiftly moving flowslide or mudflow which may give a disaster with many casualties.

The failure of tailings dams, leading to the discharge of their semi-liquid contents as mudflows may be equally disastrous. Bight (1997) discusses some disasters, including one in Italy causing nearly 300 deaths. It appears that escaping tailings are likely to move as far as a mudflow if the ground on which they move is wet. On dry ground movement may be restricted.

9.4
Slopes in Rock

If the limiting vertical height of a slope in a weak limestone is calculated on the basis of its unconfined compressive strength, values of the order of 100 m may be obtained. Very few, if any, slopes of this height are to be seen in such materials and observations of slope failures in all types of rock have demonstrated their association with geological discontinuities such as faults, joints and bedding planes. Observation also demonstrates that the parameters of importance when considering the role of discontinuities in slope failures are their *strength*; their *orientation*; their *persistence* and their *frequency* and the *geometry* and *orientation* of these discontinuities in the slope.

There may also be permanent seepages from slopes, which are present most seasons of the year and indicate the 'spring-line' from a permanent groundwater table. Up thrust pressures from such a water table will influence stability. New slopes cut in the dry season may be stable when the groundwater table is low, but become unstable if, in the rainy season, the water table rises (Fig. 9.2).

9.4.1
The Strength of Discontinuities

To complete the basic slope stability equation given in Eq. 9.1 values of c and φ must be obtained. However, as described in Chap. 3, values of φ are dependant on the roughness of, and stress upon, the discontinuity surface. Samples of discontinuity surfaces may be tested in the laboratory and in the field but there will always be some uncertainty that the roughness examined on the scale of the test will be the same as that at full scale. Also there is no guarantee that a discontinuity tested is representative of all the discontinuities in that set. This is particularly the case if the discontinuity is an integral discontinuity, such as a bedding plane or foliation plane, for these are not necessarily uniform as a result of their genesis and may be selectively and irregularly weathered.

General assessments of the strength for particular discontinuities in a particular rock type may be obtained by the study of failures and by back analysis. The back analysis process is illustrated in Fig. 9.7. The investigator must first find a rockslide that has occurred along a discontinuity (perhaps of a particular type) in the rock mass. The existing profile is surveyed and the profile before the landslide is reconstructed by surveying adjacent slopes. The weight of the rock mass which moved is estimated from the geological profile and from measurements of unit weight on rock samples.

Fig. 9.7. A profile for back analysis

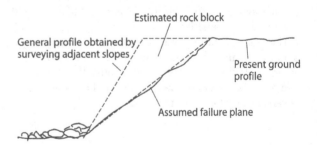

Water seepages are noted. It is then assumed that the estimated rock block had a factor of safety of 1.0 at failure and that there was no cohesion on the failure plane. The standard analysis procedures are then used to determine φ. This value of φ (which may contain 'i' or even some allowance for cohesion) can then be used for the analysis of slopes resting on discontinuities of a similar geological character.

This technique is difficult to use for examples of slope failures may be scarce and the exact circumstances surrounding the actual failure (whether the tension crack was water-filled or not, for example) are almost impossible to establish. However the method can be very powerful in the case of open mine slopes for there the original ground profile before failure and the circumstances around the actual failure should both be known. A mine is often a long-term project and the faces are often designed for minimal stability. It follows then that as the mine progresses a number of failures will occur in the temporary faces. If these are back analysed, quite realistic values of strength parameters may be calculated and then used for the design of slopes for which long-term stability is required. If necessary, slope failures can be induced by toe excavation and then back analysed. However, this is a rather hazardous procedure.

Another approach is to also analyse rock slopes, or components of rock slopes, that have not failed, but assuming that they are about to fail. Such an analysis may give conservative values of strength parameters to be compared with those from laboratory test results and back analyses of failures.

It is usually assumed that strength parameter values are uniform and independent of discontinuity orientation. This is not necessarily true, particularly in the case of shear planes, which may be present in rock and soil masses as the result of past or other shearing motions (such as shearing induced by moving glaciers). These shear planes may be slickensided with ridged lineations which follow the direction of shearing. In such cases the strength of the discontinuity will be less in the direction of the lineations than in other directions and the direction of maximum weakness of the discontinuity will not necessarily be that of this maximum dip (Fig. 9.8).

9.4.2
Modes of Failure

Three common modes of failure may be observed in rock slopes, namely plane, wedge and toppling failure (Fig. 9.9); each involves sliding on a discontinuity as a failure mechanism.

Plane Failure

Slope failure by the movement of a mass of rock along a single discontinuity is described as '*plane failure*' or '*translational failure*' as such a mass on such a plane translates from its origin to its resting place. This is the case of the totally isolated rock block sitting on a discontinuity, a situation seldom met in reality. Sliding failure from a long continuous rock slope is often contained by '*release surfaces*' which may allow the mass to move down dip along the basal plane. Figure 9.10 shows release surfaces which by their orientation may allow, or inhibit, plane failure. Even if orientated to allow movement they may be bonded to the potential moving block and influence its stability, but this influence is almost impossible to assess in practice.

Fig. 9.8. On a slickensided discontinuity the direction of least resistance to sliding may not be that of maximum dip

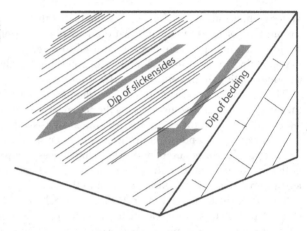

Fig. 9.9. Main modes of failure for rock slopes

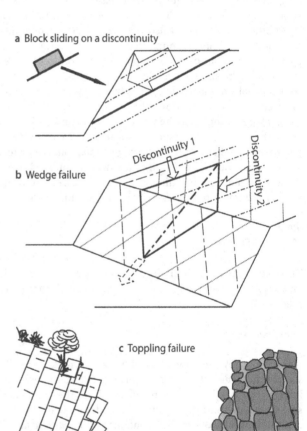

a Block sliding on a discontinuity

b Wedge failure

c Toppling failure

1. With root wedging

2. By weathering of coarse igneous rocks

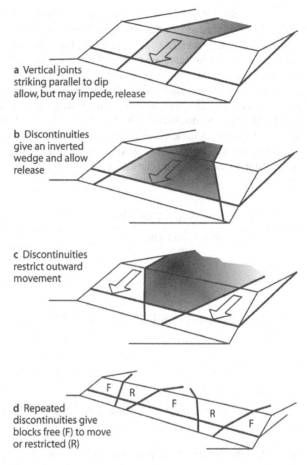

Fig. 9.10. Release surfaces govern which parts of a long slope may give plane failure

a Vertical joints striking parallel to dip allow, but may impede, release

b Discontinuities give an inverted wedge and allow release

c Discontinuities restrict outward movement

d Repeated discontinuities give blocks free (F) to move or restricted (R)

In order that sliding should occur on a single plane in dry conditions, the following geometrical conditions must be satisfied:

i   The plane on which sliding occurs must strike parallel or nearly parallel (within approximately ±20°) to the slope face.
ii The failure plane must '*daylight*' on the slope face; this means that its dip must be smaller than the dip of the slope face and that its position in the slope causes it to intercept the slope face.
iii The dip of the failure plane must be greater than the angle of internal friction on this plane.
iv Release surfaces which provide negligible resistance to sliding must be present in the rock mass to define the lateral boundaries of the slide.

Plane failures are analysed two-dimensionally; the rock block that may fail is considered to be built up of a series of slices of unit thickness taken normal to the slope direction. The stability calculation is the same as that for the cuboid block, except that the weight of the block is rather more difficult to calculate because of its irregular shape (Fig. 9.11).

If the slope angle, height and the dip of the basal discontinuity are known (Fig. 9.11) then it is quite easy to calculate the weight of the potential slide. Hoek and Bray (1981) have produced charts to allow calculation of the stability of slopes of various dimensions and geometries. Figure 9.11 also shows the true profile of an existing slope as surveyed, as opposed to the simplified profile which might be utilised to calculate slope weight. Whether or not such a simplification is justified depends on the irregularity of the true profile. Should there be a tension crack in the slope (Fig. 9.11) the calculation must include the possible condition of the crack being water filled and water pressure acting on the crack, and the underside of the slope block. In this case, for a line of section slice of unit width,

$$F_s = \frac{cA + (W\cos\beta - U - V\sin\beta)\tan\phi}{W\sin\beta + V\cos\beta}$$

(9.10)

where

$$V = \text{Force of water in tension crack} = \gamma_w \frac{1}{2} D_w^2$$

(9.11)

and

$$U = \text{Force of water on sliding plane} = \gamma_w \frac{1}{2} D_w L$$

(9.12)

where L is the length of the potential sliding discontinuity from tension crack to its outcrop.

The simple equations developed above illustrate the significance of water-filled tension cracks and the importance of draining these cracks when they occur along or behind the slope crest. The development of a tension crack may be a consequence of sliding movement along a basal plane but could also be the result of a general redistribution of stress following the withdrawal of support to the rock mass from which the slope has been formed. *It is important to determine, if possible, the cause of tension crack development* for if movement on a basal plane has occurred then its strength may have been reduced from peak to residual values. Whether the appearance of tension cracks may be considered as implying that the slope is near failure or not, their presence constitutes an extra hazard because they make it easier for surface run-off water to enter into the slope.

Wedge Failure

A block of rock may be defined by two discontinuities, dipping towards each other, whose line of intersection dips out of the rock slope (Fig. 9.9b). The most common

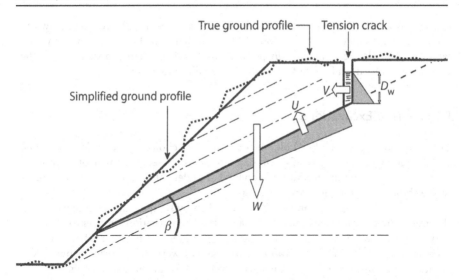

Fig. 9.11. Profile for analysis of plane failure

combination of discontinuities giving a wedge is that of a bedding plane and a joint. Analytical techniques have been developed to assess the stability of such wedges (Hoek and Bray 1981), assuming that the wedge is supported by friction only and that φ is known for both discontinuities.

Toppling Failure

In rock masses where the discontinuity pattern gives tall rock columns it has been observed that columns may fall when sliding movement on their steeply inclined sets of discontinuities is sufficient to bring the centre of gravity of the block outside its base. Such failure is described as toppling (Fig. 9.9c). The toppling block must rest on an outwardly, and commonly shallowly, inclined plane. In some coarser igneous rocks, particularly dolerites and gabbros in sills, tall columns may suffer weathering, concentrated around joints, and exfoliation. This leaves rock blocks separated by rotten rock which may readily weather away and promote falls (Fig. 9.9c).

9.4.3
2D Theory versus 3D Reality

Most of the drawings in the preceding part of this chapter show sections of slope in two dimensions to aid explanation of stability theory but slopes are, in reality, three dimensional so that all stability problems must be considered three dimensionally. Thus a true cross-section may be as in Fig. 9.11 but, if the slope is on a promontory, the tension crack may outcrop and thus never fill with water. Rock slopes may have general inclinations and directions that are favourable for stability with regard to the overall discontinuity pattern but any gullies, clefts and promontories will locally change slope orientation and allow failure to take place. These and similar problems can be reduced by analyzing the stability of a slope using a number of 2D sections drawn in different directions through the slope. These will help identify directions in which failure is possible (i.e. *kinematically* possible) and of these, define those on which the forces operating are sufficient to actually cause failure to occur.

9.5
Engineering Excavations

Before a slope can be made a hole must be dug. The choice of excavation technique will depend upon the nature of the material to be excavated, the shape of the hole to be made and whether the excavation is on land or underwater. Figure 9.12 gives a guide to the choice of techniques for excavation.

In soils on land excavation may be undertaken by scrapers and bulldozers when the area to be excavated is large or long, as in road construction. Smaller excavations and closed excavations (such as for foundations) are mostly dug by mechanical shovels and buckets. Very small excavations (for drains, sewers etc.) are dug by bucket excavators. Specialised machinery has been developed for pipeline trenching.

Underwater excavation is done by dredging. Some types of dredger are illustrated in Fig. 9.13. There are many different designs of dredger, most of which are adaptations of surface excavation techniques and equipment for use underwater. However, there are major differences in the methods used to remove excavated material from site. Thus, for example, material cut out by the cutter head in Fig. 9.13a is sucked up into the dredger and may be pumped through floating pipelines to shore.

9.5.1
Surface Excavations in Rock

In rock the suitability of a particular excavation technique is determined by the properties of the rock mass which consists of materials cut by discontinuities. The properties of main importance are:

- the strength of the rock material,
- the size of the rock blocks (as determined by the three dimensional spacing of discontinuities),
- the openness of the discontinuities,
- the orientation of the discontinuities.

The success or failure of many a rock excavation contract has depended on the decision, made at the time of tendering, regarding any pre-treatment that the rock mass should receive before excavating. Possible pre-treatments to loosen the rock mass before excavation are:

- ripping, in which a massive steel tooth is driven into, and dragged through, the rock by a bulldozer (Fig. 9.14);
- blasting to loosen or 'lift' the rock mass;
- blasting to fracture and loosen the rock mass.

a

Rock

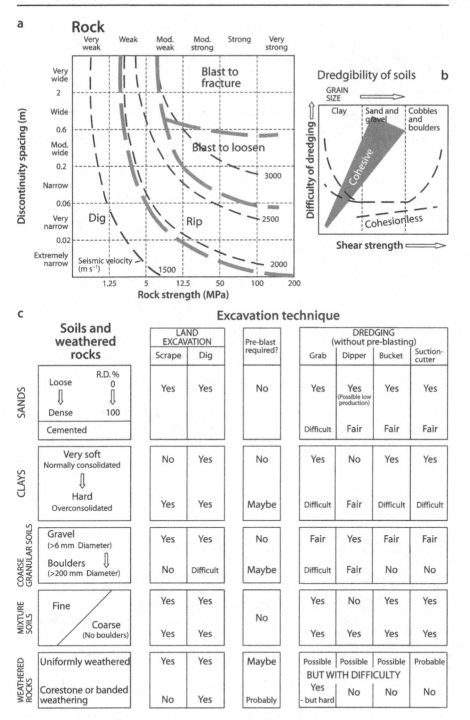

Fig. 9.12. A guide to the applicability of excavation techniques

Fig. 9.13. Some types of dredger (in each case the vessel is shown only schematically); **a** suction cutter dredger, which with the cutter head illustrated may be used to cut rock; **b** bucket dredger, often used to remove soil from channels which may be silting up; **c** dipper dredger, which is an underwater shovel

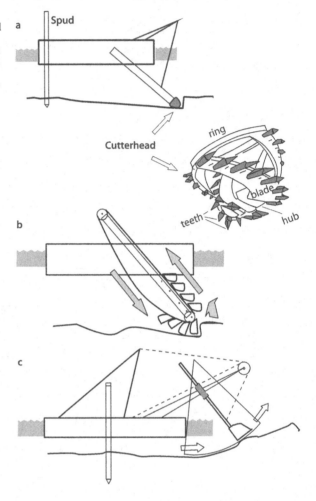

Fig. 9.14. Ripping a steel tooth towed behind a bulldozer. Whether or not a rock mass can be ripped will depend on its characteristics and the power of the ripper. In significantly anisotropic rock rippability may depend upon direction of attack. In direction *b* the tooth plucks out blocks and is more effective than ripping in direction *a*

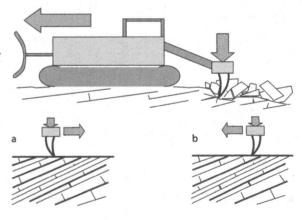

How much the rock has to be loosened before being dug out will depend on the size and power of the digger used. Assessing which technique should be applied to loosen the rock mass is of considerable importance in terms of the cost per unit volume for excavation. This rises considerably if blasting is necessary.

One of the first approaches to establishing the technique suitable for excavating particular rock masses was made by Franklin et al. (1971), in which the two parameters of rock strength and discontinuity spacing were utilised. Figure 9.12b gives approximate limits to digging, ripping, blasting to loosen and blasting to fracture based on the simple idea that the bigger the rock blocks and the stronger the rock, the greater the effort to be applied to excavate. The rock mass characteristics of discontinuity spacing and rock material strength are also reflected in measurements of the seismic velocity of the rock mass, for velocity through the material generally increases with increasing strength while it decreases with increasing discontinuity frequency. Very approximate velocity boundaries have been established, for particular rock types, to indicate the suitability of ripping or blasting as *pretreatments* before digging (Fig. 9.15).

Fig. 9.15. Approximate velocity boundaries between rippable and non-rippable rock masses. The generalised strength and spacing indicates the relationship between rock strength, block size and rippability

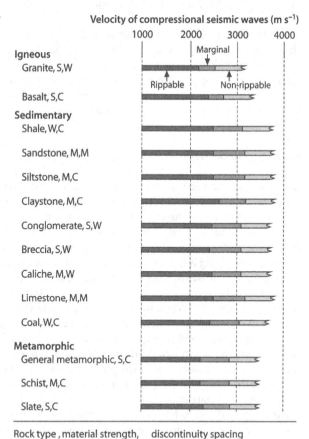

Rock type , material strength,   discontinuity spacing
    S = strong   W = wide
    M = moderately strong M = moderately wide
    W = weak   C = close

The velocity measurements would also be influenced by such rock mass characteristics as discontinuity openness, the position of the water table and discontinuity
anisotropy. This suggests that velocity measurements combined with discontinuity
spacing and rock material strength may provide a satisfactory guide to any excavation pretreatment required. However, it should be noted that the seismic velocity in a
layered rock mass may be an average velocity. If geophone spacings are large, high
velocity layers in a low velocity matrix may give an intermediate velocity which may
seem to indicate rippability although a few high velocity (i.e. strong and intact) bands
may make this impossible. Accordingly, whatever measurements are made to assist
estimates of rippability, there must be a thorough knowledge of the geology of the site.
Much will also depend on the power of the machine used and, for a given machine, on
whether the machine is in good mechanical condition and the driver is skilled in his
work.

If the rock mass has a strongly anisotropic structure, the rippability of the rock will
depend much upon the direction of the principal discontinuity in relation to the direction of attack of the ripper; Fig. 9.14 illustrates the differences. If the dip follows
the length of the excavation the direction of ripping may be chosen to give the most
favourable result.

Another way to examine the excavatability of a rock mass is by use of a rock mass
classification. Table 9.1 gives that of Weaver (1975) as an example. Since appraisal by
rock mass classification can take account of more factors than other simpler methods
it is likely to develop into the main tool to determine excavatability. However, for most
major civil engineering work excavation trials are held using a variety of excavation
techniques; the limits for ripping can then be determined on site.

9.5.2
Excavation Underwater

Under water excavation is more difficult than on land where the operator of the excavating machine can see what he is excavating and can adjust his attack to suit minor
variations in rock structure to get the maximum use of the available power.

This is not the case underwater where the operator has little flexibility in the use of
the cutting and digging mechanism and cannot see what he is doing.

Also, in dredging it is as much a problem to bring material to the surface as it is to
dislodge it from the sea or river floor. Thus core stone boulders from a corestone weathered mass could be dislodged by the cutter of a suction-cutter dredger but not necessarily brought to the surface by the suction pipe. To remove such dislodged material
another dredger, say a grab dredger, might have to be brought on site at great expense,
entailing extra costs and perhaps loss for the contractor. It follows then that *investigations for dredging contracts, particularly rock dredging, should be of higher standard
than those on land* and should incorporate not only good quality sampling, geophysics and so forth but also adequate examination of on-shore and underwater outcrops.

A problem of particular importance is that of lightly cemented granular soils. In
dredging granular soils by suction-cutter the cutter undercuts slopes which fail and
then the slumped and disturbed material is sucked to surface. If the soil is lightly cemented the slope will not fail, all material has to be cut before sucking, and much more
energy has to be expended for each unit of material excavated. Unfortunately materi-

Table 9.1. Rock mass classification for excavatability (from Weaver 1975)

Parameter and rating	Rock class				
	I (Very good rock)	II (Good rock)	III (Fair rock)	IV (Poor rock)	V (Very poor rock)
Seismic velocity (m s⁻¹)	>2 150	1 850–2 150	1 500–1 850	1 200–1 500	450–1 200
Rating	26	24	20	12	5
Rock hardness (MPA)	Extremely hard (>70)	Very hard (20–70)	Hard (10–20)	Soft (3–10)	Very soft (1.7–3)
Rating	10	5	2	1	0
Rock weathering	Unweathered	Slightly weathered	Weathered	Highly weathered	Completely weathered
Rating	9	7	5	3	1
Joint spacing (mm)	>3 000	1 000–3 000	300–1 000	50–300	<50
Rating	30	25	20	10	5
Joint continuity	Non-continuous	Slightly continuous	Continuous – no gouge	Continuous – some gouge	Continuous – with gouge
Rating	5	6	3	0	0
Joint gouge	No separation	Slight separation	Separation <1 mm	Gouge <5 mm	Gouge >5 mm
Rating	5	5	4	3	1
Strike and dip orientation	Very un-favourable	Un-favourable	Slightly un-favourable	Favourable	Very favourable
Rating	15	13	10	5	3
Total rating	100–90	90–70	70–50	50–25	<25
Rippability assessment	Blasting required	Extremely hard ripping and blasting	Very hard ripping	Hard ripping	Easy ripping

als of this nature may break up during sampling by boring so that a routine inspection of the sample reveals nothing of the cemented nature of the material in situ. Detailed study of seismic records may help to identify the problem, for cementation may raise velocities to values higher than would be expected for non-cemented material. Carbonate deposits (reef debris, carbonate sands etc.) should always be regarded with suspicion, particularly if they were once, in the geological history of the area, above sea level allowing duricrust formation to take place. The fall of sea level during the Pleistocene and its subsequent rise makes this a possibility in many areas.

9.6
Investigations of Slope Stability

Slope stability investigations fall into three broad categories. These are:

1. investigations of ground masses into which slopes will be cut in order to design that slope;
2. investigations of existing slopes in order to determine their stability;
3. investigations of existing landslides.

9.6.1
Investigations for the Design of Excavated Slopes

The objective of these investigations is *to build up a three dimensional geotechnical model of the ground* in which the slope is to be cut. The model must include such quantitative data as is necessary to put numerical values into slope stability formulae on material and discontinuity properties, the distribution of materials, the position and orientation of discontinuities, and the position of the groundwater table(s). Seasonal variations in groundwater levels should be known together with such environmental influences as may act upon the completed slope.

Investigations of slopes in soil are mostly undertaken by standard boring and sampling techniques. If the slope will be such that soil will overlie rock in the slope then it is essential that the topography of the bedrock surface be known, in detail, for there may be a danger that soil will slide on the bedrock surface. Such detail cannot be found by boreholes alone; additional data should be acquired using geophysical methods; seismic or ground penetrating radar methods are usually appropriate.

If the slope is in rock then discontinuity orientations and strengths must be known, in order to make a stability analysis. While such data may be derived from boreholes it is best obtained by mapping existing outcrops or man-made excavations (tunnels, shafts, pits, or large diameter boreholes).

Information on discontinuity orientations may be displayed by plotting on a stereonet. Figure 9.16 outlines the basic principles of the construction and use of stereonets. More detailed descriptions may be sought in Hoek and Bray (1981) and Phillips (1968). The objective of this is to establish a limited number of dominant orientations which may then be used in stability analysis and slope design.

Experience has shown that within a particular volume of rock mass, which may be described as a *domain*, the discontinuities will fall into sets with more or less uniform

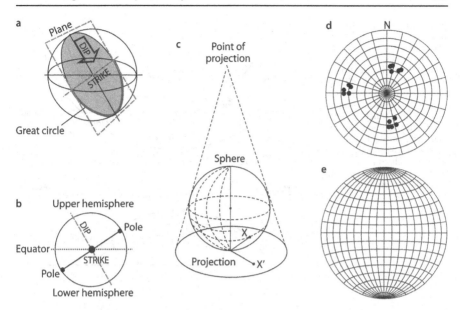

Fig. 9.16. Stereonets. **a** A dipping plane enclosed in a sphere intersects on a great circle. **b** The sphere is divided into upper and lower hemispheres. A line drawn from the sphere centre at right angles to the plane intercepts the sphere surface at a pole. **c** Lines of latitude and longitude on one half of the sphere may be projected onto a plane surface to give a projection. A pole on the surface of the sphere (X) may be shown on the projection (X'). **d** A projection showing the poles of the sphere is a 'polar projection', as in (c) The *dots* represent plots of individual discontinuities which fall into three groups. **e** A projection from the side showing the equator an 'equatorial projection'. Projections vary depending on the position of the point of projection and may be designed to show true angles (as in (e)) or equal area. Usually poles are plotted on a projection of a lower hemisphere

orientations (Fig. 9.17). How big a domain is will depend upon the complexity of the geological structure. In general, the older the rock mass the more complex the geological structure and the smaller the domains. If, in a given rock mass, discontinuity orientations are plotted on the stereonet irrespective of the mass structure, the stereonet for a mass with many domains will display a cloud of data points. Accordingly one of the first tasks in preparing orientation data is to select domains within the mass. This is not easy if the data to be plotted comes entirely from boreholes.

If the data is plotted on a stereonet then the cloud of *poles* may be contoured to show where the greatest density of poles lies. Methods of contouring pole density are discussed in Hoek and Bray (1981). The orientation of the point of greatest density is then considered to be the orientation of the discontinuity to be used for stability analysis. In such a plot it is prudent to indicate which type of discontinuity is being plotted, whether, for example, it is bedding, a joint or a shear zone.

It must be understood that the contoured stereogram is only a representation of the information recorded. It is not a type of statistical analysis. Indeed, many engineering geologists prefer to select dip directions for use in stability analysis by eye in the field, rather than at the desk through contouring.

Fig. 9.17. The plan shows strikelines on a bedding plane surface in an anticline. It is proposed to make a road cutting through the anticline. Directions of dip and dips will continuously vary around the anticline, but three domains are chosen within which the bedding orientation is considered to be uniform. Stable slope angles are calculated using these values in each domain. Joint sets in each domain could also be considered uniform

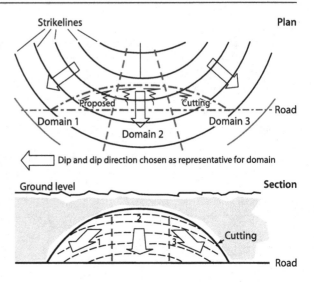

One problem associated with contouring poles on a stereonet is that this displays only orientations and in this respect treats all surfaces equally. *Factors of equal importance in slope stability are discontinuity persistence and frequency.* These factors, together with orientation, must be considered to select those discontinuities, in a domain, which will influence the design of a new cut slope; the poles to persistent surfaces can be highlighted to help them stand out from the background of poles to less persistent surfaces. The choice of domain boundaries may also be influenced by the φ value assigned to the discontinuity which dominates stability. If this is close to the dip angle, then perhaps domains may be chosen relative to quite small variations in dip. Clearly there are many factors to be considered in constructing the geotechnical model in which slope design may be undertaken.

9.6.2
Investigations of Existing Slopes

Some slopes in soils, for example those cut for motorway construction, may exhibit signs of distress and potential or limited movement and have thus to be examined to determine their actual stability. In such a case the investigation should gather adequate data for a satisfactory analysis by conventional methods.

Rock slopes may exhibit rock falls and their examination in the field is necessary to determine the extent of the danger and, if necessary, to gather sufficient data to allow design of remedial works. Mapping may be done on various scales. On the large scale the first task is to determine the general structural geology and to see how this relates to existing slope stability. Edinburgh Castle rock is a basalt plug, once a feeder pipe to a volcano, which dominates the centre of Edinburgh (Price and Knill 1967). This has been subject to glacial and other erosion, and to some degree modified by man. The plug exhibits three main joint sets. These are:

1. dome joints, which dip downwards and outwards radially and quite shallowly from the plug centre to its periphery;
2. radial joints, very steeply or vertically dipping, striking along radii from the centre; and
3. circumferential joints, which dip very steeply and srike parallel to the plug periphery.

Joints from each set are present in every exposed rock face, but vary in frequency and persistence. Where dome joints are dominant slopes tend to be relatively gently inclined due to past rock slides on this joint set. Where circumferential joints are dominant slopes are steeper and show signs of potential or actual failure by toppling. Where radial joints are dominant, slopes are steep and relatively stable; the basic situation is illustrated in Fig. 9.18. It should be noted that joint dominance may vary with height as well as position in plan.

Such slope assessments may be done on a background topographical map prepared with contours drawn relative to a vertical datum plane (Fig. 9.19). If the general topography gives uninterrupted views of the rock slope and allows stereopairs of photographs to be taken horizontally, such a map may be prepared by terrestrial photogrammetry or its digital equivalent. Measurements taken from them establish discontinuity orientations. Alternatively, if the intention is to stabilise the rock face by bolting, maps may be prepared using the front face of scaffolding as a datum plane and taking horizontal distance measurements from this (see Fig. 4.9). Discontinuities may be plotted on the resulting map which is thence interpreted to define rock blocks and set target depths for rock bolting.

Photogrammetry or scaffolding allows the whole slope to be examined, but neither may be possible, thus limiting the observer to the lower slope that can be reached. An engineering geologist should be equipped not only with conventional compass/clinometers which must be placed on a discontinuity surface to measure dip and strike, but also with compasses and clinometers which allow dips and strikes to be measured by sighting along them. With such instruments orientations can be measured at locations which cannot be reached. Slope profiles may be constructed using compass, clinom-

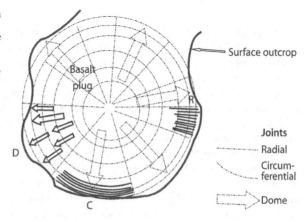

Fig. 9.18. Schematic illustration of principal joint sets in Edinburgh Castle basalt plug. Where dome joints dominate (*D*) falls may be by plane sliding, where circumferential joints dominate (*C*) by toppling and where radial joints dominate (*R*) slopes are relatively stable

Fig. 9.19. A simple rock slope survey. Contours and strike lines are drawn relative to a vertical datum plane. Contour and strike line values are thus horizontal distances from that plane

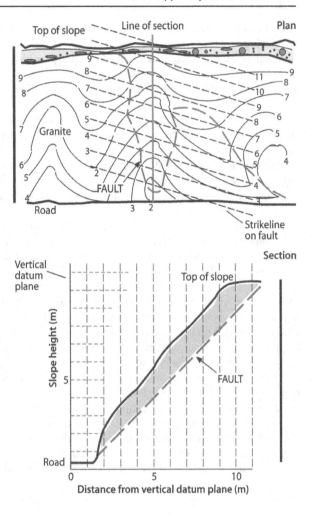

eter and tapes (to measure baselines); portable range finders, sextants and binoculars are also useful additions to simple field survey equipment.

Discontinuity data may be obtained by observations along a 'scan-line'. This is nothing more than a tape stretched along a rock face, usually at its base, and the orientation, character and tape length location of each discontinuity is recorded. Computer programs exist which perform statistical analyses of scan-line data to aid assessment of slope stability. Scan-line data may be conveniently recorded on a 'discontinuity data survey sheet' or a form linked to the entry of data into the computer program.

If the purpose of the work is to assess the stability of a slope in detail then, following the survey, the rock slope is divided into rock blocks which have failure potential dependant on the dip of basal planes, discontinuity strength and so forth. This requires an understanding of the three dimensional structural geology of the rock slope. Such an understanding is absolutely essential if remedial works to improve the factor of safety of the slope are to be designed. Alternatively the data may be entered into a rock mass classification designed to assess the general stability of a rock slope as described in Sect. 4.9.

9.7
Design of Slopes

All excavated slopes are designed to a chosen factor of safety. This is generally not lower than 1.5 and not greater than about 2.5, but in every slope design there may be 'hidden' safety factors obtained by choosing conservative values of strength, or pessimistic views with regard to water pressure. It is important to decide whether the slope should have "*long-term*" or "*short-term*" stability. Most road, railway, canal etc. slopes are designed for long-term stability in which case account must be taken of the possible reduction in strength of materials and discontinuities as the result of weathering and erosion of the slope by gullying from rain wash. Slopes with short-term stability could be some open mine slopes and the slopes for foundation excavations. In excavations in soil over rock the critical factor may be the slope of the bedrock topography on which the soil rests.

In slopes in rock where stability problems usually occur by sliding over discontinuities the slope can be designed on the basis of the strength of these discontinuities. This is done using the same basic formulae as in the example of the sliding block of rock (Eq. 9.1). The only complication is that the shape of the block is no longer neatly rectangular, but an irregular shape whose weight is difficult to calculate. However if the geological situation is favourable it may be possible to design a rock slope so that any unfavourably inclined discontinuities do not outcrop on the slope. Thus, in rock slope design one of the simplest design techniques is to arrange for the slope direction and angle to be such that no steeply dipping discontinuity outcrops from the excavated slope (Fig. 9.20a).

This very simple approach is often the best for road cuttings and similar permanent excavations when space and orientation of the works permits such freedom of choice, for it avoids having to assign values of c and φ to discontinuities. Calculating safe slope angle based on c *and* φ values is possible in fresh rock masses such as, perhaps, open pit mines, but in civil engineering excavations, which are often within the weathered zone, it is very difficult to assign a single value to c and φ that the designer will be confident will apply to the whole slope. Also civil engineering excavations are permanent and long-term weakening of discontinuities by weathering after excavation has to be considered.

In almost any excavation there must be some discontinuities which outcrop on one slope. Thus in the Fig. 9.20b discontinuity 'a' outcrops and could allow sliding on the east slope if 'a' is a very weak discontinuity, particularly prominent and frequent, or liable to long-term deterioration. If so the slope could be flattened to the inclination of 'a'. Alternatively, it is possible that local support works could be planned for areas in which, after excavation, it seemed likely that local failures on 'a' could occur.

The discontinuity set on which sliding may take place need not be continuous through the slope for movement to occur. In Fig. 9.20c short discontinuous 'a' joints serve for sliding surfaces for small instabilities but could also link along 'x-y' to give a whole-slope failure.

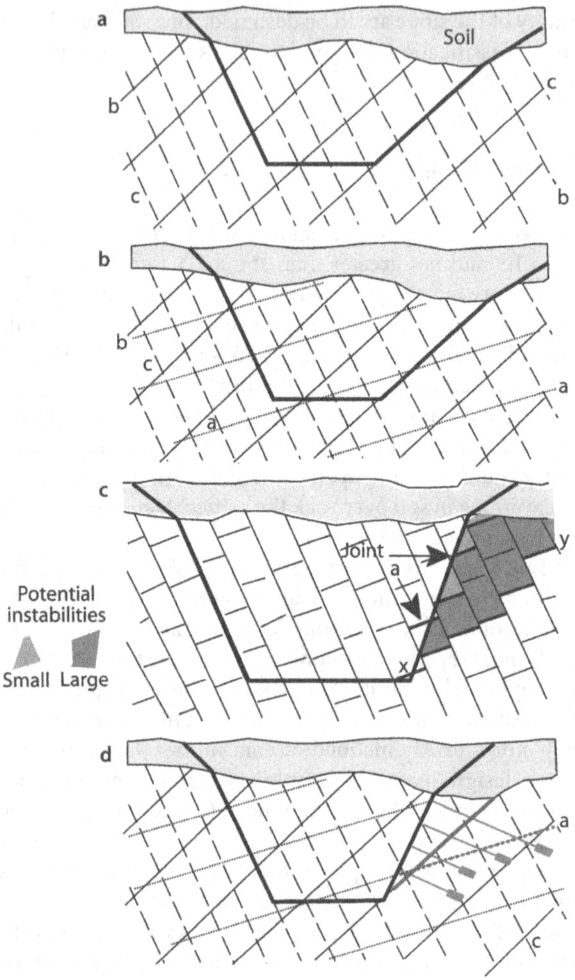

Fig. 9.20. Aspects of slope design in rock

If the inclination of the slope produced by this method is not acceptable then the slope may be designed using values of c and φ for the discontinuities which seem appropriate. This may require flattening the slope to give the required long-term factor of safety. However, if for reasons of land purchase or adverse environmental impact the plan area of the excavation should be kept as small as possible, then the slope might have to be supported by rock anchorages which, in Fig. 9.20d are placed to resist collapse along discontinuities 'c' and 'a'.

9.7.1
Benching on Slopes

Many excavations are dug out in layers, because of their height or for convenience of excavation, so that the resulting slopes have a stepped appearance. In open mines these

Fig. 9.21. Benched rock slopes. The overall stability of slope *AB* is adequate but falls from potentially unstable slopes of component benches could block bench access and bounce on accumulated debris to reach road. Slope *AC* has stable component slopes and is totally stable but occupies more space and entails more excavation

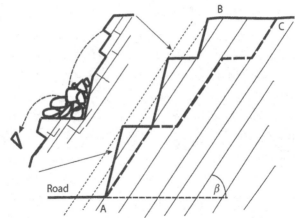

steps or benches may be used as temporary or permanent haulage roads. In slope design it is important to decide whether the benches should lie inside or outside the safe slope. Figure 9.21 shows two possible slope designs in a rock mass containing dominant discontinuities dipping at β degrees. In design AB the overall slope is stable but the benches are potentially unstable. In AC both slope and component benches are stable but both excavation volume and land acquisition are greater.

It is sometimes considered that benches serve to trap rock falls from plunging down the whole slope. However, in slope AB the rock falls onto a bench may help promote its collapse and rock falls that fall on the bench may accumulate until they form a launching ramp for falls that come from above. If benches are used as rock traps they must be regularly cleared of fallen debris.

9.7.2
Drainage

Figure 9.2 demonstrates the adverse effect of water pressures in a slope. It is thus good practice to insert drainage holes into rock slopes to allow any water in the slope to drain away as rapidly as possible. In most rock slopes water movement will be along discontinuities. Accordingly the direction of the drain holes should be such as to intersect the maximum number of, or at least the dominant, discontinuities. To do this requires an understanding of the discontinuity pattern in the slope in three dimensions. Well maintained and sealed drainage channels at the top of rock slopes can intercept and drain away any surface runoff that would otherwise enter the slope.

9.7.3
The Effect of Excavation Technique on Slope Stability

The stability of a slope depends to a great extent upon discontinuities (Fig. 9.22). Even in a carefully excavated slope stress relief may induce joint opening which allows ingress of water and plant roots to induce long-term decay (Fig. 9.22b). Rock excavation often requires blasting to break the rock before it can be excavated. If this blast-

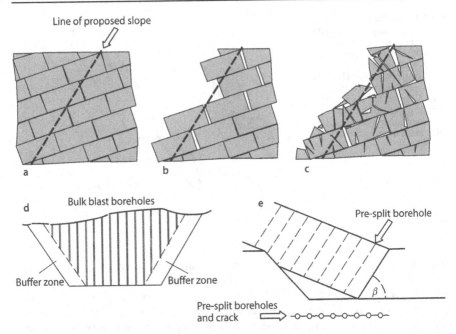

Fig. 9.22. The influence of excavation techniques on rock slope stability. After excavation of a slope in rock mass (**a**) stress relief may bring about joint opening (**b**) which could be a source of long-term deterioration. Excessive blasting (**c**) can open existing discontinuities and create new ones so that any slope design based on the original rock mass properties is no longer valid. This may be overcome by blasting with a buffer zone (**d**) or, seen in perspective, pre-splitting (**e**)

ing fractures the rock from which the final slope is to be made, than these fractures will reduce stability (Fig. 9.22c). Most engineering geologists have seen slopes in which the rock mass structure was basically favourable for stability but, once excavated, became unstable due to excessive and poorly executed blasting. For permanent slopes care should be taken that the blasting for excavation does not disrupt the slope. While much disruption may be avoided by employing experienced and able blasting contractors working to a clearly written and closely defined contract, various working techniques may be employed to reduce rock mass disturbance.

The fracturing that may be induced in the final slope as the result of excavation blasting is minimized by establishing a '*buffer zone*'. No blasting takes place within this zone which is partly fractured by blasting from the area above. The fractured rock in the buffer zone is pulled down using excavating machinery to form the final slope. The buffer zone may be 2–3 m wide (Fig. 9.22d). Damage to the designed slope from the bulk blast may also be prevented or reduced by the technique known as '*pre-splitting*'. In this a row of boreholes is drilled along the plane of the design slope. The holes are parallel, about 1 m or so apart. They are charged with suitable explosive which is then detonated. This causes a crack to run between the holes so that the plane of required slope is marked by a continuous crack (Fig. 9.22e). The rock in the excavation is then removed by normal drilling and blasting techniques; fracturing caused by the normal blasting is found not to pass through the boundary fracture plane.

The method produces very fine, clean rock faces. However, it must be noted that the inclination of the pre-split fracture must be that of the stable slope whose stability is determined by the discontinuities within it. Pre-splitting does not make a slope stable but prevents it becoming unstable as the result of fracturing following blasting. The method works best in uniform rock conditions. It may not work at all if the rock mass is particularly anisotropic as the result of geology or weathering, for in such a case the boreholes may not be parallel and success in crack-forming would vary because of the varying response to shock of the different rock materials.

In a rock face that has been well pre-split the observer should see parallel lines of half boreholes on the face. The efficacy of the pre-split may be assessed by comparing the length of boreholes drilled to the length of borehole seen in half section. Thus a 'Pre-splitting Index' can be easily defined as:

$$\frac{\text{Length of halfsection boreholes seen}}{\text{Length of pre-split bore holes drilled}} \ (\%) \tag{9.13}$$

The nearer the index approaches 100%, the better the pre-split. If, however, the design slope follows a discontinuity plane then a good 'split' may be obtained without a high index for the blast would tend to open the pre-existing discontinuity rather than create another one. On some contracts very good slopes may be formed only to be damaged by later minor works. Thus blasting out a drainage trench at the foot of a face could, if carelessly done, damage the toe of the slope above.

While techniques may be applied to reduce bulk excavation disturbance in the rock mass forming the final slope, some disturbance is inevitable. Some problems may be caused by error. Thus, for example, infrequent shear planes, near impossible to detect in investigations, may give local stability problems. Completed slopes must be examined to discover such features that may give rise to short- or long-term stability problems and remedial measures applied. It is important, particularly in road excavations, that, for reasons of safety and economy, that remedial works be implemented before the road is opened.

However, experience suggests that the greatest source of excavated rock slope instability is poor workmanship in the use of explosives. It is noticeable that many very old slopes excavated by hand and by blasting by gunpowder in short boreholes are stable, while their modern counterparts in similar geological conditions but excavated using much more powerful explosives and deeper boreholes are unstable. There is little point in spending time in investigation and elegant rock slope design if the excavation technique employed destroys any possibility of achieving a stable slope.

9.8
Existing Landslides

9.8.1
Recognition and Identification

Existing landslides, some originating many thousands of years ago, may pose major engineering problems. Those which are presently dormant may be reactivated inadvertently by engineering works while those which are presently active, continuously or intermittently, pose a constant problem.

There are many different types of *mass movement* (a term covering all types of slope disturbance) ranging from simple rock falls from a cliff to rock slides or topples (Fig. 9.9), debris avalanches, rotational failure (Fig. 9.23) to very large complex movements such as may be found in most recent mountain chains. The larger landslides usually involve rock and produce a hummocky landscape with ponds or swampy areas. An example profile is given in Fig. 9.24. Most slides are betrayed by ground fissures and, if supporting vegetation, by trees showing bent trunks.

Perhaps the first most important step in dealing with landslides is to recognise that one exists, for it is generally not a good idea to attempt to undertake engineering works on, in or through landslides. Landslides are best avoided.

Many authors have attempted to classify mass movements, perhaps the most widely cited being that of Varnes (1958). However, while all landslides are dangerous, those which move more rapidly than people can escape may be considered the greatest hazard. The classification of Hutchinson (1988) includes comments on the rapidity of movement of mass movements. Those generally recognised as the most dangerous involve the movement of saturated debris and include such phenomena as mudflows, peat bog bursts, lahars (from volcanoes) and also dry debris flows from scree slopes.

Fig. 9.23. A rotational landslide or slump

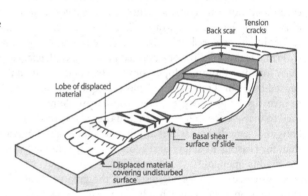

Fig. 9.24. An example of a profile of a landslide in rock; **a** before the landslide, the porous jointed rock releases water on to the underlying argillaceous rock, promoting softening and weathering; **b** when the soft rock is too weak to support the overlying material a complex slide occurs, flowing downhill and displacing the blocks of softened rock to give a hummocky landscape

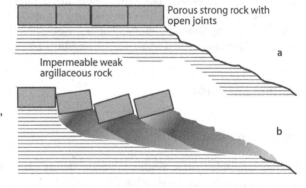

Mining waste tips may fail and flow if saturated, as may saturated tailings if a tailings dam bursts.

9.8.2
Rates of Movement

The Fall of Threatening Rock

On 22 January 1941, Threatening Rock, or Braced up Cliff as it was known to the Navajo Indians, finally fell. This 30 000 ton monolith of Cliff House Sandstone, which rose behind the ruins of Pueblo Bonito in Chaco Canyon National Monument, New Mexico (Fig. 9.25), was of concern to the National Park Service Personnel, as it had been to the inhabitants of Pueblo Bonito in the 11th century, who had attempted to wedge the rock in position with tree trunks.

Because of the concern that fall of the rock would severely damage Pueblo Bonito, National Park Service personnel began, in 1935, to measure the distance between the rock and the cliff from which it had been detached. The measurements were recorded until the fall of the rock in 1941. In the figure cumulative rock movement and (Fig. 9.22) cumulative precipitation are plotted as a function of time (Schumm and Chorley 1964). Although precipitation was well above average during the latter part of 1940, the precipitation does not show the progressively greater increments with time that is shown by the curve of rock movement. The rock appears to have been stationary for consecutive periods of several months, but then moved progressively farther as time passed. The irregularities of the curve in Fig. 9.26 suggest an intermittent application of increments of stress either causing or as a result of periodic failures.

The most rapid rates of movement occurred between the dates of hard frosts or during the winter, whereas the periods of slight movement occurred during the summer months. Seasonal frost action, although effective in promoting surface creep in soils of that region, could not be expected to be as effective in moving a 30 000 ton block of sandstone. Possibly the higher precipitation of summer, on average 15 cm for

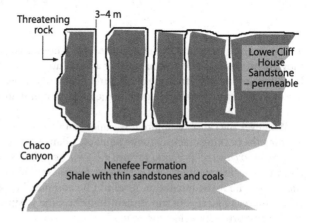

Fig. 9.25. Threatening Rock (from Schumm and Chorley 1964)

Fig. 9.26. Graph of cumulative rock movement and cumulative rainfall in the last days of threatening rock

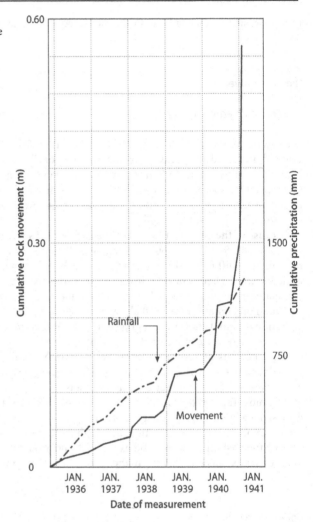

each period of least movement, in contrast to 8 cm for each winter period of greater movement, was less effective in causing a build up of water pressure beneath and behind the mass than the freezing of fractures over winter; these would prevent the drainage of winter snow that had thawed in the rock mass and normal discharge of groundwater from the mass.

Whatever the cause of failure it is important to note the acceleration of movement opening the backing crack up to the point of collapse. *Should a steep rock face, whether natural or excavated, be found to have a backing crack it is desirable to instrument the crack to measure regularly any vertical or horizontal movement of the face in front of the crack.* The instrumentation need not be complex but should the measurements reveal increasing rates of movement then the time is ripe to consider the safety of people and property below the slope.

Portuguese Bend

The active Portuguese Bend Landslide occupies about 1.5 km^2 on the south side of Palos Verdes Hills, south of Los Angeles, California. The head of the slide lies about 130 m above sea level, and the toe generally lies just seaward of the shore line in the shallow surf zone. Moderately indurated tuffaceous and sandy shales and silt stones of Miocene age dipping towards the coast within a seaward plunging shallow asymmetric syncline are interbedded with several bentonite beds, at least one of which is believed to act as an incompetent zone along which the landslide is creeping. Some more recent sediments have been discovered which are believed to have been deposited within lakes held in depressions formed by earlier landsliding.

Although the presently active landslide is a portion of an ancient landslide which is known to have been active several thousand years ago (Merriam 1960), it was inactive when a number of roads and homes were constructed on it, mostly in the early 1950s, in the area west of Crenshaw Boulevard and north and south of Palos Verdes Drive South (Fig. 9.27). The recent phase of movement is believed to have begun in

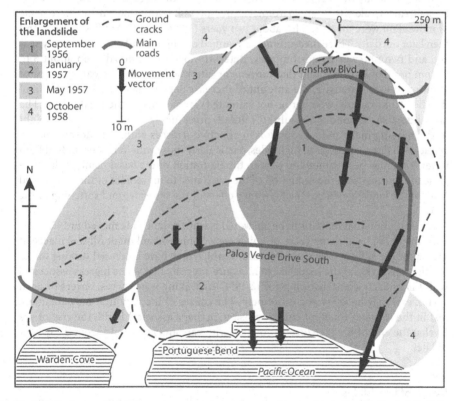

Fig. 9.27. Portuguese Bend. The movement vectors and cracks shown relate to the situation in July 1957 when the landslide was active. Only the major ground cracks are shown; there were many more smaller ones. The numbered and shaded areas show the spread of the landslide between September 1956 and October 1958 (adapted from Merriam 1960)

September 1956, and has continued at varying rates. From 1956 to 1973 parts of Palos Verdes Drive South were displaced seaward a maximum of approximately 40 m (Easton 1973). Between 1956 and 1958 the landslide spread so that in Fig. 9.27 area 1 represents its extent in September 1956, 2 its additional extent in January 1957, and 3 and 4 extensions in May 1957 and October 1958 respectively. Survey points were established in 1956 to monitor movement; some of the movement vectors recorded between September 1956 and December 1957 are shown in Fig. 9.27.

Boreholes sunk to investigate the landslide suggested that the lowest shear surfaces on which movement was taking place could be as deep as 90 m below surface. Attempts were made to stabilise the landslide by inserting precast concrete caissons, about 1¼ m diameter and 6 m long into holes drilled into rock below the presumed sliding surface. These were grouted into place, at locations near the coast, leaving about 3 m projecting upwards into the moving mass and were supposed to act as shear pins to restrain the moving mass. The movement was not stopped, although each of the caissons provided an estimated resistance of about 450 000 kgf and about 25 were installed. Some tilted, some may have sheared, but it is thought that the disrupted landslide mass may have flowed around them.

In a later study by Easton (1973), which examined the influence of such factors as rainfall, earthquakes, tides and barometric pressures on rates of movement, increments of movement were measured for about two years (1970 and 1971) at first weekly, and then later on, daily. These measurements spanned one major and one minor rainy season and two noteworthy earthquakes. Rates of movement ranged from about 5 to 20 mm per day. Easton found that movement rate was increased by earthquakes of significant local intensity, the amount of increase depending upon the condition of the slide (for example, whether or not saturated) at the time of the earthquake. The landslide also responded rapidly to rainfall, presumably due to increased weight of the moving mass and a reduction in effective stresses within it. Movement also accelerated during periods of high tides, particularly those with longer periods of high water. This may be attributed in part to the saturation of the basal plane of the slide close to the ocean and the reduction of the resistance to movement of the submerged front of the landslide due to the buoyancy effect of the sea water and softening of the clay.

One question that remains to be answered is why the landslide moved in 1956 after what may have been many years of quiescence. Some embankment fill was placed at the top of the slide at Crenshaw Boulevard, which may have increased driving forces but the sensitivity of the landslide to moisture suggests that water ingress associated with the housing construction prior to 1956 (by leaking water pipes, sewers or lawn sprinklers) may have revived movement. The causes of the very first movement are lost in the mists of time but coastal erosion, perhaps associated with the rise of sea level at the end of the Pleistocene, coupled with earthquakes, would seem likely candidates.

9.8.3
Extent of Landslides

The occurrence of a rotational landslide on a long uniform engineered slope raises not only the question of why it happened but also "*why in that particular spot?*" The answer to the second question lies in a variation in geological and hydrogeological conditions giving a weaker ground mass within a particular volume of slope. When the landslide has moved back scars and side scars usually present slopes which are steeper than the slope that has failed. If the changes in geological and hydrogeological conditions which occur at the scar locations are abrupt then the slide may stop at the scar. If not, then the steepness of the scar slopes may ensure continuing failures so that the landslide gradually spreads. Such spreading extension of landslides is relatively common in the larger more complex slides. In any case weathering will act on scar slopes and the slid mass and bring about further instability. Steward and Cripps (1983) have described the chemical weathering of the pyritic shale of the Edale Shale and indicated the part that this plays in phases of reactivation of the Mam Tor landslide in Derbyshire, England.

9.9
Remedial Works for Slope Instability

Stabilisation works may be undertaken on existing landslides or other forms of clearly unstable slope, such as a rock cliff giving rise to rock falls, or on slopes shown by calculation to have a factor of safety lower than that deemed satisfactory.

Existing landslides may be either active or dormant. In both cases the most important part of the investigation is to recognise that the landslide exists. Such landforms may be recognised by ground surveys but are generally much more easily seen on aerial photographs. Such photographs will not only show the form of the landslide but may detect changes in vegetation which are a consequence of old landslides. False colour photographs may indicate changes in moisture content which indicate the presence of the slipped mass. On the ground bends in tree trunks may show that movement occurred within the lifetime of the tree. *Recognition of old landslide terrain is not easy*, particularly on the ground and often requires that combination of intuition and knowledge which comes only after long years of experience.

Once the landslide has been recognised it is then necessary to determine its extent in depth as well as in area and, if possible, to understand why the landslide occurred in the first place. This is particularly important with old landslides because they may have originated as a consequence of *circumstances which no longer exist*. In the northern hemisphere many old landslides are believed to have occurred at the end of the Pleistocene period following the retreat of the ice, when many slopes were rapidly destressed and the climate was very different to that of today. Earthquakes, a consequence of isostatic re-adjustment after the Pleistocene, may have triggered some of the slides.

To understand the present stability of the slide *the various shear surfaces in the slide must be discovered*. These may be much deeper than expected and since many landslides are complex, containing slides within slides, the first shear surface encountered in a borehole is not necessarily the one controlling the present stability. Accurate mapping of the extent of the slide and the types of mass movement within it may lead to some assessment of the location of surfaces of movement.

Groundwater conditions must always be established. Since groundwater table levels may be seasonal long-term observations are necessary and thus piezometers must be installed. Rainfall records should include intensities and some idea of rate of infiltration and run-off should also be gathered. If the slide is moving then the rate of movement should be monitored. Inclinometers set into boreholes may indicate the shear surface on which the mass is moving.

The remedies for slope instability are well known and proven. However, there may well be as many examples of unsuccessful application of the remedies as of successful application. This is generally the consequence of a lack of understanding of the cause of the particular problem. *Thus while the choice of remedy might be correct, its application to the wrong depth or in the wrong place or at the wrong time will not lead to a satisfactory conclusion.* There must be adequate investigation to recognise the cause of the problem before the solution can be applied. This may take some time. If immediate action is required no engineer can go far wrong if he initiates the construction of slope drainage works.

9.9.1
Methods of Stabilisation

The methods of stabilisation that are most commonly applied fall within three groups. These are: (1) changing the geometry of the slope, (2) drainage, and (3) giving support by reinforcement or force; less commonly, protecting the slope against deterioration and changing the character of the material in the slope. Some of these techniques are illustrated in Fig. 9.28.

Fig. 9.28. Methods of improving stability of slopes

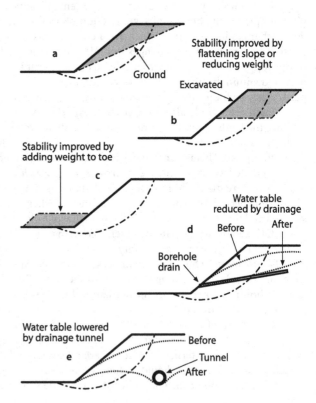

Changing the Geometry of the Slip

The stability of a slope depends upon the inclination of the slope in relation to the strength of materials, the strength and orientation of discontinuities and the groundwater conditions within the groundmass. The driving force for landsliding is the weight of the mass above any potential failure surface and it is thus clear that a reduction in the angle of the slope (Fig. 9.28a) or the removal of material from the top of the slope (Fig. 9.28b), will improve the factor of safety of the slope.

Although removal of material seems an easy way to improve stability it may pose considerable practical problems. In the case of slope flattening this should not be applied as a routine measure to improve stability without first determining the location of the failure surface on which movement may be taking place. In the case shown in Fig. 9.29 the slope is flatter and the total weight of material potentially in movement is less. However, the soil removed has reduced the forces resisting movement more than it has reduced the forces promoting it. It is better to remove from the top of the slope, although this should be done with caution for the weight of excavating machinery could cause failure. Hutchinson (1988) proposed a "*neutral line*" theory to help define the position of OO' and hence the zone from which weight should not be removed. Small landslides that could enable larger movements to develop upslope are often removed rather than stabilised and replaced by dense granular permeable rock fill.

Drainage

Slopes may be drained by the introduction of shallowly inclined drainage holes (Fig. 9.28d). The drainage holes are usually drilled into slopes by rotary drilling methods. The hole is lined with a slotted tube and often this tube is itself surrounded by a solid ceramic filter material if the borehole is in soil. Occasionally drainage may be undertaken by pumped wells but these are seldom used as a permanent means of slope stabilisation because of the operating costs. However, they may be used as a temporary remedy, modifying the pumping rate to find the correct reduction of the water table level to stop slope movement. In rock, driving a drainage tunnel (Fig. 9.28e) to

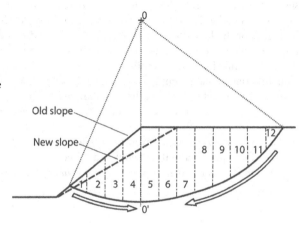

Fig. 9.29. Flattening a slope. In the case shown slices of ground to the left of *OO'* resist rotational sliding while those to the right promote it. Flattening the slope as shown will reduce the forces resisting movement more than it reduces the forces promoting it and make the slope less stable

achieve this required reduction on a permanent basis (Sherrell 1971) is possible. This is a particularly useful technique in large open mines where drills are readily available to drill wells and tunnels are easily constructed.

Support

In the case of potential or actual landslides support may be provided by placing weight at the foot of the slope to resist downhill movement (Fig. 9.28c). Such work was done to allow building a motorway on a landslide at Walton's Wood. The paper by Early and Skempton (1972) describes the work done in 1963, which may be considered a landmark in the development of geotechnical engineering.

At a location, known as Walton's Wood, Staffordshire the attempt to build an embankment for the M6 motorway, on what was later realized to be an old landslide, led to severe slipping. The landslide mass consisted of clay mixed with fragments of rock of all sizes derived by weathering, shearing, and perhaps solifluction, from mudstones, sandstones and thin coal seams of the upper Coal Measures (Fig. 9.30). Instability of the hillside had probably been initiated in Late Glacial times by the erosion of an ice-marginal drainage channel, and there is evidence that the slope had not yet reached a state of complete equilibrium when the engineering works began. The investigations of this failure took place in two stages. *In the first stage, the principal object was the design of remedial measures.* This involved: defining the limits and nature of the landslide by surveying, boring and trial pits, measuring the strength of the clay on these surfaces, and measuring the pore water pressures in the landslide mass

The landslide has a width of about 300 m and extends about 200 m up the hillside from its toe to the highest back-scarp. Numerous arcuate scarps existed (Fig. 9.30a), some being comparatively fresh and others subdued and weathered. The landslide mass, or 'colluvium', attains a maximum thickness of 10 m. It consists of clay, with numerous fragments of mudstone, sandstone and coal. The average inclination of the slope is 11°. In many cores from the borings, and in the trial pits, slip surfaces could clearly be seen. They were polished and striated, and the clay for several millimetres above or below typically had a light grey colour, in contrast to the adjacent yellow-brown clay. This colour change is probably the result of a secondary chemical development due to reducing water percolating along the slip surface. The piezometric observations demonstrate the existence of a perched water table (Fig. 9.30b) in the colluvium, at a higher level than the water table in the rock. This is due to the higher permeability of the landslide allowing rainfall to seep down within it above the almost impermeable underlying mudstone. It is the perched water table which controls stability of the hillside. From a study of the observed shears in the trial pits and in cores from the continuously sampled boreholes it was possible to reconstruct the most likely pattern of slip surfaces within the landslide mass. The stability analysis of these surfaces showed that values of φ' between 13.5° and 15° were required for limiting equilibrium.

This value of φ' was extraordinary low; according to the ideas current at that time, a value of φ' equal to about 20° would have been expected. *The second stage of the investigation had as its objectives a scientific study of the reasons for this discrepancy in*

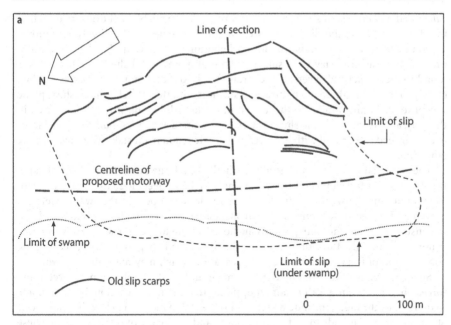

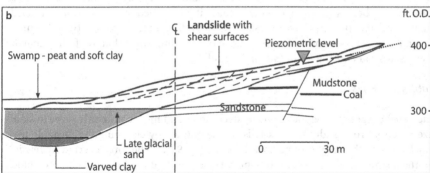

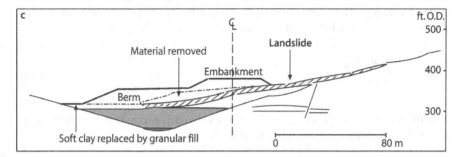

Fig. 9.30. The Walton's Wood landslide (from Early and Skempton 1972); (**a**) and (**b**) show the landslide in plan and section respectively and (**c**) shows the remedial works

calculated and measured strengths. A clue already was available in the observation that the clay particles on the slip surface were highly orientated in the direction of movement, in contrast to an almost random orientation elsewhere in the clay. Samples of previously unsheared clay were subjected to multiple reversal shear box tests. It was found that after several reversals the strength fell to a constant value corresponding closely to the strength on the natural slip surfaces, and that a slickensided shear plane was produced quite similar to those seen in the landslide. The strength measured in this way, and as found on the slip surfaces, was defined as the *residual strength*. Moreover, it agreed rather well with the strength as calculated from stability analysis of the landslide.

The remedial measures consisted principally of a large buttress or 'berm' of earth fill at the toe of the slope and trench drains 3 to 4.5 m deep at 20 m centres beneath the rebuilt embankment (Fig. 9.30c). The soft clays and peat in the swamp were removed and replaced by granular fill. An alternative to mass loading would have been to construct a retaining wall, perhaps anchored, at the foot of the slide. However, this is not so commonly done for a landslide that has already begun to move, because the construction of the wall involves some excavation which may make matters worse.

Support can also be given by reinforcement and by the application of a retaining force, when a retaining wall is not an option. Attempts have occasionally been made to drive piles through landslide material into the stable material below the shear zone at the bottom of the slide to add reinforcement and a measure of retaining force. These have seldom been successful, usually because it has proved to be very difficult to calculate the forces the piles have to resist and the sliding material has continued to move around and between them. The resistance of retaining walls that are failing may be increased by anchoring them to firm ground.

Cables and Bolts

Such anchorages may be rock bolts or steel cables and most commonly serve to stabilize potential rock slides or rock falls. These operate by clamping the unstable rock block back to the stable sound rock behind (Fig. 9.31). The force exerted by tension on the anchor can be oriented to oppose the forces that tend to push the rock block down the slope. Figure 9.31b shows a steel cable anchorage used to improve the stability of a rock block resting on a sloping discontinuity. Without the anchorage and assuming $c = 0$ (Fig. 9.1) the factor of safety (F_s) of the block is given by:

$$F_s = \frac{W \cos \beta \tan \phi}{W \sin \beta} \qquad (9.14)$$

If the cable anchorage is tensioned so that the block is pressed against the backing discontinuity with force F the factor of safety becomes:

$$F_s = \frac{\left[W \cos \beta + F \sin(\beta + \alpha) \right] \tan \phi}{W \sin \beta - F \cos(\beta + \alpha)} \qquad (9.15)$$

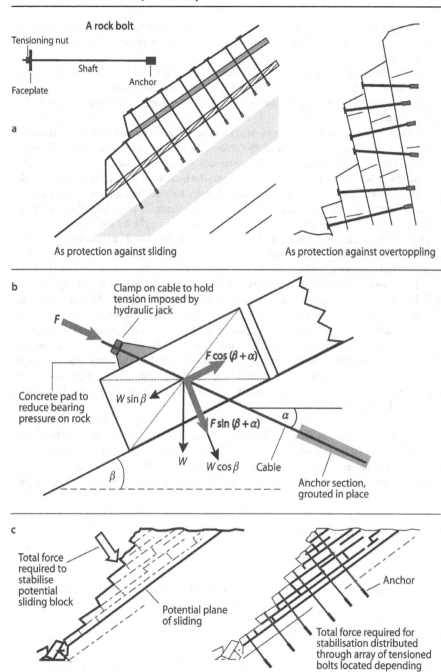

Fig. 9.31. Rock bolts (**a** and **c**) and cable anchorages (**b**) used to stabilise rock slopes

The angle at which the ground anchor is inclined to the possible sliding plane is critical; anchors must be inclined against the dip of the plane of sliding so that they give no down dip component of force.

Ground anchorages may be grout anchored steel cables or steel bars or mechanically or resin anchored rock bolts. Both are widely used for many engineering purposes but generally cable anchorages are used for deep long anchorages and bolts for shallow work.

For both types of anchorage successful installation and operation requires that, the rock is strong enough to withstand the stresses imposed under the surface bearing plate, the rock will not creep under these stresses and thus reduce anchorage load, the rock near surface will not weather away around the bearing plate, and the steel of cable or bolt shaft will not creep or corrode and thereby reduce the tension in the anchorage and thus the force applied. Figure 9.31a shows examples of potential slope instability and indicates how tensioned ground anchorages might be used to make some slopes safe. Both the long cable anchorages (10 to 40 m+) and rock bolts (3 to 6 m) must be protected against corrosion, especially if they are used as part of a permanent engineering structure.

Anchorages should not be randomly distributed on a slope but used to support key rock blocks. Thus, in Fig. 9.10, anchorages installed to support the blocks which are free to move (F) will automatically support the intervening blocks. It is quite easy to identify major rock blocks and to calculate the force necessary to support them to a given factor of safety. However, the major blocks will also be divided by discontinuities which might cause the block to spring apart if subjected to the force from a single massive anchorage. Accordingly the total load required to stabilise a large block may have to be applied by many smaller anchorages located so as to ensure no disruption of the total block (Fig. 9.31c).

It may be that problems of corrosion, creep or weathering are such that the application of tensioned anchorages seems undesirable. In such a case an alternative treatment would be to install untensioned rock pins which would resist potential rock movements in shear. They may be considered to increase the cohesion across any potential sliding plane. Thus the equation for the factor of safety in the dry condition (F_s) becomes:

$$F_s = \frac{cA + sn + W \cos \beta \sin \phi}{W \sin \beta} \qquad (9.16)$$

where s = the safe shear strength of a steel rock pin and n = number of pins.

9.9.2
Remedial Works for Rock Falls

Natural or constructed rock slopes which are, on the scale of the total slope, inherently stable may, through a process of natural decay or as a consequence of poor construction, give rise to rock falls. While any rock slope anywhere may pose such problems most of them that have required attention are, in the author's experience, road and railway cuttings, coastal cliffs overlooking or supporting structures e.g. lighthouses, hotels and cliffs underlying historical monuments, castles, cathedrals, earth mounds and the like. While the rock falls may not be particularly large, perhaps no more than a few tonnes weight, they pose a hazard to traffic and structures below them and to the stability of structures built near the edge of the cliffs. The remedial works that may be undertaken are briefly described in Table 9.2.

Stabilisation Works

The purpose of these is to prevent rock falls occurring by preserving and supporting the rock face. Preservation always involves cleaning down loose debris, plants, trees etc. the sealing of joints and cracks by grouting and surface coatings and the installation of drainage works, usually sub-horizontal boreholes inclined to the rock face thus acting as drains. It is essential, if surface coatings are applied, such as cement based *shotcrete* or *gunite*, that *drainholes* are installed to drain off any groundwater that might build up behind the surface coating.

The support works for unstable rock slopes involve the installation of tensioned rock bolts and untensioned reinforcing pins, underpinning of overhangs and the installation of buttresses. Both underpins and buttresses are usually anchored into place. The first task is to choose the location and type of works to be undertaken. Usually it follows the procedure given below:

1. A sketch topographic map of the slope is prepared using simple survey methods (Fig. 9.32). This may serve to delineate the limits of the slope on which work has to be done. A map of the slope area chosen is then prepared (using methods suggested in Chap. 3) showing contours to a vertical datum plane.
2. Geological features (major joints, faults etc.) are plotted on the map. Orientations of discontinuities are measured.
3. Potentially unstable blocks are recognised by the geology they present. Weights of blocks are calculated, using the surface outcrops and orientations of the joints bounding them, to determine their shape in three dimensions.
4. The amount of force or reinforcement required to raise the factor of safety of the slope to a given value is calculated.
5. The total force or reinforcement required is divided by that which can be applied by the chosen design of bolt or pin.
6. The number of bolts and pins needed for stabilisation is distributed within the block area with regard to the internal jointing in the block, working space limitations etc.(Fig. 9.31c).

In the design of the works careful consideration must be given to safety. Bolt, pin, drain and grout holes are usually drilled by rotary percussive methods using compressed air as a flushing medium. If any rock block is truly and dangerously unstable then drilling a borehole through it could, for the driller, become nothing more than an elaborate method of self destruction. If any large block should become dislodged on to the scaffolding it could well cause the scaffolding to collapse, endangering those working on it or below it. Accordingly the works should be planned to either allow dangerous blocks to be supported by cables and nets anchored to more stable areas or to fall through prepared gaps in the scaffolding. The scaffolding should also be

Table 9.2. Remedial works for rock falls

Stabilisation works (made to prevent falls occurring)		Defence works (prevent falls reaching sensitive areas)	
Preservation works	Support works	Close defence works	Forward defence works
These reduce the action of the processes that cause rock falls. 1. Cleaning down 2. Grouting – to reduce permeability, prevent plant root entry, bond blocks together 3. Drainage – to relieve water pressure 4. Surface coatings – to prevent water entering a rock face, reduce erosion and weathering	These increase the safety of the face by applying force or adding reinforcement. 1. Force – rock bolts (or cable anchorages) 2. Reinforcement – rock pins, and • Underpinning • Buttresses • Cables and chains (usually only temporary works)	Retain rock falls on the face or allow them to fall close to the face. Usually exposed steel and require constant cleaning out of caught debris and maintenance. 1. Heavy gauge nets (such as submarine or torpedo nets) 2. Fences at foot of face 3. Absorbent beds	Situated well away from the face giving falls and sited to protect particular areas. Location and design critical. 1. Rock traps and ditches 2. Fences – mostly flexible to absorb impact

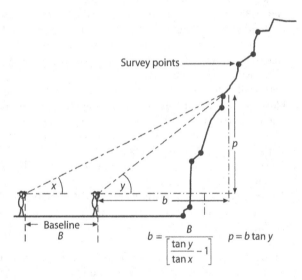

Fig. 9.32. A rough slope profile established by clinometer and tape. Baseline B is measured by tape

Survey points

$$b = \frac{B}{\left[\dfrac{\tan y}{\tan x} - 1\right]} \qquad p = b \tan y$$

anchored to safe parts of the rock face and, if the works are being undertaken above trafficked areas, fences or barriers should be erected to prevent any small rock falls reaching them.

Grouting should be undertaken in small stages, each stage, once grouted, thence being supported by bolts or pins before proceeding with a higher stage. In Fig. 9.33, which shows part of a rock slope, grouting is undertaken in stages from the scaffold-

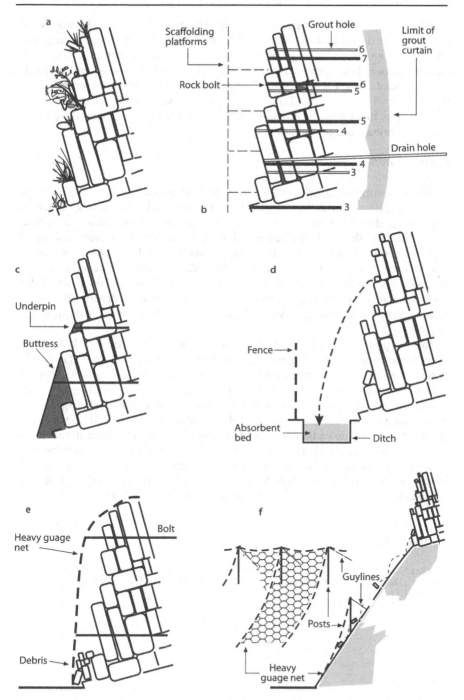

Fig. 9.33. Stabilisation and protection works; **a** part of a rock face before cleaning down; **b** bolting and grouting in stages, with drainage holes; **c** underpins and buttresses; **d,e** close defence works, which could be used in combination; **f** forward protection works

ing platforms so that stage 4 grouting is limited to that stage (Fig. 9.33b) and is undertaken after stage 4 bolting has secured the face below. Thence stage 5 bolting precedes stage 5 grouting. Grouting, usually cement grout, should not be undertaken with high pressures and any open surface cracks should be sealed before it begins. The grouting is intended to prevent the uncontrolled passage of rainfall or groundwater through the face and fill voids that could enable blocks to move and possibly fall or slide. However drainage holes must be provided, otherwise water flowing around the grouted area could bring damage to adjacent faces.

Protection Works

Stabilisation work is expensive and it may be preferable and cheaper to protect sensitive areas against damage from rock falls by the installation of ditches and fences (Fig. 9.33d). The fence is to catch any rocks that may bounce from either the bottom of the ditch, or other rocks already in it. Filling the ditch with an absorbent bed of sand helps reduce bounce. Nets may be draped over the face. In Fig. 9.33e the nets are held away from the face by spacer bolts so that falls are guided to be bottom of the face. Nets in close contact with the rock or fastened to the face are not desirable for while they will, for a time, stop debris reaching sensitive areas the debris will build up so that, as the net deteriorates, eventually the net and debris may come down in one major event.

Forward protection works seek to deflect or catch rocks falling from rock faces high above target zones from rolling or bouncing down the underlying slope (Fig. 9.29f). Fences or walls may prevent this movement if correctly sited. Correct siting depends upon an appreciation of the shape and frictional properties of the slope below the rock face (Hongey Chen et al. 1994). Such fences or walls may be severely damaged when struck and it may be better to utilise flexible net catch-traps (Fig. 9.33f) in which the weight of net slows down boulder velocity without imposing too much strain on the structure. *Whatever the character of the protection work, they require maintenance.* Apart from replacing any damaged or deteriorated nets, fences, posts etc. any rock collected in ditches, traps or nets must be cleaned out and removed. Some thought should be given to how this may be done at the time of construction of the works.

9.10
Further Reading

Bromhead EN (1992) The stability of slopes, 2nd edn. Blackie Academic, London
Simons N, Menzies B, Mathews M (2001) A short course in soil and rock slope engineering. Thomas Telford, London

Chapter 10

Withdrawal of Support by Underground Excavations

10.1
Introduction

Withdrawal of support by underground excavation is done each time an animal makes an underground hole for its food and shelter. Mankind encountered caves and could see that this removal of material was possible on a large scale, and eventually became able to extract minerals for tools and materials for construction, and to create voids for storage and space. The latter has become increasingly important due to the growing pressure on the use of surface space. Extraction of material from the subsurface implies that the existing equilibrium between stress and strain in situ, and its configuration in the ground, will be disturbed and redistributed due to the deformation of the ground into the space left by the extracted material. Support may be required to maintain the void created. Extraction of material may cause surface subsidence with implications for surface structures. In many textbooks, a distinction is made between underground structures in soil and rock masses. The author considers this differentiation artificial because the boundary between what is considered a soil or a rock mass is often not sharp; excavation and support methods originally developed for rock are later often found also to be suitable for soil masses and vice versa, and seldom does a project deal with rock or soil masses only. Hence, the engineering geologist should be able to deal with underground excavations both in soil and in rock masses, even if only for judging different location alternatives. This chapter describes the extraction of solid materials, by methods common to civil engineering (mainly tunnelling) and mining, and the influence geology can have upon the ground response associated with this withdrawal of support.

10.2
Stress

The stress configuration in the sub-surface is one of the most important parameters governing the withdrawal of material from the underground, the behaviour of an underground opening, and the resulting subsidence at ground level.

10.2.1
In Situ Stress Field

The in situ, or virgin stresses, are the stresses in the subsurface before any extraction of material. Stresses in the subsurface are caused by the weight of the overlying material (the *overburden*). The vertical stress (S_v) at depth H is:

$$S_v(H) = \int\limits_{h=0}^{H} W(h)\,dh \tag{10.1}$$

in which $W(h)$ are the unit weights of the ground above depth H. Deformation of the mass under the vertical stress causes horizontal stresses. If the material would be water, the vertical stress would equal the horizontal stresses (an hydraulic stress configuration) because water cannot sustain shear stresses. Other masses do sustain shear stresses and therefore the material will not completely transform the vertical stress into equal horizontal stresses, but only a certain fraction of it depending on the deformability of the mass. The fraction is depicted by the ratio between horizontal (S_h) and vertical (S_v) stresses, without any natural (e.g. geological) or man-made influences being considered and for an area of flat topography. This is known as earth pressure at rest, K_0. In an isotropic, homogeneous, linearly-elastic material the value depends on the Poisson's ratio (v) of the mass and $K_0 = v / (1 - v)$. For many rock masses, v is about 0.25 and, thus K_0 is about 1/3 and the horizontal stresses are both about equal to 1/3 of the vertical stress. K_0 is higher (normally between 0.4 and 0.5) for loose, non-cemented sand *and under certain geological circumstances can be greater than 1.0.*

Soil and rock masses deform with time so as to minimise the shear stresses within them. The rate of time dependent deformation is governed by factors such as the material, the amount of discontinuities, the temperature, and the magnitude of confining stresses. Some masses may deform over millions of years while others deform within a couple of days or years. This time delay is the reason for many in situ stress conditions being more representative of the geological *past* than of the present. In particular, the overburden stress in the past has been higher than at present because the surface of the earth has been eroded. The magnitudes of horizontal stresses are however still largely reflecting the high overburden stress of the past; hence K_0 can be large. A similar effect is known from areas that have been glaciated during the Pleistocene; the weight of the ice is still reflected in horizontal stresses that are higher than those calculated from the present overburden. Note this mainly applies to the horizontal stresses as the vertical stress reduces soon after unloading – a fact that can be observed during deep engineering excavations. Topography and tectonics can also cause the stress *orientation* and magnitude to be very different from what could be expected based on overburden thickness. Values of S_h near mountains, plate margins, and active faults can be up to 10 times the local vertical stress and in the direction of present day stress.

The parameter K is used for the ratio of average S_h over S_v, without concern to the origin of these stresses. Average K-values as a function of depth for various areas in the world are shown in Fig. 10.1; here average values of S_h are used, but it should be noted that considerable differences might also exist between the two horizontal stresses at the same location (i.e. the maximum and minimum horizontal stresses), particularly near tectonic active areas.

If no other information is available, some rules of the thumb for the in situ stress configuration to expect are available (Table 10.1); these should be used with care as the stress field may be completely different, as Fig. 10.1 shows. For a new project, it is

Fig. 10.1. Measured virgin stress field (data from Bieniaswki 1984, except Costa Rica, which is from Lopez and Pirris 1997)

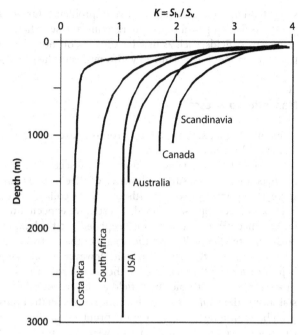

Table 10.1. Magnitudes of virgin stress field (rules of the thumb)

	Depth (m)	
Vertical stress	< 50	Stress due to weight overburden
	50 – 2 000	Overburden weight, but large scatter has been measured between 0.5x and 2x overburden weight
	> 2 000	Overburden weight
Horizontal stress	< 50	$K = 0.4$–0.6 for cohesive material, e.g. clay, 0–0.4 for discontinuous rock masses, and 0.25–0.45 for loose, non-cemented material
	50 – 500	$K = 0.4$–3.5
	500 – 1 500	$K = 0.5$–1.5
	1 500 – 2 500	$K = 0.7$–1
	> 2 500	$K = 1$

Note:
The direction of $S_{h(max)}$ is not given by this rule of thumb, only its possible magnitude.

Note: The above are the values as used by the author if no other information is available; the values should be handled with great care ($K = S_h/S_v$); S_h are the average horizontal stresses ($S_{h(max)} + S_{h(min)} / 2$).

advisable to investigate other underground workings (mines, tunnels, shafts.) that may be nearby. The miners, consultants, or contractors may provide measured data for the stress field or give personal accounts of problems during construction due to the virgin stress field. If possible go underground and see whether there is evidence of anisotropic stress fields, such as heavier support, or more fracturing in a particular direction, that cannot be explained by differences in the ground.

10.2.2
Man Induced Stresses

In many situations, the site for a proposed excavation is to be made in a stress field disturbed by other excavations.

Consider Fig. 10.2, which shows a vertcal section through a synclinal ore body that has been mined from surface downwards. The 'crown pillars' made to allow mining of stopes (large 'mining rooms') without causing collapse of the hanging wall, were left in place after the stope was mined. These were expected to break under the increased stress concentrations as the mine deepened and the hanging wall was supposed to collapse unto the footwall. Then the weight of the central core of the syncline would have been spread over the whole footwall. However, the crown pillars did not collapse, but kept transferring the weight from the central rock mass to the footwall. This resulted in a strongly anisotropic stress field at the crown pillars, created problems with the stability of the haulages (tunnels), and stresses on the bottom part of the syncline that prohibited mining. A solution of the problem was found in blasting the crown pillars causing the collapse of the hanging wall thus spreading the weight of the core over the footwall. A modified stress field can be caused by an existing excavation or surface structure; foundations of high-rise buildings may influence the magnitude and orientation of the stress field underground where tunnels for transport may later be excavated.

10.2.3
Stress Measurements

Stress measurements are the only reliable way to determine a stress field. Stress measurements are, however, very sensitive to local variations in the ground and many measurements may have to be done before the obtained values can be assumed reliable. Most standard methods for measuring stresses are based on the same principle: a volume of the ground is allowed to deform, expand due to unloading or contract due to loading and is then either re-loaded or un-loaded until the original volume is obtained; the loads required to do this give the original stress condition. Examples of stress measurements are: overcoring, flat jack and dilatometer tests, and induced fracturing.

10.3
Stress around an Underground Excavation

In general, the ground around a subsurface excavation supports itself by arching as the ground deforms in the direction of the opening. This occurs because as the ground moves towards the excavation the volume of space it collectively occupies decreases. This

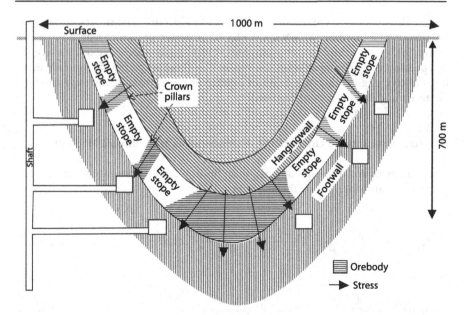

Fig. 10.2. Man-made induced stresses due to poor mine planning (a number of stopes has been left out for clarity of the figure)

Fig. 10.3. The roof does not support the overburden stress

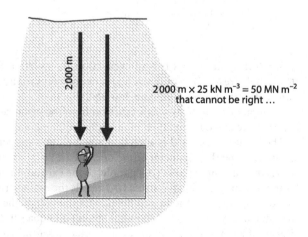

2 000 m × 25 kN m^{-3} = 50 MN m^{-2}
that cannot be right ...

causes the blocks of rock to lock against each to produce an arch. That arching exists can easily be shown by a simple calculation. Assume that a tunnel of 5 wide and 3 m high has to be made at a depth of 2 000 m, and the unit weight of the ground is 25 kN m^{-3} (Fig. 10.3). The vertical stress is then 2 000 m × 25 kN m^{-3} = 50 MN m^{-2}. The force on the roof per metre length of tunnel is: 5 m × 1 m × 50 MN m^{-2} = 250 MN. Hence, the support per metre length has to take up 250 MN. A concrete pillar system is used for support, with a compressive strength in the order of 20 MN m^{-2}. The surface of concrete required is thus 250 MN / 20 MN m^{-2} = 12.5 m^2. However, per metre length of tunnel there is only 5 m^2 available. Hence, support is just simply impossible if the overburden pressure had to be taken up by the support. Yet there are plenty of tunnels in mining and civil engineering at a depth of 2 000 m or more, without any support at all because the ground is supporting itself.

10.3.1
A Circular Opening in a Linear Elastic, Homogeneous and Isotropic Medium

The stresses around a circular opening in an linear elastic, homogeneous, and isotropic medium are described by the following:

$$\sigma_r = \left(\frac{S_h + S_v}{2}\right)\left(1 - \frac{r^2}{d^2}\right) + \left(\frac{S_h - S_v}{2}\right)\left(1 - \frac{4r^2}{d^2} + \frac{3r^4}{d^4}\right)\cos 2\theta \tag{10.2a}$$

$$\sigma_\theta = \left(\frac{S_h + S_v}{2}\right)\left(1 + \frac{r^2}{d^2}\right) - \left(\frac{S_h - S_v}{2}\right)\left(1 + \frac{3r^4}{d^4}\right)\cos 2\theta \tag{10.2b}$$

$$\tau_{r\theta} = \left(\frac{S_h - S_v}{2}\right)\left(1 + \frac{2r^2}{d^2} - \frac{3r^4}{d^4}\right)\sin 2\theta \tag{10.2c}$$

S_h = horizontal stress before excavation, S_v = vertical stress before excavation, σ_r = radial stress, $\tau_{r\theta}$ = shear stress, r = excavation radius, d = distance from excavation centre, θ = polar coordinate, horizontal = 0°.

Examples of the stresses as a function from the distance perpendicular to the wall and roof are given in Fig. 10.4.

The in situ stress field is a vertical stress S_v, and perpendicular to it a horizontal stress S_h. Along the perimeter of the opening ($d/r = 1$) the major principal stress (σ_1) equals the tangential stress (σ_θ) parallel to the excavation perimeter and perpendicular to it the minor principal stress (σ_3) equals the radial stress (σ_r). If the in situ vertical stress (S_v) equals the in situ horizontal stress (S_h), the tangential stress (σ_θ) has the value of 2 S_v at $d/r = 1$. The radial stress (σ_r) equals 0. The radial stress at the perimeter of an excavation that is not supported has always to be 0, as otherwise the perimeter would move until σ_r is 0. If the in situ horizontal stress is lower than the in situ vertical stress, the tangential stress in the roof may become negative, which here means tensile (see stresses for $\theta = 90$°). Figure 10.4 also shows the magnitude of the stress concentrations to be independent from the size of the excavation and further shows the influence of the presence of an opening to be limited to about 4 times its radius. The stress concentrations along the perimeter of an opening show a large increase (Fig. 10.5) for non-circular openings or angular openings.

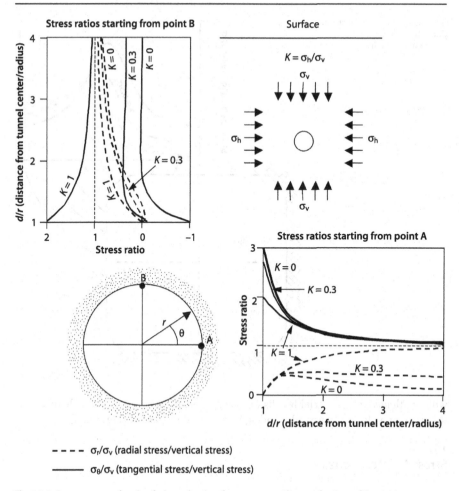

- - - - σ_r/σ_v (radial stress/vertical stress)

——— σ_θ/σ_v (tangential stress/vertical stress)

Fig. 10.4. Stresses around a circular opening in a homogeneous, linear-elastic, and isotropic mass

10.3.2
A Circular Opening in an Elasto-Plastic or Brittle, Homogeneous and Isotropic Medium

Unfortunately for the numerical modellers who employ relationships of the sort described in Eq. 10.2a–c, real intact rock *material* behaves only partially as a linear-elastic medium whereas soil and rock *masses* are not linear-elastic at all. Figures 10.6a and c show the stress-strain behaviour of ground after reaching the elastic limit: Fig. 10.6a for an elasto-plastic mass and Fig. 10.6c for an elasto-plastic mass with brittleness. Figures 10.6b and d show the corresponding stresses around underground openings. The mass at the perimeter may deform plastically (Fig. 10.6b) and could squeeze into the opening if the elastic limit is reached. If the elastic limit is reached for a material

Fig. 10.5. Stress concentration along the wall and roof of a rectangular opening

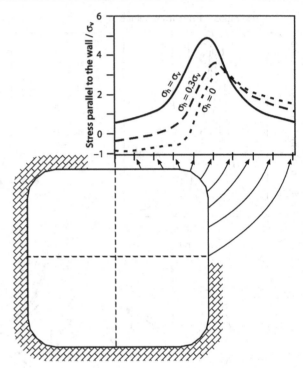

with brittleness, the material breaks (*spalling*), creating a '*broken zone*'. The mass in this zone may deform plastically and squeeze into the opening (Fig. 10.6d).

10.4
Stress Related Issues

The ground has many aspects that are not reflected in the analyses of stress outlined above and which cause it to behave in ways that are not easy to predict quantitatively.

10.4.1
Excavations in a Discontinuous Medium

In a discontinuous medium such as the ground, the deformation, strength and failure relationships are far more complicated than in a continuum (Hoek and Brown 1994). In a discontinuous mass, the ground may not need to fail anymore to create discontinuity planes because these are already present. Movements can take place fully or in part along these discontinuities. New discontinuities are created only when the stresses exceed the strength of the intact material. Squeezing or flowing of the material will occur either if the material is soft and cohesive (for example, clay) or if the block size of the material is small compared to the excavation (for example, shale or non-cemented sand). The relation between block size and stability causes the size of an opening to become important. The more discontinuities intersected by the excavation the

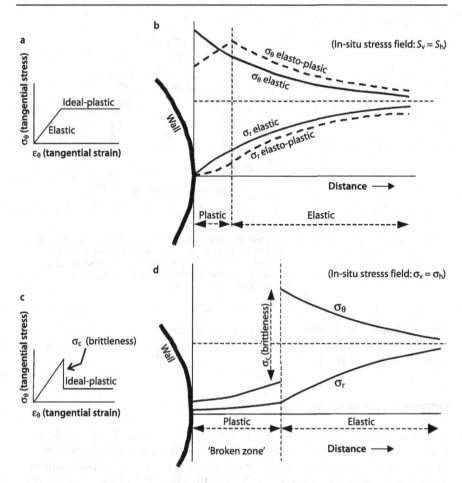

Fig. 10.6. Radial and tangential stresses around a circular opening in various media; **a, c** stress-strain curves for elasto-plastic and elasto-plastic with zero-brittleness material; **c, d** show the corresponding stresses around a circular opening (after Kastner 1971)

more chances there are of blocks moving or falling (Fig. 10.7). In addition to these the discontinuous ground is also likely to be anisotropic.

10.4.2
Swelling Materials

Some materials swell if exposed to water, even that in the atmosphere. Swelling material will deform into the excavation and change its shape and size. Swelling stresses can be extremely high and it is often impossible to counteract them. Therefore, such material should be sealed from either free water or the atmosphere, for example, by *gunite* or *shotcrete* (see below), immediately after excavation. Swelling conditions can be expected in materials containing swelling clay minerals, in particular, montmoril-

Fig. 10.7. Size matters!

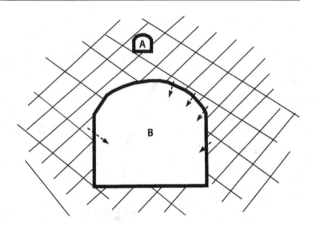

In A few discontinuities are intersected by the tunnel wall and no options for movement along existing discontinuities exist; in B many discontinuities are intersected and consequently many options for movement, and rock fall or squeezing of blocks exist.

lonite, and in shales, claystones, mudstones, weathered pyroclastic deposits, and fault gouge. Anhydrite will swell because when exposed to water it slowly converts to gypsum, which has a larger molar volume.

10.4.3
Dynamic Stresses – Earthquakes

Dynamic stresses may be superimposed on the virgin stress field by earthquakes. It is generally recognized that underground structures are less sensitive to earthquake motions than surface structures. One of the reasons for this is that the resonance frequency of most underground structures is well above the range of frequencies common in earthquake waves. During the passage of an earthquake wave the underground excavation deforms under a combination of axial compression and extension. Tension cracks may develop and normal stress on discontinuities may be reduced during tension phases of an earthquake wave, allowing movement of blocks in the surrounding ground, with loss of integrity of the mass and its consequent collapse. Support systems such as bolts or wire mesh reinforced shotcrete give reinforcement to discontinuous ground and support, such as concrete segments, can be connected by nuts and bolts so that the integrity of the support they provide is not lost. Much here can be learnt from case histories.

10.4.4
Failure Modes and Need for Support

As the forgoing sections show, an underground excavation can fail in different ways:

- *Spalling:* the stresses at the perimeter of the excavation are such that the intact material breaks (rock or firm cohesive or cemented soil); spalling of intact material can be expected if σ_θ is more than about half the unconfined compressive strength of the intact material.
- *Gravity failure:* material falls into the opening due to gravity, this may be either rock or soil.
- *Squeezing failure:* the surrounding material deforms so far into the opening that the opening is no longer serviceable; clay, shale, or masses with many discontinuities such as schist are susceptible to this.
- *Swelling failure:* when materials swell on exposure.

Combinations of failure modes are normal; for example, a unit is squeezed in the opening and this allows de-stressing of other units and discontinuities, which permits gravity failure.

Support is installed in an underground excavation to keep the ground next to the excavation walls, roof and floor, in place. Support, except for shallow underground excavations where horizontal stresses are usually manageable, never counteracts the virgin stresses present because arching effects will deflect most of the stresses around the excavation. In a strong rock or cemented soil mass (e.g. strong in relation to the virgin stress field) the radial stresses at the perimeter of the excavation are nil and the tangential stresses are taken up by the material; the excavation is 'self-supporting'; failure is not reached. In weaker masses, the magnitude of the tangential stresses cause the material in the perimeter of the void to fail and material will move into the excavation causing it to deform. The material behind the collapse forms the new perimeter of the excavation and may subsequently also fail. This is *progressive failure* of the excavation and is prevented by providing support. In shallow underground excavations with either no or small horizontal stress, arching effects may not be sufficiently developed to enable an arch to be self supporting, therefore a shallow underground excavation may need support which actually carries the ground above the tunnel.

10.4.5
Stresses around Portals

The major and minor principal stresses along a tunnel are normally oriented tangential (around) and radial (perpendicular) to its perimeter. The intermediate principal stress is parallel to the axes of the tunnel, much influenced by the virgin stress field and very important for the normal stress on discontinuities perpendicular to the tunnel axes. Near a portal of a tunnel the stress parallel to the tunnel axes will be low and possibly zero. Hence, near the portals there is little to no normal stress on discontinuity planes oriented perpendicular, or there about, to the tunnel axes. In addition to these it is likely that the quality of the ground near portals and the strength of discontinuities there will be lower due to weathering. *The combination of low stress and weathered ground often causes major problems with excavating the tunnel portals.* Many tunnels in the world have become more expensive than expected due to portal problems (Fig. 10.8).

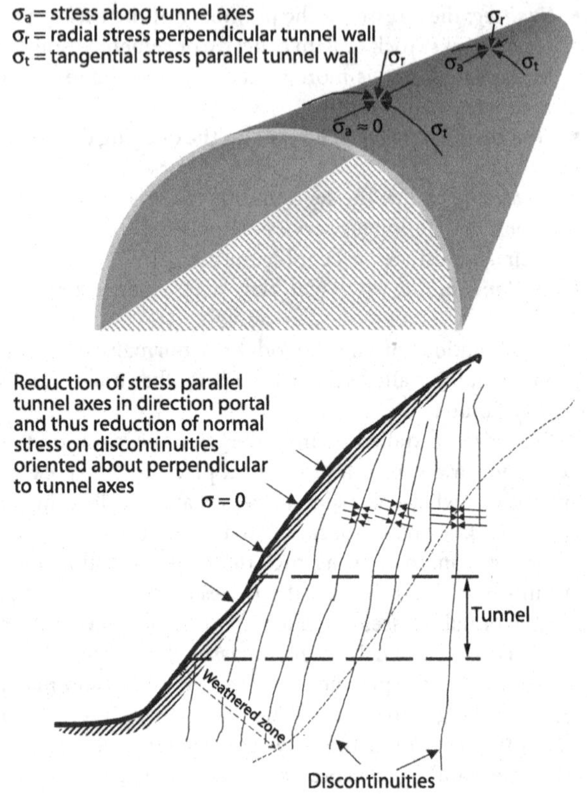

Fig. 10.8. Stress parallel to tunnel axes becomes null at portal

σ_a = stress along tunnel axes
σ_r = radial stress perpendicular tunnel wall
σ_t = tangential stress parallel tunnel wall

Reduction of stress parallel tunnel axes in direction portal and thus reduction of normal stress on discontinuities oriented about perpendicular to tunnel axes

$\sigma = 0$

Tunnel

Weathered zone

Discontinuities

10.4.6
Stand-Up Time and Time Effects

All ground in which the stress conditions change needs time to adjust to the new situation by deformation. When this deformation causes either expansion or shear, the ground becomes weaker as new fractures, asperities on discontinuities shear, discontinuities that were mated are now displaced and cement within the ground is broken. Therefore, the sooner support is installed after excavation, the stronger the ground remains. The time between excavation and failure, for underground excavations, is called the '*stand-up time*'. Stand-up times range from negligibly short (minutes) to thousands of years. Stand-up time allows an excavation to be supported before failure mechanisms develop to the point where collapse commences. Most of the mechanisms governing the time-dependent movements are only partially known and available mathematical expressions are not reliable. One way to assess stand-up time is by *experience*, which may be sustained by empirical classification systems (Chap. 4).

Time-effects cause many serious accidents underground. Many engineers, even those working underground, do not appreciate the nature of time dependent deformation. Often an excavation is considered safe because it has been standing for "*some time*". This can be a fatal error underground. It requires experience to recognise an imminent collapse; everybody should be aware that the time an underground excavation has been standing is of little value to assess its safety. *Most collapses are preceded by warnings* such as small movements in the surrounding ground, sounds from the mass, or small pieces of material falling from roof or walls. These should be recognized as signals that a potential danger might be at hand.

10.5
Excavation Issues

Water and ease of excavation are common issues for most underground excavations of any great size as considerable costs can be incurred if predictions relating to either of these are wrong.

10.5.1
Water and Underground Excavations

Water flowing into the excavation can be not only a nuisance but also a hazard. Water pressures reduce effective stress and subsequently the shear strength of the ground. Water pressures and flow may also cause material to be flushed into the excavation causing *overbreak* and instability (Fig. 10.9). Flowing water can prevent the application of shotcrete from good contact with the ground. Therefore, water flowing into an excavation in large quantities has normally to be prevented. Draining is feasible if permeability of the ground is not too high or if sealing (impermeable) layers are present that limit the quantity of water to be drained. If the rock mass is highly permeable drainage will lower the groundwater table above an excavation. This is normally unacceptable for social, environmental, and geotechnical reasons. Under these circumstances the excavation has to be sealed locally with, for example, by grouting. Even if the water inflow is no geotechnical problem, it is not regarded as good engineering to have large water inrushes into an excavation.

It is not easy to estimate water inflow into an underground excavation before the excavation has been made. The theoretical background can be found in many books, but to determine the nature of the groundwater regime and the permeability of the strata is very difficult. In a homogeneous mass without discontinuities, the water will flow via the pores between the grains. For this, reasonable estimates can be made based on theoretical calculations. Most masses are, however, not homogeneous and not with-

Fig. 10.9. Water flow and pressures from localized high permeability zones may break through to a tunnel before the tunnel reaches the zone

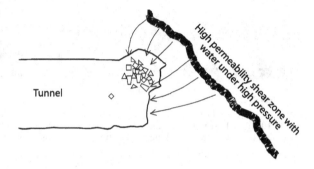

out discontinuities. Reliable estimations of the quantity of water can then only be made from boreholes to locate these zones, packer tests in these zones and from pilot tunnels. Portals of tunnels may in particular be at risk because these are near to the surface and the mass near portals is often more permeable because of a higher number of discontinuities. Rainfall may then directly flow into the portal area. This problem can be reduced by building the portal in a dry season if such a season exists.

10.5.2
Excavation

Excavations can be made in many different ways, from digging by hand to using highly mechanised tunnel boring machines (TBMs). Excavation can be made by any type of method whatever the ground; prisoners made escape tunnels in rocks using their bare hands and spoons, however in commercial practice the excavation method chosen has to be suited to the ground, available skilled labour, and the constraints on the project from time and economics. Generally, two types of excavation method can be distinguished: mechanical and blasting (Fig. 10.10); both methods can be further divided (Table 10.2). Blasting techniques were used extensively in the past, but nowadays mechanical methods are more popular, having various advantages over blasting methods especially for smaller works and for reducing vibrations at ground level. 'Specials' noted in Table 10.2 include methods that make use of the expansion characteristics of wood or chemicals, the force of water under high pressure, used for jetting to erode the rock mass, or use of sawing techniques. They are seldom used in underground excavations.

The method of excavation has a considerable influence on the quality of the perimeter of an excavation especially if high levels of stress exist. Table 10.3 gives values

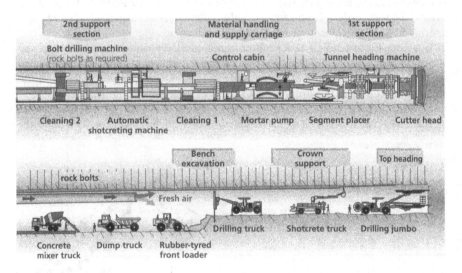

Fig. 10.10. Tunnel excavation and support organisation for the St. Gotthard Tunnel; top mechanical excavation by TBM, bottom: drilling and blasting (© AlpTransit Gotthard AG, after AlpTransit 2002)

Table 10.2. Various means of excavation

Mechanical	Digging	Man-made/shovel/excavator
	Cutting and grinding	Borehole
		Road header
		Trench cutter
		Tunnel boring machine (TBM)
		Raise borer
	Hammering	Jack hammer
		Hydraulic/pneumatic hammer
Blasting		Pre-splitting
		Smooth wall
		Conventional tunnel blasting
		Conventional large hole blasting
Specials		Wood
		Chemical expansion
		Water (high pressure breaking or jetting)
		Sawing (blade or steel cable)

Table 10.3. Excavation damage factors for a rock mass (the factors are multiplied with the classification results of *SSPC* or *MRMR* to account for the method of excavation)

Method of excavation		*SSPC* factor	*MRMR* factor
Natural/hand-made		1.00	
Boring			1.00
Pneumatic hammer excavation[a]		0.76	
Pre-splitting/smooth wall blasting		0.99	0.97
Conventional blasting with result	Good	0.77	0.94
	Open discontinuities	0.75	0.80
	Dislodged blocks	0.72	0.80
	Fractured intact rock	0.67	0.80
	Crushed intact rock	0.62	0.80

SSPC factors after Hack (1998), *MRMR* after Laubscher (1990).
[a] This value is based on hammer sizes up to 5 m with a diameter of 0.2 m.

for the damaging influence of methods of excavation on a rock mass. Natural, hand-made, and bored excavations show fewer new mechanical discontinuities than excavations made by blasting in the same rock mass, as they have not been subject to the transient loading blasting imparts to the ground. Large excavations may not be stable long enough for support to be installed; in these cases it is often necessary to complete the excavation in sequence starting with a small part of the excavation where loads are least, at the top, or crown, of the tunnel (Fig. 10.10 bottom).

10.6
Machine Methods of Excavation

Most civil and mining excavations underground are created by mechanical cutting and grinding. Such methods use a mechanical device onto which drag, disc, roller, or tri-cone bits are mounted that cut and grind into the ground (Figs. 10.11 to 10.13). These cutting tools may be made from steel, hardened steel, tungsten, or diamond and their damage and wear are kept to a minimum by using low forces for excavation. Excavation is thus a relatively slow cutting and grinding of the ground that breaks small pieces from the mass. The diameters of excavations are typically circular holes up to about 8 m for raise bores, and 15 m+ for tunnel boring machines. Cross-sections of shape other than circular can be formed using a *road header*, which consists of a rotating cutting wheel mounted on a modified excavator (Fig. 10.11); this can excavate any size and form of excavation in all but the strongest of rock masses.

As all mechanical cutting and grinding methods apply low energy, and consequently low stress levels, at any time, they generate little damage to the rock mass beyond the perimeter of the excavation. Good control over the excavation is possible, resulting in minimum overbreak and a smooth perimeter.

Mechanical cutting and grinding methods will excavate any ground, soil and rock, so the question is whether the method is economical. If a rock mass consists of larger blocks and the intact rock strength is high it will take a long time before a mechanical cutting and grinding device has excavated the space. Secondly, the wear on cutters and

Fig. 10.11. Road header with spherical cutters

machinery may be high and consequently the costs. Until some years ago mechanical cutting and grinding excavation techniques for large excavations were restricted to soil and moderately strong rock masses with intact rock strength not exceeding 50 MPa. Techniques are advancing rapidly and mechanical cutting and grinding methods are becoming economically viable in rock masses with higher intact rock strengths and larger block sizes. They are often used for other reasons, for example, if blasting is not allowed for fire risk or vibrations.

10.6.1
Tunnel Boring Machines

Tunnel Boring Machines (TBMs) are mechanized, more or less automatic machines that excavate the tunnel, provide support to the ground temporarily during excavation if needed, and install permanent support where necessary (Fig. 10.10 top). The excavated material is transported out of the tunnel by pipes, if a watery slurry, conveyor belt, or trains. TBMs can be thought of as long stiff machines; they have difficulties going round corners so they experience sever limits with vertical and horizontal gradients. This means they cannot go around an unexpected geological problem if encountered; they have to go through it; *good geological predictions are required for TBM design and use.* That is why TBMs are often specially designed for one project in a particular type of ground. The various parts that may be included in a TBM, are: (1) ripper or excavator, (2) cutting wheel, (3) face plates, shield support or pressure shield, (4) grouting facilities, (5) steering assembly between cutting wheel and shield, (6) permanent support installation facilities, and (7) foam and grease injection installations.

   In a well organized project, just one person may be able to handle and oversee the full operation. The operator can be located in the TBM, however this is not required. The author has visited a site in Japan where the entire TBM operation for an 8 m diameter concrete segment supported tunnel was run by remote control from a nice sunny surface office by a tiny Japanese lady. Obviously, a better working environment makes it easier to find skilled employees and keeps the salaries low, hence, improving economic viability. Disadvantages of TBMs are the relatively (very) high initial costs and the very costly consequences if the wrong type of TBM has been chosen. *Many examples exist of TBM projects heavily delayed and going over budget because the subsurface conditions were different from those initially expected;* some TBMs have been left in the ground and the already built tunnel *abandoned.* TBMs have become increasingly versatile so that one TBM can cope with different subsurface conditions.

Dimensions

A TBM may have a length up to 50 m or more. The start of a tunnel to be made with a TBM requires a considerable working space; not only for assembling the TBM but also for its reaction block, against which it will jack itself ("*launch*") into the ground. Generally, a space is required double the length and height of the heading machine (Fig. 10.10). TBMs can be dismantled into fairly small parts for transport, except for the main bearing of the cutting wheel which has a maximum dimension of about 2/3 of the diameter of the TBM. Access roads or shafts should thus allow this size to pass.

Stress around TBM

An opening excavated in the ground will deform with the largest deformation in the direction of the maximum stress of the virgin stress field. For most excavation methods this is no problem as long as the deformation does not reduce the minimum required excavation dimensions. For a TBM this is different. Even relatively small deformations (in the order of centimetres) may cause a TBM to become stuck, or parts of the TBM to deform. In particular, deformation of the shield may be a major problem as the sealing of the shield against the ground is reduced.

Balancing Pressures

To resist ground pressures an *earth pressure balance* shield (EPB) can be employed if it is thought that the face will not be stable and sag, so permitting too much material into the TBM and facilitating subsidence at ground level. A nearly closed cutting wheel may prevent this. If this is not enough, for example, if the mass easily squeezes or if water pressures are high, it is necessary to use a *closed shield* (using a *bulkhead* installed vertically behind the cutting wheel, if present), in which only a relatively small opening allows the ground to pass into the body of the TBM. The space between the bulkhead and the face will then be under pressure, as it supports the face, and counteracts water pressures. The shield may be a pressure shield, bentonite shield, hydro shield (Fig. 10.12), or earth pressure balance shield (EPB) depending on how the pressure is maintained.

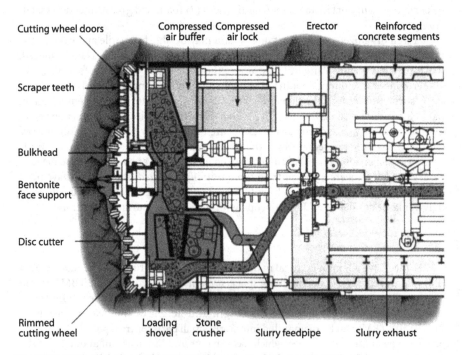

Fig. 10.12. TBM with hydro shield (©Voest-Alpine Bergtechnik, Austria 2000)

Cutting Wheels and Heads

Cutting wheels and heads (not shaped like wheels) are required to carry the cutting tools that will excavate the ground. These are adjusted to the type of ground to be excavated. In soft materials, such as clay, the cutter is more a type of knife or tooth, cutting slices of the ground. In harder ground such as rock, the cutter will be a drag-tooth or disc bit (Fig. 10.13). Cutting wheels may act as face support if mainly closed. Water jets may be mounted on the cutters to help excavation or to ease transport of broken material.

Jacking Systems

A TBM propels itself through the ground by the thrust from a jacking system mounted between the TBM shield and the tunnel lining that has been assembled behind it (Fig. 10.12). For short tunnels e.g. under railway embankments, the jacks can be positioned between a tunnel shield and a concrete installation outside the tunnel often located in a "thrust pit". The jacks push the support lining with the shield into the ground ('jacked tunnel'). Friction between the ground and support lining must be kept to a minimum. The backfill of a concrete segment lining can be grouted. Grease or foam may be used to reduce friction between shield and ground. Different foams can also be used to make the excavated ground easier to handle, or to increase the stability of the face when mini-TBMs are used, e.g. for jacking pipes.

Percussive Methods

Jackhammers and hydraulic or pneumatic hammers excavate by percussion. The devices are normally mounted on machines. The length of the hammers can be up to

Fig. 10.13. TBM cutting wheel with disk cutters; diameter 8.89 m; St. Gotthard Tunnel (photo: © AlpTransit Gotthard AG, AlpTransit 2002)

5 m with a diameter of about 25 cm. The force on the hammer may be substantial; large hammers may cause considerable damage to the rock mass. Note that in Table 10.3 the adjustment factor for pneumatic hammering is in the same order as good conventional blasting.

10.7
Blasting

In its most simple form, blasting consists of drilling of a series of holes that are filled with explosive that is then detonated. The holes are drilled and blasted according to a blast pattern that describes the location of the holes, the quantity and type of explosives, and the order and delay times between the holes, as they are normally not all blasted at the same time but with small time intervals (10 to hundreds of milliseconds). The first holes to be blasted are normally smaller holes in a triangular or pyramidal pattern (Fig. 10.14) in the middle of the excavation (the *wedge or burn-cut*) to create space for the rock to move to when blown from the following rounds of larger production holes. After the blast the broken pieces of the rock mass are removed ("*mucking*") and the process can start again. Diameters of holes are in order of 2.5 to 10 cm and length in the order of 0.5 to 7 m, depending on the quality of the rock mass and type of explosives used. Blasting in karstic rock masses requires special measures. The quantity of explosives should be adjusted to the fact that the rock mass consists partly out of open space or soft karst-hole filling material, and the blasting holes should be lined as otherwise the karst holes will be filled with explosives too leading to excessive over blasting. Conglomerates should also be blasted with care as the matrix to pebbles and boulders may disintegrate on blasting so enabling the blast to hurl the equivalent of cannon balls across any open space nearby.

10.7.1
Smooth Wall Blasting and Pre-Splitting

In smooth wall blasting the number of holes along the perimeter of the excavation is increased, and the diameter and quantity of explosives per hole decreased (Fig. 10.14). These holes will be blasted simultaneously as the last round. If properly executed the

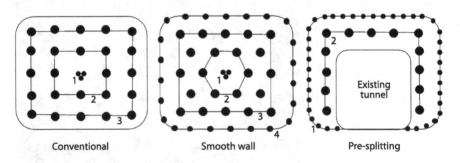

Conventional   Smooth wall   Pre-splitting

Fig. 10.14. Simplified blasting patterns (the numbers indicate the order of the rounds of blasting)

excavation will be relatively smooth and half of each hole will remain visible. Over-break will be low.

In pre-splitting the number of holes along the perimeter may be further in-creased and the diameter decreased (Fig. 10.14). The pre-split holes are simulta-neously blasted *before* the main holes and a '*split*' along the perimeter is created from one pre-split hole to the next. The split will act as an open discontinuity that reflects shock waves from the main blast. Hence, the shock waves of the main holes do not enter the mass outside the perimeter and the reflected energy helps break the middle part of the excavation. If properly done half of each pre-split hole will remain visible and the damage to the mass beyond the perimeter will be minimal. Pre-splitting for full-face tunnel blasting is not very useful as stresses in the face tighten the rock and can prevent the good development of a split. However, for sec-ondary blasting such as widening an existing tunnel, pre-splitting may give good results.

10.7.2
Conventional Large-Hole Blasting

Large-hole blasting makes use of blast holes with diameters of about 15 to 25 cm. The quantity of explosives in each hole is large and the rock mass will normally be shat-tered. It is used for underground or surface mining were damage to the surrounding rock mass is of less concern. It is used in civil engineering when large quantities of a rock have to be removed and vibrations are not an issue for concern.

10.7.3
Advantages and Disadvantages of Blasting

Conventional tunnel blasting is characterised by damage to the rock mass outside the designed perimeter of the excavation. Generally, the control on the excavation size is poor and the resulting roof, walls, and floor will not be smooth. On detonation new discontinuities are formed and existing ones are widened; gasses discharged by the explosion increase the loosening of the mass. The roughness of discontinuities may be affected as asperities are sheared or crushed as rock blocks are displaced, giving less rough and non-fitting discontinuity planes that result in lower shear strength. The effects from blasting are reduced if special techniques are used like pre-splitting or smooth wall blasting. These methods, however, can increase the costs considerably. In many countries, extensive and expensive security measures are also required to pro-tect the explosives from theft by villains.

So what are the advantages of blasting? Simply, it is an excavation method that can be used in hard rock when other techniques are either impossible or uneconomic. Conducted properly, blasting produces good results with minimal damage to the ground. The quality of the blasting engineer is vital. Contracts for blasting should be carefully drafted; often the mere fact that the contractor is paid for the quantity of explosives or the quantity of blasted rock, lures the contractor into using maximums of explosives to do the job as fast as possible with maximum profit, but without much concern for the results.

10.8
Ground Improvement and Support

Sometimes the ground needs some help to carry itself and it is strengthened by techniques of *ground improvement*. Ground improvement can be thought of as providing "*internal*" support to a volume of ground whilst "*support*" is something applied externally to that volume.

Thinking on support has changed; in the past support was installed to carry the ground but nowadays it is realized this is not always necessary, because the mass should carry itself as far as possible. Support is only needed to help the mass carry itself. Often a thin layer of shotcrete acting as glue to keep small pieces of rock in place is sufficient for arching (see Sect. 10.3) to occur whereas in the past heavy steel sets with lots of wood timbering would have been installed to obtain the same result. Movement allowed in ground that cannot support itself will normally result in a weaker mass. Therefore, the faster support is installed the less strength is lost. It should however be realized that some relaxation of the ground is always necessary to obtain the arching effects that can be generated without failure and which enable the ground to carry itself. Generally, the time required to install support is time enough to obtain the arching effects. In some situations, it can be necessary to delay support installation or allow additional deformation by other means (as making special de-stressing slots) because it is simply impossible to withstand the stresses if not done. This normally happens if large stresses and deformation occur, as happens often in underground mining but can also be encountered in deep tunnels through mountains in which there is poor ground. Depending on the time available in relation to the excavation process, it is normal to divide the types of support into: (1) pre-excavation support, including drainage, (2) support during excavation, and (3) permanent support (Table 10.4). *The design of ground treatment and support requires the very best ground models that can be generated.* The engineering geologist needs to appreciate what is expected from the ground by these techniques and for this reason the main techniques are briefly described.

10.8.1
Ground Improvement

Freezing

Ground is frozen by pumping a cooling agent through a double-tube pipe system in boreholes either from surface, a small diameter service tunnel, or from so-called feeder holes drilled from the tunnel (Fig. 10.15); the cooling agent is pumped in the borehole through the inner tube and returns through the outer tube. Water in the ground is frozen and binds the particles and blocks together, making the mass impermeable. It is absolutely essential that the ground contains water. Without water the mass cools, but there is no gluing effect because ice cannot be formed. Even if only some minor layers do not contain water, the method is questionable. The unfrozen layers will deform due to the expanding surrounding frozen mass and subsequently de-stress when the surrounding layers are thawed. This may lead to stress relief throughout the whole perimeter of the excavation and subsequent instability. The expansion of the ground

Table 10.4. Support types

		Pre-excavation	During excavation	Permanent
Freezing		X	X	
Grouting		X	X	X
Drainage		X	X	X[a]
Forepoling	Jet grouting			
	Spiling	X	X	X
	Pipe roof			
Shield (with or without face plates, face shield, pressure, hydro, bentonite, or EPB)			X[b]	
Caisson			X	
Shotcrete (with or without wire mesh)	On face		X	
	On walls, roof and floor		X	X
Rock bolt, dowel, anchor	In face		X	
	Radial in wall, roof, or floor		X	X
Rock bolts/anchors with straps			X[c]	X
Steel				X
Concrete				X
Cut-and-cover		Not a support type but methodology of working		
Timber			X[d]	X[d]

[a] In some situations it is feasible to maintain drainage for the lifetime of the tunnel (e.g. mining).
[b] See TBM, Sect. 10.5.2.
[c] Generally very time consuming and, hence, mostly not effective during excavation.
[d] Questionable as support means.

Fig. 10.15. Freezing or grouting from surface, service tunnel, or feeder holes (note that the freezing and grouting holes, if cored, will also give information about ground conditions ahead of the tunnel face)

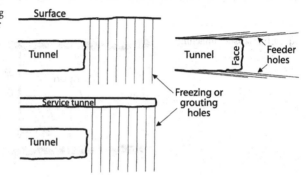

may displace other nearby structures or foundations of surface structures, and surface heave may be a problem if applied for shallow excavations. The method is expensive.

Grouting

Grouting pumps a cement milk or resin into the ground (the ground is "*grouted*"). This grout can be designed to provide mechanical strength to the ground (by bonding particles together or filling their voids with a strong material) or simply reduce its permeability by filling its voids. Often grouting is undertaken to achieve both these ends. The grout is pumped into the ground from ground level, or a small diameter service tunnel, or from feeder holes (Fig. 10.15). A large advantage of the method is that the support is also present as permanent support after excavation. Knowledge of the permeability of the ground and groundwater flow is essential. If the ground has too low a permeability, grout will not reach all locations and some areas will not be treated. Generally, the method works very well to improve the strength and deformation characteristics of the ground. Whether it will also make the ground impermeable is questionable. In many cases the milk or resin will not be homogeneously spread through the ground and the ground will not be completely impermeable after one round of grouting. Additional grouting rounds are then necessary to stop leakages. Special care should be taken that the grouting is not done with too high a pressure as this will cause the ground to break and permeability after the grout process may have increased rather than decreased. *It is important to supervise this work closely.*

Grouted feeder holes (Fig. 10.15) can also be used for inserting bolts, or pipes of steel or fibre glass into weak ground ahead of the tunnel as *pre-excavation support* called *Forepoling*.

Caisson (or Compressed Air) Tunnelling

This is used to support water saturated weak ground and to reduce water inflows in wet tunnels. A bulkhead is installed separating the face of the tunnel from the remainder of the tunnel and the air within it pressurized to counteract the hydraulic gradient that is driving water to the tunnel (Fig. 10.16). Labour and machinery necessary for the excavation pass through an air lock in the bulkhead and work under compressed air. The method works best in cohesive or cemented ground where pore and fissure sizes are small (sand and silt size). Large voids and open fissures will allow the air to

Fig. 10.16. Caisson tunnelling

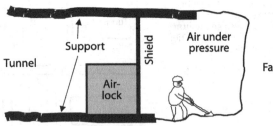

escape and pressure to be lost. Working under compressed air is unpleasant, unhealthy and expensive.

Gunite and Shotcrete

Gunite is a mixture of sand/cement/water sprayed onto a rock surface for waterproofing. It is usually sprayed in thin layers on wire mesh that is held to the rock by bolts and serves to give structural strength to the gunite and prevent it cracking. *Shotcrete* is a quick setting mortar sprayed on the ground. It is applied in layers about 4 cm thick at a spray velocity of about 150 m s^{-1} and closely follows the excavation contours. Additives can be used that shorten the setting time allowing thicker layers to be sprayed in one layer; fibres of plastic or steel can be added to increase the tensile strength of the gunite. Multiple layers can build up to a final layer of (in principle) unlimited thickness; however, the usual thickness is about 4 to 6 cm and the maximum is around 40 to 50 cm. Wire mesh, mats of steel, and/or steel beams may be installed between layers of shotcrete as reinforcement to give strength. Bolt support can also be added. Shotcrete supports the mass by keeping the smallest pieces of rock in place by gluing the mass together. This prevents small deformations becoming large movements. The first layer can be sprayed directly after excavation to minimise deformation. Shotcrete is presently one of the most popular forms of temporary and permanent support.

10.8.2
Support

Bolts, Dowels, and Anchors

These are three similar types of support that work by keeping the ground together and providing reinforcement in the mass. The naming is not clearly defined, generally, an anchor has a flexible cable and is long (ranging from metres to many tens of metres), whereas a bolt is stiff and relatively short (0.5 to 10 m). Bolts and anchors can both be tensioned to create a compressive stress in the ground between their nut and clamp; this increases the shear strength on surfaces so strengthening the ground (Fig. 10.17). Bolts and anchors can be used in soil or rock masses, but are most effective in rock masses with reasonable intact rock strength and a not too small block size. If the intact rock strength is too low, the rock mass will squeeze or shear along the bolt or anchor, and if the block size is too small the small blocks will be squeezed out between the bolts or anchors. In these types of ground bolts and anchors are normally combined with netting and grouting (to be done before the bolts or anchors are installed).

The engineering geologist may be asked to advise on the length of bolting or anchoring to be used and the tensioning to be applied. If the ground deforms it will naturally tension an untensioned bolt and anchor that has a plate on its nut, but this may take time. If the deformation is to be controlled then tension may be applied at an early stage. The installation of these devices is therefore divided into *"tensioned"* and *"untensioned"* Advice may also be required on the likely corrosion of the anchors and bolts in the natural groundwater of the mass.

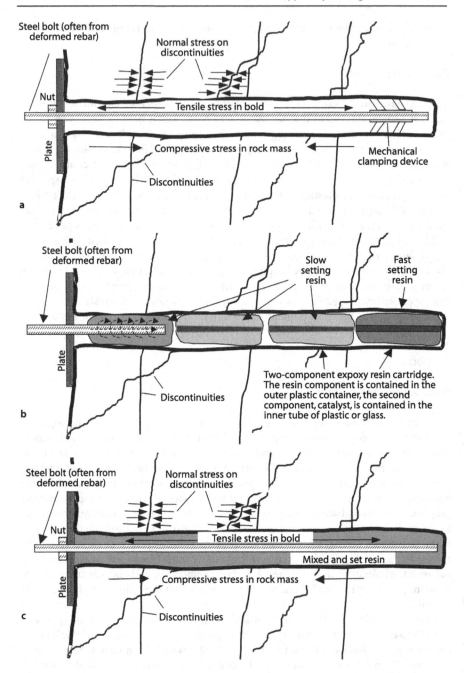

Fig. 10 17. Mechanical (**a**) and resin (**b**) bolts. A borehole is filled with cartridges containing two-component resin (**b**). The rebar is forced in the hole while rotating and breaks the cartridge and mixes the resin and catalyst components. When the deepest installed fast-setting (hardening), cartridge has set, the rebar is tensioned. The resin contained in the other cartridges sets later, gives additional strength, and protects the rebar from corrosion (**c**)

Bolts are fastened in the mass by cement or epoxy resin, by clamp systems (Fig. 10.17) or by the elastic expansion of steel. Anchors are normally fastened with cement or resin. Anchor cables have a protective sheathing against corrosion.

Bolting Patterns

Bolts can be installed in soil or rock masses as an umbrella, inclined from the tunnel over the face as forepoling, where they serve as a pre-excavation support. Bolts can also be applied to stabilise the face of an excavation if required. For radial bolting different opinions exist for how bolts should be installed: either perpendicular to as many discontinuities as possible (Fig. 10.18a) or just perpendicular to the wall, roof, or floor of the excavation (Fig. 10.18b).

The first is often difficult to execute and requires skilled labour. The later is easy, even by unskilled labour. Bolts should be installed perpendicular to the discontinuities where one clear discontinuity set exists. Most masses contain more than one set and it will be impossible to install the bolts perpendicular to all discontinuities. The installation of bolts perpendicular to the walls and roof is easier and cheaper so that more bolts can be installed and the final support is normally effective. Bolting installed perpendicular to discontinuities where a limited amount of mass is unstable is called 'spot bolting' (Fig. 10.18c). Shotcrete and concrete can be used as required (Fig. 10.18d).

Bolts with Straps, Cables and Nets

Metal straps, tensioned cables and nets can be placed between bolts to keep intervening bolts in place; cables force the blocks of the mass together to create a tighter struc-

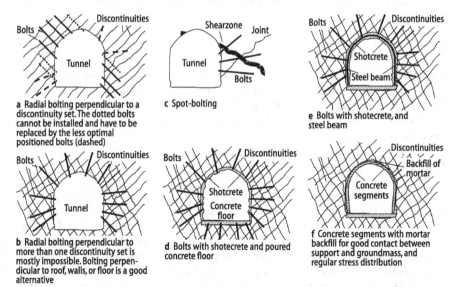

Fig. 10.18. Various support types and combinations

ture and nets prevent severely broken rock masses from unravelling. They tend to be used with strong but broken rock. Their installation follows the excavation perimeter as far as possible, is labour intensive and time consuming.

Steel Support

Steel support exists in two forms: flexible and fixed. The flexible steel support, normally in the form of so-called '*yielding steel arch sets*', are able to deform and therefore release the stresses in the rock mass. Bolts keeping the steel parts together can regulate the yield strength of the arch. The support is suitable for environments were large stresses are expected that cannot be stopped economically; mostly in mining. The stress on the steel arch should everywhere be about the same. If not, bending and buckling of the arch occurs and the support fails. Small pieces of timber or another flexible material between the mass and the steel are used to spread the load uniformly to the steel. When the major part of the stress relief and deformations has occurred the arch sets can be covered with shotcrete or concrete (Fig. 10.18e).

Concrete Support

Reinforced concrete support is used if large stresses are expected and movements are not allowed. Pre-fabricated plates and segments (Fig. 10.18f) or concrete formed by pouring behind shuttering, may be used. The first will act as support immediately after installation whereas the second will only start working if the concrete develops its strength within 28 days after pouring. It should be noted that the space between the concrete segments and the ground has to be filled with grout or concrete, however, this too will need some time to obtain strength. Behind many TBMs are systems for installing pre-cast concrete tunnel lining segments and backfilling them to give the support required.

Cut-and-Cover

Cut-and-cover first makes a trench into which the tunnel is then built, and finally covered with the excavated material. A trench with deep walls can be made by installing large size overlapping concrete piles and diaphragm walls, excavating the ground between them, installing the floor and roof of the tunnel, and backfilling with the excavated material.

Timber

This has been used extensively as support in the past; nowadays timber support in civil engineering is limited. Timber is used in the form of beams, or as blocks of wood in combination with steel or other support. Timber is very compressible and can deform greatly before it will break. This makes timber very suitable in locations were large displacements (and forces) need to be controlled, for example, in mining. This compressibility makes timber a very poor form of support from a geotechnical point of view. The surrounding ground can move thus causing loss of structure and reduction of shear strength along discontinuities. Timber rots with time and has to be replaced regularly; this allows relaxation of the ground and further movement and loss of strength. When rotting, it uses the oxygen in the excavation; a hazardous atmosphere can exist in poorly ventilated underground spaces.

10.8.3
New Austrian Tunnelling Method (NATM)

The New Austrian Tunnelling Method (NATM) has become very popular in recent years and been applied for parts of the Channel Tunnel between France and Britain, and for many tunnels in Germany and Austria. The NATM (Müller 1978; Pacher et al. 1974; Rabcewicz and Golser 1972) requires the designer to encourage the rock to carry as much of the load as it safely can be allowed to do providing support only where deemed to be necessary. It involves characterisation and classification of the ground, rock mass modelling, and construction adjusting in real time to the ground deformations monitored. This is done by using semi-flexible support with a closed invert (tunnel floor), that may consist of steel arches or rock bolts and wire mesh with sprayed concrete installed directly after excavation to minimise movements in the rock mass. The time of installation may be delayed to allow maximum arching effects to develop. It is an extremely sophisticated method of excavation that can go seriously wrong because the support should be adjusted to the circumstances encountered while tunnelling is in progress. Therefore, the system requires contracts to be drafted so that contractors are able to immediately install adjusted support (e.g. without time consuming contract negotiations with the client and consultants) while the work is ongoing. In practice NATM is a combination of methods and methodologies that already existed but have been grouped together as a 'tunnelling method' (Kovári 1994). Various modifications, adjusted to local circumstances, have been developed worldwide. Many claim to follow the NATM system without exactly knowing what the system incorporates, as witnessed by the collapse of the Heathrow Express Link underground railway line in 1994. The investigation after the collapse showed that support installed was at some locations considerably less than prescribed in the design. This has raised questions on the correctness of the NATM philosophy for arranging the installation of support in such a way that the client has no direct control during the work. The method also makes severe demands upon engineering geologists because the ground model required is NOT that of the ground "as investigated" but as investigated *and allowed to deform a little*. Every measured geotechnical parameter changes with deformation – so what IS the character ground through which these tunnels are being built? Probably nobody knows!

10.9
Site Investigation for Underground Excavations

The object of site investigations for underground excavations must be to recover information adequate enough to anticipate problems of excavation, water inflow, support and future behaviour of the materials. *Extensive use should be made of conventional geological survey methods applied to a large area surrounding the site in order to understand the geology and to view the strata likely to be encountered in the excavation en-mass rather than only in samples taken from boreholes.* Geophysical methods may be useful, particularly cross-hole shooting between investigation boreholes and, for excavations underwater, continuous seismic profiling. Considerable attention must be given to portals as these pass through the weakest and most permeable ground conditions likely to be encountered and often involve going through landslides. It is extremely useful to visit other underground workings nearby, talk to miners or consultants and contractors involved in the workings, and read publications relating to them. An underground work nearby is in fact a large true-scale test of what is going to be done and can give priceless information on possible virgin stress field problems, types of support and excavation methods, problems with support and stability. The author has noted that such visits are seldom done, probably because junior engineering geologists are afraid to show their lack of experience or are afraid that the information will be refused because of company confidentiality. In general, however, the atmosphere of openness and non-secrecy is remarkably larger in underground works than in surface works

The factors to be studied in a site investigation may be summarized below:

1. geological conditions with particular reference to lithology and structure, including faulting and the orientation and spacing of discontinuities;
2. engineering properties of the materials to be excavated and supported, with particular reference to strength and deformability of intact material and discontinuities, to abrasion on cutting tools, and squeezing and swelling;
3. hydrogeology;
4. stability of portals and access shafts;
5. natural hazards such as earthquakes, temperature underground, dangerous gasses, avalanches over portal areas and the like.

A factor often forgotten is the mobilisation of equipment and material to the project site and the availability of appropriate skilled labour. It may be satisfying to choose a state-of-the-art ground investigation, but if it and its labour have to travel far to reach site or if a new road has to be built to access site, transport and labour costs may well rule out the preferred methods. It is therefore a good idea for the engineering geologist to check available local resources before the "equipment stage" of site investigation is decided; a telephone directory, a few telephone calls, and Internet may be very informative.

10.9.1
Initial Estimates

The ground should be divided in geotechnical units, based on the geotechnical properties and the cost per unit of the work done in that unit should be approximately calculated, if required for different alternative sites or routings. Choice of excavation method and type of support is largely based on experience, but choices should be justified. Rock mass classification systems supply predictions for suitable support in various ground conditions (Barton 1988, 2000; Bieniaswki 1989; Carranza-Torres and Fairhurst 2000; Laubscher 1990; Laughton and Nelson 1996) and as discussed in Chap. 4. The key-block theory (Goodman 1995) may be used to predict likely bolt and anchor loads and lengths. It is advisable to use more than one system or method of prediction, as these do not always come to the same recommendations. The differences, however, may indicate which system or method is the most suitable for a particular project and ground. Some of the rock mass classification systems are linked to a "*stand up time*" that depends upon rock class and tunnel span and will indicate how quickly support has to be applied after excavation.

An indication of the type of excavation method to use can be based on the graphs published in the literature (e.g. Franklin et al. 1971), but should be refined by data from the manufacturers of excavation equipment and explosives for the project design and construction phases. Choice of explosives is often limited to the explosives manufactured locally as long transport or importing explosives is cumbersome, time consuming, costly, and often just impossible. Ground properties required for the choice of TBM design can best be derived from the various manufacturers of TBMs, all of whom have websites showing the latest developments and are quite prepared to provide additional information (e.g. Fig. 10.19). The designers will want to know whether the ground needs to be supported, the water pressures to be expected and the strength of the ground to be excavated. An estimation of production rates in rock may also be made with the QTBM system, which is based on Barton's Q-system rock classification system (Barton 2000, 2005). The International Tunnel Association (ITA) provides useful publications on TBMs.

10.9.2
On-going Investigations

Site investigations do not stop once tunnel construction has begun. It is now standard good practice to have continuous recording of geological conditions encountered as the tunnel progresses. Information is continually re-interpreted to give a better idea of what lies ahead. In some tunnels, '*probing ahead*' is undertaken to establish the nature of the ground to be excavated in the next phase of tunnelling. Figure 10.20 shows

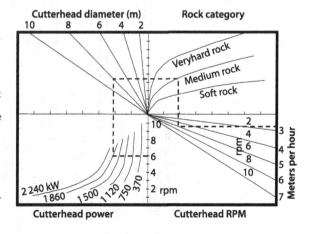

Fig. 10.19. Production monogram (after Terratec 2002). Explanation: (1) Select cutterhead rpm and draw a horizontal line across until it intersects the available cutterhead power. (2) Continue the line vertically upwards from this point until it intersects the required cutterhead diameter. (3) Continue the line horizontally across to the line which most closely represents the local rock strength. (4) Continue the line vertically down until it intersects the selected cutterhead rpm. (5) Finally, continue the line horizontally across to read off the predicted machine excavation rate. Example: 6 rpm + 1 860 kW for 8 m diameter in medium rock gives 2.7 m hr^{-1}

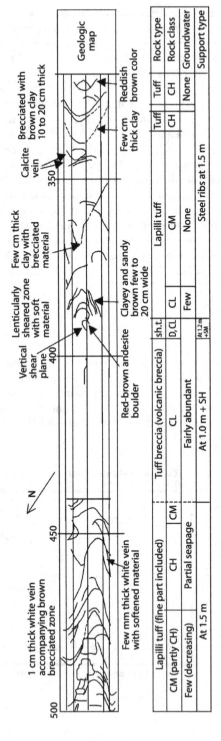

Fig. 10.20. Example of a tunnel record with geology, rock type and class, hydrogeology, and support details (simplified after Mitani 1998)

a traditional way of recording tunnel geology encountered. The top part shows the geotechnical units and the lower part shows the main properties for the tunnel support design. This type of section gives a fast overview of the whole tunnel, however, it is only a section and provides only details along one particular line. More sophisticated means for visualising a tunnel, the properties of the ground, and calculating and visualising costs are possible with the use of 3D-GIS programs (Fig. 10.21).

10.9.3
Site Investigation for Surface Effects of Tunnelling

Tunnels in soft ground will give rise to surface subsidence if crown settlement is allowed to proceed unchecked. Accidental inflows of water bearing non-cohesive ground into the tunnel can cause surface subsidence. In hard ground blasting in shallow tunnels may cause damage to buildings at surface. Blasting tests and vibration measurements should be carried out before the tunnel is too far advanced to obtain the best blasting system and pattern possible for the least surface disturbance compatible with good driving progress. It is prudent to have an inspection (with photographic records) of surface property *before* tunnelling commences and to plot surface effects onto the geological map of the tunnel route *as tunnelling proceeds*.

10.10
Subsidence

This is an extensive subject and is treated here in two parts; subsidence over tunnels and subsidence over mines.

10.10.1
Subsidence due to a Single Tunnel

Tunnels may cause surface subsidence. The amount of subsidence encountered depends on the volume of ground that is excavated *in excess of* the volume of the void created

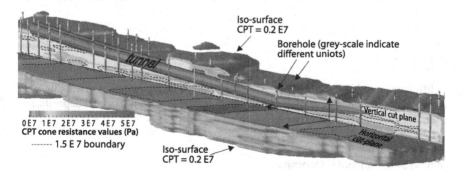

Fig. 10.21. Example of 3D-GIS visualisation of proposed tunnel alignment in a solid volume model of distribution of CPT cone resistance values, with boreholes showing geotechnical units and two cut-planes to show the distribution of CPT values (Heinenoord Tunnel, Netherlands; after Ozmutlu 2002)

(*ground loss*), relaxation of the ground surrounding the excavation, the depth of the tunnel below surface, and the horizontal distance of the point of subsidence measurement from the tunnel centreline. Surface subsidence due to a single tunnel can be relatively small because modern types of excavation and support are designed to minimise ground loss and relaxation. A first approximation of subsidence over a tunnel can be made with an inverted error function (Peck 1969):

$$S_x = S_{max} \exp\left(-\frac{x^2}{2i^2}\right) \tag{10.3}$$

in which S_x is the surface subsidence at a horizontal distance x from the centreline, S_{max} is the maximum subsidence above the crown of the tunnel, and x is the distance from the centreline. $i = KH$ in which K is dependent on the ground characteristics, generally 0.4–0.6 in cohesive material and 0.25–0.45 in non-cohesive material, and H is depth of the centreline below surface. S_{max} can be approximated (Leca et al. 2000) from $S_{max} = V_s / (i\sqrt{(2\pi)})$ and $V_s = V_l\pi r^2$ in which r is the radius of a circular tunnel and V_l is the factor of ground loss. V_l is to be estimated based on type of support and workmanship.

10.10.2
Subsidence due to Mining

While the contribution of mining to the development of society may be demonstrated readily, mining has also had unfortunate consequences. Early mines in bedded deposits removed the deposit leaving pillars of ground to support the mine from collapse. Later techniques removed all the material, allowing the mine roof to collapse in a controlled and, for the miners, a safe way. Mines working mineral veins extracted these from the surrounding ground by a variety of techniques. Whatever the technique used the immediate or later collapse of the mine opening generally disrupted the rock mass and produced some degree of surface subsidence. As well as this, most mines have shafts or adits to allow access to the material to be worked and these too collapse once abandoned.

10.11
Present Mining

A difference should be made between mining bedded, almost horizontal deposits, such as gypsum and coal, and mining of ore, which may also be in horizontal deposits, but is often in more irregular or steeply dipping deposits.

10.11.1
Mining Sub-Horizontal Deposits

Most of the problems associated with present and recent mining of sub- horizontal bedded deposits stem from the extraction of coal in so-called '*long wall*' mining. Coal is extracted by digging machinery in *"panels"* that may be two to three hundred metres wide and kilometres long (Fig. 10.22a). Extraction may advance from either tunnels at the access end of the panel or between the tunnels driven to the full panel length. Total extraction permits the strata above the seam to collapse into the mining void and this produces surface subsidence. Long wall mining methods produce almost immediate subsidence at ground level, which is usually complete within a short time after mining has finished; there seems to be little long-term significant residual subsidence associated with this form of extraction.

The extent of this surface subsidence is greater than the area mined. Its limits are defined by the angle of draw (Fig. 10.22b), which depends upon the geology above the seam, but in rock is commonly about 35°. If rock is overlain by a substantial thickness of soil overburden, the angle of draw in the soil will depend upon soil properties, but will be greater than that in rock. The subsidence trough resulting from working a horizontal seam under a flat landscape is illustrated in Fig. 10.22b and c and shows a complex waveform. The maximum vertical subsidence is usually less than the height of the mining void, perhaps about 90%, the remainder being taken up by the opening of

Fig. 10.22. Subsidence from total extraction (long wall) mining of a horizontal coal seam under a level landscape

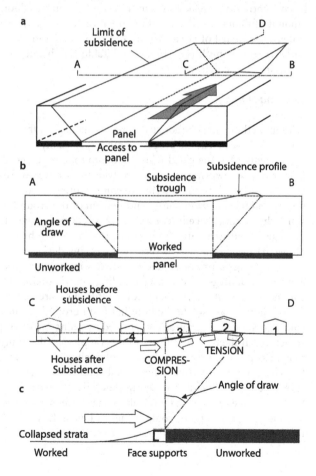

discontinuities within the collapsed mass (a phenomenon called *bulking*). It has been found that the maximum subsidence is only achieved if the ratio of panel width (w) to panel depth (d) exceeds about 1.4. If w/d is less than 1.4 it is described as sub-critical, if $w/d \approx 1.4$ as critical, and if $w/d > 1.4$ as super-critical.

Subsidence Damage

Figure 10.22c shows the effects of the subsidence wave passing under a row of houses. House 1 has yet to be affected. House 2 has risen slightly and its foundations are in tension. House 3 is tilted and its foundations are in compression, while House 4, assuming it has survived tension, compression and tilting, is now vertical and on level but subsided ground. Mining must stop somewhere and should that be at the point shown in Fig. 10.22c, Houses 2 and 3 would be permanently in the condition shown. The damage to structures depends much on the amount of strain suffered by the structure as the result of the foundation ground lengthening or shortening. For uniform strain the change in length will depend on the original length of the structure (the longer the structure the greater the change), but that change, while small, will tend to be concentrated at weak points in the structure such as doors and windows. Thus, the amount of damage will depend upon strain, length, and design of structure. The National Coal Board of Great Britain has published a graph relating damage to strain and structure length (Fig. 10.23) and Table 10.5 briefly details the type of damage occurring in each class of damage.

Reducing Damage

Mining methods may be adapted to reduce damage. If the mining void is not allowed to completely cave but, in the process of mining, is wholly or partially backfilled, subsidence will be reduced. If the panel width to depth ratio is arranged to be less than 1.4, maximum subsidence will not be achieved. If two seams, one above the other, are being worked at the same time it is possible, by *harmonic mining*, to so arrange mining progress that the ground tension from one is cancelled out by the ground compression from another. While the mining procedures outlined may reduce damage, there may well still be some damage. In old structures, appropriate reinforcement before mining commences may reduce damage. In new structures, engineering design will aid damage reduction.

　Most designs aim to isolate the structure from ground strains and reinforce it against tilt damage. In ordinary buildings, construction on rigid raft foundations is common and trenches, with a flexible infill, sunk around the building to below foundation depth may isolate the building from ground strains. Long structures may be split into smaller structural units linked by joints filled with flexible materials. Large structures, if likely to be subject to several phases of mining over a number of years, may incorporate jacking pockets in split cellular raft foundations to compensate for tilt. There are designs to ameliorate subsidence damage for almost all types of structure. Pipelines, for water, gas, sewage and oil are particularly important for if they are broken in zones of tension their leakage may have major consequences in the long term. Thus leaking water pipes could initiate a landslide, while slight gas leaks into an infrequently used cellar could be a source of explosions.

Fig. 10.23. Damage to structures from subsidence related to strain and structure length (after National Coal Board 1975ab)

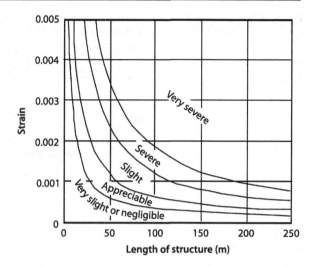

Table 10.5. Subsidence damage

Class	Change in structure length (mm)	Description
Very slight	< 30	Hair cracks in plaster
Slight	30 – 60	Slight fractures outside; redecoration required
Appreciable	60 – 120	Slight fractures outside; doors and windows stick; service pipes may fracture
Severe	120 – 180	Open fractures; door and window frames distorted; floors slope, walls lean or bulge; if compression, over thrusting of brick work
Very severe	>180	As above but worse, requiring rebuilding of structure

Subsidence Forecasting

The National Coal Board has produced data charts and calculation techniques, based on observation and experience, to allow accurate forecasts of subsidence from proposed mine workings to be made (Whittaker and Reddish 1989). Forecasting subsidence, however, does require accurate geological knowledge of the area to be mined. The presence of faults is particularly important, because their reactivation above mines, promoted by mining, may considerably modify the subsidence profile (Fig. 10.24). The presence of such faults should be detected before mining begins, for they would influence mine design. Any area likely to develop a subsidence scarp should be avoided for new construction and the engineering geologist should find the fault outcrop at rock head. This would be influenced by the thickness of overburden over rock head and any change of dip of the fault between mine and surface.

Fig. 10.24. The influence of faulting on the subsidence profile

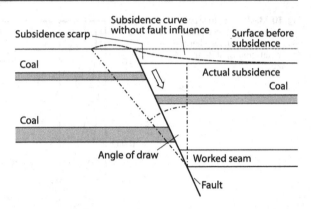

10.11.2
Mining Sub-Vertical Deposits

Hydrothermal deposits of ore often occur as veins and lobes that have a steep dip. They are extracted by many methods and subsidence due to such ore mining depends on the type of mining method used (extensive descriptions of different mining methods can be found in Hustrulid 1982). Block caving or undercutting methods, where a hole is created at depth and the ore excavated into it and transported to the surface, will give similar results as long wall mining because the ground above the ore collapses onto the footwall, however, the extent of the area is mostly far smaller than that in sub-horizontal mining and hence the surface subsidence will be less manifest than in long wall mining of, say, coal. Cut and fill techniques reduce subsidence generally to acceptable quantities with low subsidence gradients. Open stoping or sub-level stoping creates *stopes* (caves) with dimensions that can be over 100 m high, 50 m wide, and 30 m thick, can give serious and uncontrollable subsidence problems. If these stopes remain open after mining a large amount of open space remains in the subsurface, which is without any form of control. The crown pillars and subsequently the stopes themselves collapse with time, often with a domino effect. This can cause a series of local earthquakes and very large surface displacements (up to tens of metres in a few minutes for near-surface stopes). Even if the crown pillars are destroyed after mining so that the *hanging wall* has collapsed, movements in the ground above the stopes may continue for a long time because the total rock mass above the stopes need not settle immediately; after collapse of the hanging wall, cave like open spaces can migrate upwards. Hence, local earthquakes with large surface movements can be expected for a long time after mining has ceased. Later backfilling or grouting from surface is generally not economic because of the large volumes involved. Areas undermined with open stoping are therefore to be avoided for development and open stoping, without immediate backfilling of the stopes after mining, is no longer permitted in many countries; however, in some countries, notably in Africa and South America, working and abandoned open stope mines, without backfill, may be encountered.

10.12
Past Mining

Mining methods used at the present were also used in the past and much of what is said in the previous sections is just as relevant to former mines that are now abandoned. However, mining is as old as man and very old workings of different kinds remain awaiting the unwary engineering geologist. The subsidence caused by these old mine workings is now considered.

10.12.1
Bell Pits

Historically, coal mining began with the selective working of the most attractive seams at their most accessible positions (normally their outcrops) and gradually extended to greater depths and to thinner seams with the progress of technology. After the surface excavation of coal at the outcrop, the first attempt at underground mining was the 'bell pit', a shallow shaft (seldom more than 10 m deep with a diameter usually about 1 to 2 m.) expanded at its base by undercutting into the seam once it had been reached to give a bell-shaped cross-section (Fig. 10.25a). Radial mining from the bottom of the shaft was undertaken until the area of extraction was such that natural or artificial support was not feasible. Sometimes about three to six short radial headings were driven from the base of the pit (Fig. 10.25b). Stone age bell pits are believed to have been originally put down to exploit flint stones and the method subsequently adopted for the mining of coal as well as other materials; it was commonly used until the seventeenth century. A whole progression of bell-pits may often delineate the coal seam outcrop and cones of waste around the shafts may still be visible in open country and on air photographs. The surface instability problems arising from this form of working can be compared to those of old mine shafts. It must be assumed that bell pits were not backfilled when abandoned. No records of the location of bell pits were kept and often only the irregular ground surface betrays their presence.

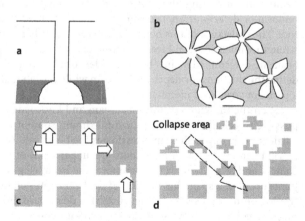

Fig. 10.25. a Bell pit; **b** bell pit 'swarm' with occasional links between galleries; **c** room and pillar (pillar and stall) working; **d** Pillar robbing as final phase of room and pillar working; arrow indicates direction of retreat from mine.

10.12.2
Room and Pillar Workings

The oldest form of true underground mining, with the emphasis on driving headings in the seam, rather than on sinking shafts, is that in which the mineral (most commonly coal or limestone) was removed from a grid of intersecting headings, leaving more or less regularly spaced pillars to support the roof, at least during the working life of the mine. The *room and pillar* method consists basically of two sets of roadways driven in the seam, roughly at right angles to each other (Fig. 10.25c). For coal extraction, room widths of between 1.8 and 4.6 m appear to have been the norm before the last century and prior to the mid-nineteenth century, the room and pillar method was almost universally employed in Great Britain. In later and more developed variants, on reaching the economic or legal limits of the mine, the pillars were *robbed* or removed in succession by pillar robbing (Fig. 10.25d) to such an extent that they crushed under their overburden load unless there was a requirement to prevent surface subsidence. Often *illegal pillar robbing* took place without concern for the consequences.

Collapse

Room and pillar mines are subject to *long-term* deterioration of both pillars and roof as a consequence of weathering and creep. Collapse may take place hundreds of years after abandonment of the mine. Where mining was active, it was usually outside town boundaries, but the abandoned mine may now lie beneath areas newly developed for housing and industry associated with the expansion of the town or city. *Thus, many towns and cities whose initial growth was a consequence of active mining now face the hazards from the potential collapse of abandoned mines beneath them.*

Roof Behaviour

At the time of mining, the roof may be controlled by temporary support which, on removal or deterioration, transfers the full strata load to the rock span forming the roof. The resistance of this beam depends on a number of factors including its tensile strength, dimensions, and freedom from discontinuities. The dimensions of the strata beam are determined mainly by the incidence of bedding planes and joints. The former define the thickness or depth of the beam while the latter affect its ability to behave as a uniform structural member.

Given the thickness of the rock beam and knowledge of its tensile strength it is possible to calculate the theoretical maximum span of the beam between pillars. A single joint in any such span, transforms the beam into two cantilevers that will tend to sag from the two abutments until the upper parts of the joint surface are in tight contact and a "wedging" effect supports the beam (Fig. 10.26a). Where two or more joints occur in a span, that portion of the beam between the joints is held only by the friction of the joint surfaces (Fig. 10.26b). A downward divergence of joints allows the block to fall out easily (Fig. 10.26c); downward convergence will tend to hold blocks in place (Fig. 10.26d).

Fig. 10.26. The influence of joints on roof stability

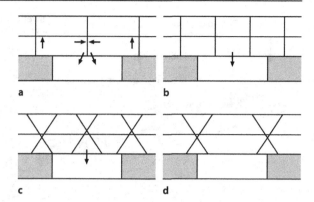

Given a set of conditions in a mine roadway where the immediate beam of roof strata is unstable, it then becomes necessary to consider the response of the overlying and succeeding beams. If the stability is largely joint-determined, the failure-process will tend to pass upwards until prevented by changing ground conditions. Where, however, the stability is largely determined by the tensile strength and dimensions of the beam, the breaks from the collapse of succeeding beams will tend to encroach into the mining opening and produce a natural arch. In practice, however, most instances of roof collapse contain an element of both beam failure and joint control and the final form will depend on the relative importance of these two processes. The maximum span length (L) for a layer of rock can be expressed as:

$$L = \sqrt{\frac{2\sigma t}{W}} \qquad (10.4)$$

where σ is the tensile strength of the rock, t is the thickness of the layer and W is its unit weight.

Void Migration

Progressive collapse of lithologically similar rocks spanning a void will lead to an upward migration of the void, which may reach ground level, giving rise to a sinkhole (sometimes known as a crown hole, *chimney*, or *sitt*) (Fig. 10.27). The height to which a void might migrate, or collapse take place, is of the utmost concern with regard to the use of the land above the workings for building construction. These processes depend for their progress on the support the fallen material can give to the roof. When this occurs, usually due to the bulking of the fallen material, the process is brought to an end and the system becomes stable.

Bulking

This can be expressed as the ratio of the volume occupied by broken rock mass to the volume occupied by the same weight of rock when intact. The bulking factor (B_f) var-

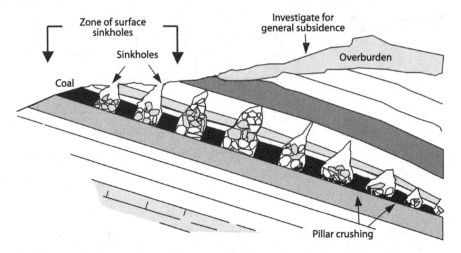

Fig. 10.27. Sinkholes and subsidence developed by void migration and pillar crushing

ies with rock type and for sandstones it may be about 1.5 and for soft shales and mud-stones 1.2. If the height of a working is h, the *theoretical* maximum height H to which a collapse can penetrate above the roof of the working is given by:

$$H = \frac{h}{B_f - 1} \tag{10.5}$$

Provided that the overlying beam of strata can carry any future ground loading, a surface instability problem will then arise only if the fallen material is removed so permitting collapse to continue. The inclination of the mine opening will influence this process for when the incline is small the collapse material will tend to rest where it falls but where the dip exceeds the natural angle of repose, the material will tend to roll away and collect at lower levels. Observations suggest that crown holes are rare when the rock cover above the worked seam is more than about four times the height of the workings. For general guidance only, a thickness of strata above a seam equal to *six times the height of the working* can probably be regarded as defining a zone to which void migration is confined in all but exceptional circumstances.

10.13
Mine Stability

When the ground above old mines is to be developed the ability of the mine to support the ground above them for another 60 to 100 years has to be assessed. If there is doubt about this ability the mined area below the development may have to be backfilled. Failure usually starts because old pillars begin to collapse.

10.13.1
Pillar Failure

A single pillar will fail when its strength drops below the level of the imposed stress. Failure may occur either by slow yielding or sudden collapse, but in either case a portion of the load the pillar has been carrying is transferred to adjoining pillars, by arching within the overburden. If the magnitude of the transferred load is sufficiently high, the adjoining pillar will also fail, causing still another redistribution of load. Successive pillar failure (the *domino effect*) and repositioning of the arch continues until the overburden is no longer able to span the distressed mine zone, and fails either by shear at the periphery of the distressed zone or by flexure. The sides of the collapse arch may then move into the depression to give a subsidence trough whose extent is related to the angle of draw (Fig. 10.22).

Pillar Strength

The strength of a pillar depends upon its size and geometry, weakness in the pillar (fractures, bedding planes, clay veins, etc.) and the intact strength of the rock itself. Many formulae have been developed to calculate the strength of cubic pillars, usually incorporating the parameters of pillar width (W), pillar height (H), and unconfined compressive strength σ_c. Pillar strength may be calculated by the formula:

$$\sigma_p = \frac{\sigma_c\, N_{shape}\, N_{size}}{F} \tag{10.6}$$

where σ_p = pillar strength, N_{shape} is a correction for the difference in shape between the cubic pillar and the cylindrical laboratory strength test sample, N_{size} is a correction relating the strength of the rock pillar including discontinuities to that of the laboratory strength test specimen without discontinuities, and F is the factor of safety. N_{shape} and N_{size} for coal are (Hustrulid 1976):

$$N_{shape} = 0.875 + 0.250 \frac{W}{H} \tag{10.7}$$

and

$$N_{size} = \sqrt{\frac{h}{h_{crit}}} \tag{10.8}$$

where h is the height of the cylindrical laboratory specimen of intact rock material used for strength testing (diameter = $h/2$) and h_{crit} is the minimum height of a rectangular cross-section specimen (like a pillar) of rock mass such that no further change in height produces any change in strength. This is clearly related to the frequency and character of discontinuities in the rock mass. For limestone pillars in the Netherlands, which may not have the same shape all over the pillar, it is suggested (Bekendam and Price 1993):

$$N_{shape} = 0.87 + \frac{A}{C}H \qquad (10.9)$$

where A is the pillar base area and C the pillar perimeter.

Estimation of Collapse Potential

Goodman et al. (1980) have developed a method to evaluate the collapse potential over abandoned room and pillar mines. This method can be used if a map is available showing the locations and dimensions of rooms and pillars, as abandoned. Stress on pillars may be calculated by the *tributary area method* (Fig. 10.28). This assumes that each pillar, with plan area Ap is loaded by a volume of ground of plan area Av bounded laterally by vertical planes extending up to surface and located at half room width around each pillar. The method is probably as good as the data available in these studies and the factor of safety of a pillar (F_s) is given by:

$$F_s = \frac{\text{Pillar strength}}{\text{Load on pillar}} \qquad (10.10)$$

If some fall below unity, or an assigned value of F_s, they may be considered to no longer carry load, which must then be redistributed to adjacent pillars by a recalcula-

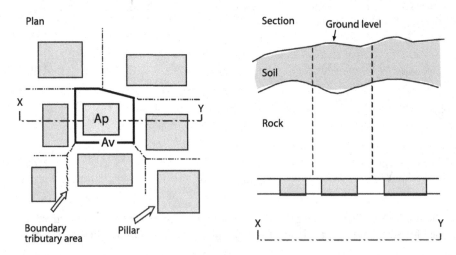

Fig. 10.28. Pillars and tributary areas

tion using the relevant areas. The process is repeated until all pillars with a factor of safety below the assigned value are identified. The procedure serves to identify those pillars that are most likely to collapse. Should the project be to build a structure on ground above the mine, the stresses imposed by that structure may be added to the ground load on the pillars to assess its influence on pillar stability.

This approach requires a plan of the mine *in its present condition*. If old mine plans are available it is often difficult to locate where the mine is relative to present surface features, the mine plan may not be accurate, and illicit pillar robbing may have taken place after formal mine closure. Consequently, before any reasonably accurate pillar stability assessment of this type can be made, the mine must be surveyed. In wet mines this is particularly difficult because most have been flooded after closure, and pumping out the mine is hazardous for changes in effective stress may initiate collapse.

10.13.2
Mine Stability Analysis

In most mines the pillars show varying degrees of deterioration evidenced by cracking, which may be used as a basis for a deterioration classification of which an example is given in Fig. 10.29a. Here pillars range, with increasing cracking, from the un-cracked pillar (Class I) to the completely failed pillar containing many cracks, lateral spalling, giving it an hourglass form, and sometimes showing evidence of compression (Class IV). Pillar cracking and shortening may be also associated with roof collapse. This may result from failure of weaker bands lying above the immediate roof beam

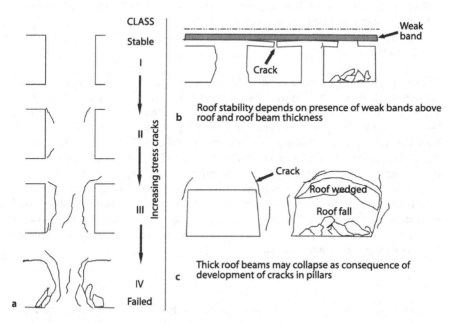

Fig. 10.29. a Pillar deterioration classification based on cracking; **b** causes of roof instability, weak layers in roof beam; **c** upward progression of pillar cracks severs beam ends

(Fig. 10.29b) and by upward progression of pillar cracks into the end of the roof beam (Fig. 10.29c); each increases the intensity of load on the pillar. The cracked pillars are progressing to failure and redistributing their overburden load to adjacent pillars, which may lead to a domino-like progression to general failure. The stability analysis of a mine that can be entered requires the mapping of much information to enable that analysis to be made.

10.13.3
After Effects of Mining

Mining breaks the ground, reduces its strength and deformability and increases its permeability. *The engineering geology of mined ground is extremely complex and can be time dependent; it is some of the most difficult ground in which to engineer.*

Loss of Structure

In all mining methods, except those where backfill has been applied directly after mining, an open space has been allowed to collapse. This causes the ground between the working and ground level to become loosened, which may affect foundation-bearing capacity and settlement. This loosening and any relict stress still present in the mass will significantly influence underground excavations in the mass. Knill (1973) has described problems encountered when tunnels were driven through coal bearing Carboniferous rocks under the River Tyne in north-eastern England. Subsidence from past mining had opened discontinuities in the rock mass with faults and steep cracks infilled by alluvium, which lead to extensive water inflows and rock falls in the tunnels and very difficult tunnelling conditions.

Rising Groundwater

Many mines have closed in recent decades and water, which was continually pumped out to keep them dry while in operation, is now returning to flood them. With flooding, rebound was measured at locations on the surface that had previously subsided; Fig. 10.30 shows the uplift experienced at a survey point above one of the areas once mined for coal in South Limburg, The Netherlands, following the rise in level of mine water. The maximum values of uplift and mine water rise on the graph are those that were expected as a consequence of a further reduction of pumping in 1994. Such mine water rises and uplifts may have serious consequences. Apart from any damage to surface structures from the uplift strains, there is a risk of collapse of once dry abandoned room and pillar mines and old mine shafts. The magnitude of uplift that may occur following mine abandonment can be estimated. For coal longwall mining this may be done using Bekendam and Pottgens (1995):

$$\Delta h = h D_m \Delta_p \tag{10.11}$$

where h is thickness of the zone of disturbed rock, taken as four times the seam thickness; D_m is uniaxial dilation coefficient (Pa^{-1}) of the disturbed rock mass, and Δp is the increase of pore pressure (Pa) within the disturbed rock mass.

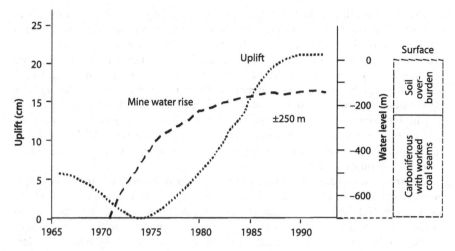

Fig. 10.30. Surface uplift as a consequence of rising mine water (after Bekendam and Pottgens 1995)

Much of this groundwater can have a very unpleasant chemistry as it has come into contact with minerals that have been exposed to decades of oxidisation in the mine air; it is often acidic and can be mildly radioactive. Its movement is hastened by the looseness of the ground above the mine where discontinuities are now open. The permeability of these is approximately proportional to the cube of their aperture; in other words a *small increase in fracture opening produces a large increase in fracture permeability*.

Foundations over Undermined Areas

There are two main options to build safely over undermined areas. These are:

- to take the foundations to below the workings,
- to fill the workings to prevent their collapse.

The first case is only feasible if one seam is worked at shallow depth; then the workings may be excavated out and standard foundations placed on strata below them. Alternatively, a structure may be supported on bored and cast-in-place piles taken to a suitable bearing stratum below the workings. In such a case the piles must be '*sleeved*' with a flexible material, such as bitumen, between the pile and the surrounding ground so that if the mine collapses (for reasons entirely unrelated to the construction) the collapsing strata will not impart extra frictional load to the piles. One hazard in the use of such piled foundations is that the piling process, or other changes, might encourage new collapse in the workings and consequential subsidence. Moreover, if the seam is dipping, collapse of the workings may laterally load the piles causing them to fail by buckling or shear.

Infilling the workings is usually considered the best option and, if the mines cannot be entered, this may be done through boreholes from the surface. Various fluid and fluidised infilling materials have been used, ranging from sand and fly ash to clay, but aerated concrete is generally favoured. If the mines are extensive, the fluid infill may spread beyond the area to be treated, so the first step in any infill operation from surface is to construct a peripheral dam bounding the infill area. This may be done by closely spaced large diameter boreholes into which fine gravel is poured to give a barrier of overlapping gravel mounds in the mine which may later be grouted. More widely spaced holes are then bored, usually on a grid pattern, for the introduction of the infill material that will support pillars and roof. To complete the work more fluid grout can be injected from the infill boreholes to fill up any opened joints or bedding planes that might impair the bearing capacity of the foundation ground.

The area of the mine filled is that which might be influenced by stress from engineering construction. This area may be estimated conservatively using a 45° angle from surface (Fig. 10.31). Such work is expensive; many multi-storey buildings constructed to give extra housing in Britain in the 1960s were chosen to be multi storey to reduce the cost of foundation treatment per dwelling when they had to be built on undermined areas.

Mine Shafts

The minimum safe distance for locating multi-storey blocks from open or poorly filled shafts depends on the nature of the surface deposits and their thickness. The minimum would be a distance equal to the depth of the superficial deposits, though this might require modification if sensitive clays or waterlogged silts were present. Such shafts sometimes can be plugged at depth but if they cannot be filled, may be made safer by the installation of a circular retaining wall around the shaft.

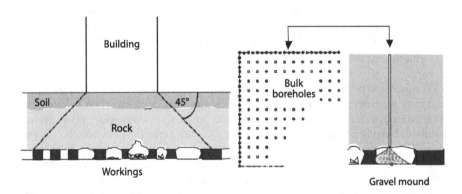

Fig. 10.31. Infilling old mine workings. The area to be infilled is that stressed by the structure, which may be defined conservatively as that bounded by a 45° angle from the foundation area and depth. The area to be bulk filled via boreholes (only a part of the borehole plan is shown in the drawing) is bounded by a peripheral dam installed via closely spaced boreholes

10.14
Site Investigations for Subsidence Areas

The objectives of a site investigation in a mining area are likely to be one or more of the following:

- detection of old mine shafts including bell pits;
- detection of old underground workings, their depth, extent, method of working and degree of extraction;
- location of zones of faulting along which subsidence due to collapse of old workings could be expected to be concentrated.

10.14.1
Detection of Old Mine Workings

Old Mine Shafts

In considering old mine shafts as a target for a site investigation, account should be taken of the possible variations, both in original construction and later history, that may be present. The principal variable factors are:

- *type of shaft:* unlined or brick lined;
- *later history:* either unfilled, open and covered with a raft (often only of wood) at ground surface, or completely or partially infilled and possibly rafted over at depth;
- *nature of filling:* material from original excavation or mine waste mixed with debris from mine buildings, etc.;
- *water:* air above water table or water below water table.

Thus the *target* for the site investigation can vary from an unlined bell-pit of 1 to 2 m in diameter and up to about 13 m deep and not backfilled, to an open brick-lined shaft 2 m diameter, 30 m deep, partly air filled and partly water filled, rafted over with wood 2 m below existing ground level and covered with made ground.

Old Workings

Old workings are also quite variable, depending on:

- thickness of the seam,
- percentage of coal extracted in room and pillar or pillar and stall workings,
- state of preservation or collapse of old workings,
- level of the water table.

Thus, the target may vary from either a water or air filled cavity of 1 to 2 m height, to a zone of collapsed strata several times that height.

Fault Zones

The importance of faults as '*safety valves*', absorbing strains in an area, has been pointed out already. It is recommended to avoid building on ground 15 m either side of known surface positions of faults. Mine records are usually the best source of information on faults.

10.14.2
Investigation Techniques

Drilling

In many cases it is necessary to penetrate the ground to seek details of working or seam depths, cavities, pillar dimensions and extraction ratios, and to examine the ground overlying workings for migrant voids, bedding plane separation and general fragmentation. A considerable number of drilled boreholes may be required but these may be provided at reasonable cost (about one quarter of that of core-drilling) and with some speed (about 80 to 130 m of drilling per day as compared to 10 to 13 m per day) by probe or open-hole drilling. The use of compressed air rather than water for circulation has the advantage of greater sensitivity in the detection of broken ground and workings, and assures the recognition of the presence of groundwater. The borehole may be advanced by rotary methods. More often the use of the tri-cone roller bit is to be preferred because the larger size of the cuttings produced helps ascertain the detailed stratigraphy.

Trial Pits, Trenches and Inspection Shafts

Trial pits and trenches are invaluable for examining very shallow mine workings or determining boundaries of infilled *opencast* workings (quarries from which coal has been taken). If documentary evidence or preliminary boreholes reveal that mine workings are open and reasonably dry they may be examined directly by sinking a shaft to seam level. Such an examination may provide more reliable information than the equivalent value of boring as the condition of the rock can best be ascertained by examination from a shaft, in which if appropriate, in situ tests may be conducted.

Geophysics

Geophysical techniques are, in principle, a powerful tool for locating old underground workings (extensive descriptions of all common geophysical techniques can be found in McDowell et al. 2002). However, there are three important difficulties:

1. The data obtained in a geophysical survey are virtually always ambiguous and therefore the results depend very much on the quality of their interpretation. *The interpreter has to be experienced in site investigations for old mine workings* as well as being an experienced geophysicist. A non-geophysicist is likely to commission an inappropriate geophysical survey that delivers disappointing results. The geophysicist can make the same mistake if he or she knows nothing of mining.

2. The second problem is that geophysics virtually never comes to a result that can be guaranteed to be correct. Other methods such as drilling and trenching have to be used as control on the geophysical results.
3. A third problem is that geophysical surveys can be expensive. This is caused by the relatively long time expensive labour is required for both the survey and the interpretation of its results. Many of the latter problems have been greatly reduced since computers became available. This has opened options as three-dimensional seismics with shear and compressive waves, and three-dimensional resistivity surveys for engineering purposes. The latter may be very useful to obtain indications where and at what depth old mine working may be present.

In all these cases there must be a measurable contrast in the physical property of the features sought for them to be "seen" by geophysics. Shafts backfilled with the same material as the surrounding ground may be "invisible" and seen possibly only by their lining (Hack 2000).

Detection of Old Mine Shafts

In those cases where documentary evidence from the location of an old mine shaft is uncertain and no surface features indicate its presence, the location of the shaft may be attempted by *magnetometry*. The method can only be used in areas where the magnetic field is not disturbed by anomalies due to buried services, buildings, deep fill or live power cables, which would mask the small anomalies likely to result from the presence of a mine shaft. Following the location of a promising magnetic anomaly it is necessary to excavate this in order to confirm that it is caused by a shaft. If an investigation is required to prove the absence of a shaft from a site it is unwise to assume that the lack of distinctive magnetic anomalies conclusively indicates the absence of the shaft. Probe drilling may be used to assess the condition of the shaft infilling, but often the necessity for introducing casing to support the hole in loose infilling so restricts the rate of drilling that probe drilling holds few advantages. Accordingly, standard drilling methods are generally more appropriate so as to conduct standard penetration tests and take samples. Samples may later be chemically analysed for soluble sulphate content if grouting the shaft is considered a suitable means of stabilization. It must be emphasized that drilling into a marginally stable shaft could initiate its collapse. Consequently, the equipment and personnel should be supported on a drilling platform that should not rest on the shaft infilling. Similarly, the potential risks arising from the possible emission of inflammable or poisonous gases should be appreciated. Down-the-borehole camera and television techniques may be used to provide visual confirmation of bedding plane separation and fissuring above seam level as well as conditions in the seam itself, although definition may be poor below ground-water level and the range of visibility is limited in old workings because of the difficulties of illumination. Down the hole acoustic methods (*echo sounders*), continuously recording seismic methods and ultra-sensitive methane detecting devices (for locating mine shafts) can all be used to assist detection.

Design of Investigations

The initial aim of the investigation should be to establish the base site geology and the number of boreholes required to do this will depend on the complexity of the situation. On undermined sites, however, they may well be more frequent than elsewhere because of the higher degree of precision of investigation required. Similarly, the diameter of rock cores may need to be greater than usual in order to ensure maximum recovery of ground broken by mining. The selection of techniques to be employed, and the probable depths, number and distinction of boreholes, should always be tailored to the site. No two investigations need be the same.

The selection of techniques to investigate mine workings will depend on the depth of working and the nature of the problem. In the case of room and pillar workings rotary probe boreholes are most often a satisfactory method unless entry from a shaft is feasible or the workings are shallow enough to be revealed by trenching. Probe boreholes are seldom of value in assessing the disturbed nature of ground broken by total extraction mining and this work is best done by coring, possibly with associated permeability testing. The initial cored boreholes should penetrate at least 14 m into bedrock or into the seams, which, if worked, might affect the stability of the site. In the case of room and pillar extraction the general principle that migrant voids are unlikely to rise to a height above roof level greater than 6 times the height of the workings suggest a maximum depth to which investigation need be taken. The number of boreholes depends upon the complexity of the problem. In some cases, the standard investigation may prove the depth of the seam and the method of working, and this may be sufficient to provide a reasonable assessment of the situation, however it is prudent to sink a number of probe boreholes to verify continuity of the condition of the workings. Attempting to establish the absence of workings in a seam by boring requires more boreholes than investigating the geometry of known workings, for the more boreholes sunk that find the seam solid, the less the probability of workings. The adoption of a grid system has obvious advantages for case of setting out and later drawing of sections etc., but to avoid the remote chance (which has occasionally been realized) of coincidence between the borehole grid and pillar distribution, it is wise to avoid locating grid axes along dip and strike of the seam, and to introduce an element of irregularity in the borehole lay-out.

Subsidence over Natural Solution Cavities

Geological units containing materials such as limestone, dolomite, and gypsum may have natural solution cavities. The cavities originated due to groundwater flow that dissolved the material, generally starting along a discontinuity and gradually forming a cavity. Natural cavities grow with time and the material in the cavity may be weakened by weathering, and creep. This causes most natural cavities to collapse after some time. The influence of natural cavities on civil engineering is similar to the results of old mine workings as discussed before. Engineering measures to be taken are similar to those taken for old mine workings. Investigations to locate the cavities are difficult. Sometimes geophysical methods may work, however, generally, the results are not reliable enough, and hence cavities may remain undiscovered. The only truly reliable method to find the cavities is to drill holes to the required depth on a grid pattern suitable for the engineering application.

10.15
Further Reading

Burland JB, Standing JR, Jardine FM (eds) (2002) Building response to tunnelling; case studies from construction of the Jubilee Extension Line. Thomas Telford, London (vol I: The project; vol II: Case studies)
Illinois State Geological Suvey (1989–onwards) Reports from the Illinois Mine Subsidence Research Program
Malone AW, Whiteside DGD (eds) (1989) Rock cavern – Hong Kong. Institute of Mining and Metallurgy, Hong Kong Section

Chapter 11

Static Loading of the Ground

11.1
Introduction

The foundation of a structure is that part which transmits the load of the structure to the ground and static loading implies a pressure from that foundation that does not change with time. However, the pressure applied by most structures will vary to some degree depending on the use of the structure. For example, foundation pressures given by ordinary dwelling and office buildings will vary as the users move in and out of the building. Such changes are not considered significant except where the structure is built merely as a container, such as a silo for grain or a storage tank for fluids. The sideways pressure of wind may be significant for tall structures and bridges, increasing pressure on one side of a foundation and decreasing pressure on the other. In the case of a piled foundation some piles may receive extra compressive loads while others could at times be in tension.

If a load is placed upon the ground then the ground will deform under the stress imposed by that load. This will result in settlement of the engineering structure that imposed the extra load. If the load is too great to be supported by the ground then the ground may fail in shear and the construction collapse (Fig. 11.1). There is a limit to the load ground can carry and this is related to the mechanical properties of the ground and to the size and shape of the engineering structure. Foundation loads imposed by structures may be of the following orders of magnitude:

Fig. 11.1. Failure of a foundation in shear

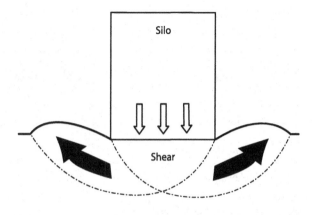

- Office and residential properties   ~10 N/m² per storey
- Compacted embankments   ~20 kN/m² per m³ of embankment
- Mass concrete (as in a dam)   ~25 kN/m² per m³ of structure

11.1.1
Types of Foundation

Civil engineering structures that impose load on the ground can be grouped into two types; mass structures and framed structures.

Mass structures made of soil, rock or mass concrete include such constructions as embankments, dams, mining waste tips and breakwaters. These are in direct contact with the ground and the magnitude and distribution of stress imposed upon the ground is related to the size and shape of the constructed mass, and the density of the material from which it is composed.

Framed constructions are buildings containing hollow spaces for dwellings, storage or industrial use. The weight of the upper parts of the structure is transmitted to the ground via the load bearing members of the structure. These load bearing members are composed of steel, concrete, bricks and wood, whose strength is generally much greater than that of the ground immediately beneath them. Contact between these load bearing elements and the ground is made via a foundation, which generally distributes the load of the structure so that the stresses on the ground are not excessive. There are four main types of foundation (Fig. 11.2).

Strip Footings

In an ordinary house or modest size building, external and internal walls carry the building load. The foundation under the wall is usually a long strip of either mass or reinforced concrete, which is somewhat wider than the wall it supports. It is very much longer than it is wide. Foundations are usually placed at some depth, normally greater than a metre below surface, to avoid swelling or shrinking associated with the ground freezing or drying.

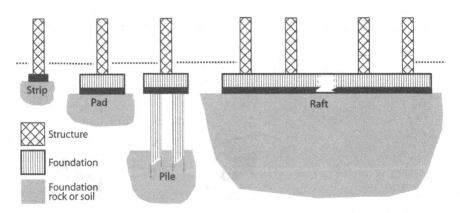

Fig. 11.2. Common types of foundation

Pad Footings

Larger buildings may be built around a framework of load bearing columns connected together by horizontal beams. Foundations at the foot of these columns are square or rectangular 'pads' of either reinforced or massive concrete. The pads are designed so as to distribute the load from the column to the base of the pad without overstressing the ground on which the pad rests.

Raft Foundations

If the foundation loads are high and the foundation ground weak, the area of pad or strip footings required to give a bearable foundation stress may be so great that the footings coalesce to form a single plate, extending under the whole area of the structure; i.e. a raft. The practicalities of foundation construction are such that if the combined area of pad and strip footings exceeds about one third of the plan area of the building, then it may be easier and more economic to construct a raft foundation. Raft foundations are almost always of reinforced concrete, perhaps stiffened by cross beams.

Piled Foundations

Piles are essentially columns of reinforced concrete, steel or timber, usually between about 0.3 and 1.5 m diameter which are inserted into the ground by a variety of methods and whose resistance to the load they carry is provided by friction on the sides of the column and reaction with the ground at their base. There are many different types of pile but they are mostly either '*driven*' or '*cast-in situ*'.

Driven piles may be square or round in cross-section and are generally from 0.3 to 0.5 m diameter. They are mostly made of reinforced concrete, but may sometimes be of timber or steel. Steel piles are often constructed of *H* or *I* section steel beams. These piles are driven into the ground using mechanical hammers until they encounter a strong layer that they cannot penetrate and/or until sufficient side friction is built up to prevent further penetration. Often piles are driven until a limit is reached defined by the number of blows required for a given increment of penetration. Driven piles *displace and compact* the ground into which they are driven. If a foundation is composed of many closely spaced equal length piles and piling begins on the foundation periphery, it is not uncommon for the piles at the centre of the foundation to encounter difficult driving conditions because they are driven into ground already compacted by the earlier piles.

Cast in situ piles are formed in place. A hole is first bored down to a bearing stratum and the sides of the borehole supported by casing. A cage of reinforcing steel is placed in the hole and concrete poured into the hole through a pipe going down to the bottom of the hole. As the hole becomes filled with fluid concrete both injection pipe and casing are pulled to the surface, leaving a reinforced concrete column in the ground. The holes are made by percussion or auger drilling machines; diameters range from about 0.4 to 1.5 m.

'Shell' piles combine driving and casting-in-place. A tube with a toe cap is driven into the ground using a hammer inside the tube and the tube is then filled with concrete.

The carrying capacity of both driven and cast in place piles is built up of '*end-bearing*' capacity (at the toe of the pile) and '*skin friction*' (on the sides of the pile). The proportion of pile load carried in end-bearing and in skin-friction varies greatly, depending on ground conditions and the type of pile.

Other Foundation Types

The types of foundation discussed above are those most commonly used; others exist which are essentially variations of the above e.g. basements, caissons, and stone columns.

11.1.2
Distribution of Stress under a Foundation

The behaviour of the ground under a foundation depends on the geotechnical properties of the ground stressed and on the magnitude of stresses imposed. The volume of ground stressed depends upon the size and shape of the foundation; the larger the foundation and the load it carries, the greater the volume of ground that is stressed. The stress imposed on the ground due to foundation loading diminishes with depth below the foundation. Mathematical analysis, based on the theories of Boussinesq, has shown that the stresses due to building loads fall within a '*bulb of stress*' beneath a foundation. Bulb of stress diagrams have been constructed for various shapes of foundation and for both vertical and shear stresses. Figure 11.3 shows the additional vertical stress passed to the ground by a foundation, expressed as a proportion of the foundation pressure. Significant additional stress extends to a depth of about 1½ the diameter (or width) of the foundation. This gives some help in deciding how deep investigation boreholes should be taken below surface.

Clearly the aim of an investigation should be to sample all the materials that are likely to be stressed by the foundation load. At least one borehole should do this unless there is satisfactory information from some other source to show that it is not necessary. If, as is often the case, the materials under the foundation become stronger with depth then subsequent boreholes need only sample the shallow, more highly stressed and probably weaker materials.

The bulb of stress diagram shows that the foundation load also stresses ground lying outside the plan area of the foundation. There is thus some reason to locate investigation boreholes outside the foundation area of a proposed structure. *In planning the locations of investigation boreholes it is always better to think of investigating the bulb of stress rather than the ground immediately under the foundation area.*

Stress induced by foundation loading is contained within a bulb of stress whatever the shape or size of the foundation, whether pad footings, strip footings, rafts or piles. However, if the total foundation is composed of a number of foundation elements then the bulbs of stress for each element coalesce to give a large bulb of stress for the total foundation. Thus a single footing loads the ground to a small depth below it and a group of footings, taken together, load the ground to a depth commensurate with the total foundation area loaded by all the footings. It is particularly important to recognise this if the ground is divided into strong and weak layers. Thus (Fig. 11.4) footings could

Fig. 11.3. Bulb of stress diagram showing vertical stresses due to pressure superimposed by a circular foundation

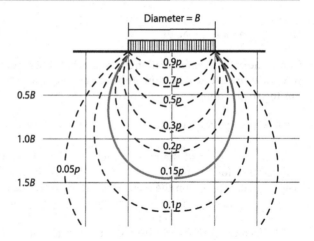

Fig. 11.4. Distribution of stress under single footing and total foundation area

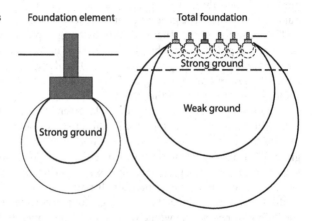

load a strong layer underlain by a weak layer. Each individual footing would be adequately supported by the strong material but the stresses caused by the total foundation could bring the underlying weak material to failure.

As it is necessary for the engineering geologist to appreciate how the foundations are analysed, an introduction to common practice is appropriate.

11.2
Bearing Capacity in Soils

There is a maximum load that can be applied to clays, silts, sands, gravels and mixtures of these, beyond which the ground will fail; this is the "*ultimate*" condition; something less than that will be the "*safe*" condition. However, safe loads may still generate a settlement that is excessive, so the safe load may have to be reduced yet further to an "*allowable*" condition.

11.2.1
Ultimate Bearing Capacity

This is the maximum foundation pressure which the ground will sustain without failure. Equations to calculate ultimate bearing capacity were formulated by Karl Terzaghi. For the strip footing the equation is:

$$Q_f = cN_c + \gamma DN_q + 1/2 B\gamma N_\gamma \qquad (11.1)$$

where c = cohesion of the underlying soil (N m^{-2}); B = width of foundation (m); D = depth of footing below surface (m); N_c, N_q and N_γ are bearing capacity factors whose value depends on the angle of shearing resistance of the underlying soil (dimensionless); Q_f = ultimate bearing capacity (N m^{-2}) and γ = unit weight of the soil beneath the foundation (N m^{-3}).

The factors N_c, N_q and N_γ are given numerically in Table 11.1. Equation 11.1 holds for strip footings only and may be adjusted for rectangular and circular footings by modifying the values of the bearing capacity factors (Table 11.1). The equation applies to cases where the groundwater table level lies deep below the foundation.

The equation shows that ultimate bearing capacity depends not only on the geotechnical properties of the ground but also on the shape of the foundation. Thus strictly speaking it is not possible, as is so often done, to talk about the *'bearing capacity of the ground'* for bearing capacity relates to both foundation shape and the geotechnical properties of the ground.

The ultimate bearing capacity formula, given above, deals with the possibilities of failure in materials at the top of the bulb of stress. Because this is the zone of most intense stress and is situated within the uppermost and probably weakest part of the ground, it is reasonable to do this. However, in shallow foundations, the near surface soils may have varying properties due to weathering and water table fluctuations, and can be stronger than ground beneath. It must also not be forgotten that weaker soils could lie below those loaded by the single foundation element, and be stressed within the bulb of stress of the whole foundation. Consequently it is sensible to begin any analysis leading to the determination of ultimate bearing capacity *by first comparing the shear strengths of the soils with the shear stresses from foundation loading*, using an appropriate shear stress distribution diagram; e.g. with depth and plotted against the geological profile. Such an approach is prudent in the case of a foundation composed of multiple foundation elements and the only approach to establish the bearing capacity of the ground under a mass foundation.

The discussion above relates primarily to cohesive soils. In non-cohesive soils the Terzaghi bearing capacity equation is modified to:

$$Q_f = \gamma DN_q + 1/2 B\gamma N_\gamma \qquad (11.2)$$

Now N_q and N_γ depend upon the angle of shearing resistance (ϕ) of the granular soil. The value of ϕ is often determined by field measurements. Because the blow count (N-value) from the standard penetration test is related to ϕ values it is not surprising that over the years graphs relating bearing capacity to foundation size and N-values

Table 11.1. Bearing capacity factors modified for rectangular and circular footings

Bearing capacity factors modified for rectangular footings
$N_c\{\text{rect}\} = N_c\{\text{strip}\}\,(l + 0.2B/L)$
$N_\gamma\{\text{rect}\} = N_\gamma\{\text{strip}\}\,(l - 0.2B/L)$
$N_q\{\text{rect}\} = N_q\{\text{strip}\}$

Bearing capacity factors modified for circular footings (mostly pile toes)
$N_c\{\text{circle}\} = 1.3N_c\{\text{strip}\}$
$N_\gamma\{\text{circle}\} = 0.6N_\gamma\{\text{strip}\}$
$N_q\{\text{rect}\} = N_q\{\text{strip}\}$

Bearing capacity factors

φ	N_c	N_q	N_γ
0	5.14	1.00	0.00
2	5.63	1.20	0.15
4	6.19	1.43	0.34
6	6.81	1.72	0.57
8	7.53	2.06	0.86
10	8.35	2.47	1.22
12	9.28	2.97	1.69
14	10.37	3.59	2.29
16	11.63	4.34	3.06
18	13.10	5.26	4.07
20	14.83	6.40	5.39
22	16.88	7.82	7.13
24	19.32	9.60	9.44
26	22.25	11.85	12.54
28	25.80	14.72	16.72
30	30.14	18.40	22.40
32	35.49	23.18	30.22
34	42.16	29.44	41.06
36	50.59	37.75	56.31
38	61.35	48.93	78.03
40	75.31	64.20	109.41
42	93.71	85.38	155.55
44	118.37	115.31	226.64
46	152.10	158.51	330.35
48	199.26	222.31	496.01
50	266.89	319.07	762.89

have been developed, and are extensively used in many countries for standard design of foundations on sands and fine gravels. In general, allowable bearing pressure increases as the N-value and foundation width increase. However, the N-value from the field test has to be corrected to N' with regard to effective overburden pressure at the test depth. Further, the allowable bearing pressure may be modified depending on the position of the water table.

11.2.2
Safe Bearing Capacity and Allowable Pressures

To determine ultimate bearing capacity there must be adequate data on ground properties and trust in the appropriateness of the calculation theory. Since neither are commonly to be found the calculated ultimate bearing capacity is divided by a factor of safety to give the *Safe Bearing Capacity*. This factor of safety is commonly between 3 and 5, and is chosen on the difficulty of the ground conditions and the importance of the structure.

Having made the necessary calculations and determined the safe bearing capacity of the ground with regard to the proposed foundation it may be concluded that the ground strength is adequate *relative to the foundation pressure*. However, while this safe bearing pressure will not stress the ground to the point of failure *it will cause the ground to deform and consolidate*. This deformation of the ground will produce settlement of the structure. The settlement of some structures may be allowed only within certain limits. Since settlement depends partly upon the pressure put on the ground it may be necessary to reduce the safe bearing pressure to an *Allowable Bearing Pressure*; this equates to the foundation pressure which will produce acceptable settlement.

If settlement of a structure is uniform there is usually little cause to be concerned about possible damage to the structure. However, if settlement is not uniform, that is, it is '*differential*' between parts of the structure, then the structure may deform beyond its limits of tolerance and be damaged. Possible causes of differential settlement are unequal loading within the structure, building parts of the structure at different times and variations in ground conditions under the structure. The first two causes can be anticipated; the last may stem from unforeseen ground conditions and is by far the most important cause of foundation problems. Examples of geological conditions that could cause differential settlement could be lenses of peat in alluvial soils and variations in thickness of compressible soil over a deepening rockhead. If the site investigation has detected unfavourable geological conditions that might give rise to differential settlement then the foundation design must be modified to overcome this problem. Thus, in the example above, piles down to relatively incompressible rock that could take the structure load with little deformation might be preferred.

If the ground is found to be overstressed at depth then an alternative foundation design may be sought. In the case of embankments, perhaps built over a long period of time (a number of years), advantage may be taken of early non-critical loadings to consolidate the weak soil and improve its strength sufficiently to resist the shear stresses developed when the embankment is higher.

11.2.3
Bearing Capacities on Boulder Bearing Soils

The comments on bearing capacity and settlement of soils given above apply to those soils which can be sampled and tested, either in the laboratory or in the field. However, there are soils, such as moraines, and fluvio-glacial deposits, which contain particles larger than any borehole diameter yet smaller than a foundation. Such materials can neither be sampled nor tested in the laboratory and even if tested by large scale field methods it is difficult to be sure that test results are representative. How these mixtures of boulders and matrix behave will largely depend upon the proportions of their components. If the boulders are surrounded by matrix, as in some 'boulder clays', mass behaviour will approximate to that of the matrix. However, if the boulders constitute the greater part of the deposit or are in contact, then the character of the matrix will have less influence. The proportions of coarse particles to finer matrix recovered in boreholes may give a clue to mass properties and this is best achieved by inspection of the material in situ with either a trial pit or natural exposure.

11.3
Settlement on Soils

Settlement is of two types, first 'immediate settlement', which results from the elastic deformation of the ground (using the word elastic in a very general sense) and then 'consolidation settlement' which is the result of the long-term compaction or consolidation of the ground under the load imposed by the structure. Once construction commences, both types of settlement begin to take place simultaneously. However it is convenient in many cases, especially with clay rich soils, to think of immediate settlement taking place during construction and consolidation settlement continuing after completion of construction.

11.3.1
Immediate (Short-Term) Settlement

This depends upon the 'elastic' properties of the fabric formed by particles and pore water between particles whereas consolidation depends on the rearrangement of the particles made possible by a reduction in pore pressure.

Terzaghi and Peck (1948) produced one of the earliest relationships between the settlement of a 1 ft^2 (= 0.3 m × 0.3 m) square plate and the settlement of a square foundation on sand:

$$S = s \left(\frac{2B}{B+1} \right)^2 \tag{11.3}$$

where s = plate settlement; B = width of foundation and S = settlement of foundation.

The layer of soil tested by the plate bearing test must be uniform for a depth of at least 4 B under the foundation. On clayey soils they suggested immediate settlement (S_i) may be calculated using the formula:

$$S_i = \frac{q2B\left(1-v^2\right)I_p}{E} \tag{11.4}$$

where S_i = immediate settlement (m); B = width of foundation (m); v = Poisson's ratio (assumed to be 0.5 for clay; dimensionless); E = the modulus of elasticity for the clay (N m^{-2}); I_p = an influence factor depending on the shape of the foundation, introduced because the volume of the stress bulb changes with the shape of the foundation (dimensionless) and q = net foundation pressure (N m^{-2}).

E is determined from the stress/strain curve produced by triaxial tests on the clay and is commonly a secant modulus to a stress level approximately equal to that produced by the foundation loading. This formula may be modified to allow for flexible and rigid foundations. Flexible foundations are those of earth structures, such as embankments; rigid foundations are of reinforced concrete, so that settlement is uniform over the whole foundation area.

11.3.2
Consolidation (Long-Term) Settlement

Consolidation depends on the water content of the soil and its permeability. Before soil grains can be brought into close contact so that pore spaces become smaller, pore contents (water and/or air) must be driven out of the pore spaces. How fast they can move out depends upon the permeability of the soil. Because the sands and gravels have mostly high permeability, almost all consolidation settlement in them is complete by the end of construction. In clayey and silty soils many years may pass before consolidation settlement is complete.

Consolidation may be predicted using the results of the *oedometer* test. The ground under a structure is divided up into layers considered to have equal consolidation characteristics, and the average additional stress imposed on each layer is calculated. Settlement of each layer may then be calculated from the formula:

$$S_c = m_v \sigma_z H \qquad (11.5)$$

where m_v = coefficient of volume compressibility for the layer (m^2 N^{-1}); H = thickness of the layer (m); σ_z = average effective vertical stress imposed on the layer as a result of the foundation loading (N m^{-2}) and S_c = consolidation settlement (m).

The total settlement (S_t) at any time t may be expressed by the formula:

$$S_t = S_i + U S_c \qquad (11.6)$$

where U is the degree of consolidation (given as a percentage) at time t, which for any degree of consolidation may be determined using the relationship:

$$t = \frac{T d^2}{C_v} \qquad (11.7)$$

where T = a time factor dependant on drainage characteristics (dimensionless), d = a thickness relating to the thickness (H) of the layer and its drainage characteristics (length) and C_v = coefficient of consolidation of the soil in the layer determined from the oedometer test within the range of pressure increase imposed by the foundation (area/unit time).

Fig. 11.5. The influence of drainage conditions on the rate of consolidation

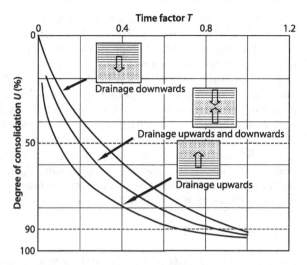

The value of d is determined by considering whether the consolidating layer will drain in one direction (upwards or downwards) or in two directions (upwards and downwards). If the layer has permeable sand above and below, $d = H/2$; if a permeable layer exists either above or below the sand layer then $d = H$. T may be read from graphs relating U (as a percentage) to T for various drainage conditions (Fig. 11.5).

The presence of permeable, usually silty, layers within clays will govern the time to reach 50% consolidation. Clearly considerable attention should be given to establishing the presence of such layers when examining cored samples and, indeed, to the design of the site investigation to recover such samples, and every engineering geologist should refer here to the work of Rowe (1972). The aid of such layers in allowing drainage of clays is demonstrated by the installation of filter layers in embankments constructed from clays: those with such layers installed consolidate more rapidly than those which do not have such aids to drainage. However, in nature, sand and silt may occur in lenses totally surrounded by clay and would thus not act as drains. Attention must be given to establishing the lateral continuity of sand layers found in boreholes.

In design practice, it is common to consider the time taken to achieve 50% consolidation and 90% consolidation. While this depends on soil characteristics it is not uncommon that in normally consolidated clay 50% consolidation may take about 2 to 5 years to achieve and 90% consolidation more than 10 years to achieve.

11.4
Bearing Capacity on Rock Masses

A large number of major civil engineering works are constructed on rock. The reaction of rock to static loading is different from the reaction of soil because of the generally much greater strength of rock as a material, the presence of discontinuities in the rock mass and the type of anisotropy of the rock mass.

One of the first questions to be asked in considering the reaction of a rock mass to static loading is "*Is it all rock?*"; all rocks, when weathered, reach a stage in their weathering when their geotechnical properties are those of soils. *The engineering geologist must be aware that within a rock mass there may be bands and zones that have the properties of engineering soils*; these need not all be weathered zones. Some so-called 'shales' are, in terms of their properties, over-consolidated clays, with the added complexity of marked discontinuities. It is thus particularly important to check on the engineering properties of argillaceous rocks in layered sedimentary sequences, such as alternating limestones and shales or sandstones and shales.

Establishing the allowable bearing pressure to be used for a particular project may begin by considering the ultimate bearing capacity (Q_f) of a rock mass. Assuming that the discontinuities in the mass have no influence on bearing capacity, Q_f is approximately four times the uniaxial undrained compressive strength of the rock material. Thus,

$$Q_f \approx 4U_c \tag{11.8a}$$

and safe bearing capacity (Q_s) is equal to U_c/x, where x is the factor of safety; a value of $x = 3.0$ has been suggested for weak gypsum-cement type rock though in most cases values less than that are adequate (Serrano and Olalla 1996). If, however, the mass is jointed and the joints are open, so that joint bounded columns of rock are not restrained by adjacent rock then clearly the ultimate bearing capacity of the mass under a particular foundation is equal to the compressive strength of the rock columns. Thus,

$$Q_f \approx 4\,U_m \quad \text{and } Q_s \approx U_m/\,3 \tag{11.8b}$$

where U_m is the uniaxial undrained compressive strength of a *column through the rock mass*. U_m may not equal U_c because the column may contain frequent and unfavourably orientated internal discontinuities in it, so that $U_m < U_c$.

If, for example, it is assumed that a rock material is weak with an unconfined compressive strength of 5 MPa and that U_m/U_c is 0.5, then Q_f would be about 10 MPa and Q_s about 0.8 MPa.

The influence of discontinuities on bearing capacity has been studied by tests on model rock masses. Goodman (1980) concluded that the bearing capacity of a homogeneous, discontinuous rock mass under a footing could not be less that the compressive strength of that *rock mass* and proposed the formula:

$$Q_f = U_{cm}(N_\phi + 1) \tag{11.9}$$

where $N_\phi = \tan(45 + \phi/2)$, U_{cm} = unconfined compressive strength of the rock mass and ϕ = angle of shearing resistance of the rock mass.

Berkhout (1985) examined the behaviour of modelled rock masses with various degrees of joint density. The rock masses were modelled from variously dimensioned cubes of artificial rock and loaded by model foundations to measure deformation and failure. Joint apertures were equal. As might be expected, clear relationships emerged between bearing capacity and joint density and between deformation modulus and joint density. Both decreased as joint density (expressed as joint surface area per unit volume) increased (Fig. 11.6).

The approaches examined above consider a single foundation element while most structure foundations are made up of many elements of different sizes covering a significant area. The results of the model studies and the Goodman equation clearly reveal the factors which are of importance in establishing the bearing capacity of rock masses; whether or not these factors can be measured in the course of site investigation is another matter.

A ground investigation might aim to divide the rock mass into one of two classes; those in which the properties of the rock mass approximately equal those of the rock material, and those in which they do not.

The classification of the mass can be achieved by comparing measurable mass and material properties, the former being influenced by the frequency, aperture, orientation and persistence of discontinuities and the latter by weathering. Such properties might be: E_{static} for the mass versus E_{static} for the material, or mass permeability (k_{mass}) versus material permeability ($k_{material}$), or compression wave velocity V_p(mass) versus V_p(material), or any other property of both mass and material which would be significantly changed by the presence, intensity and condition of discontinuities and weathering.

The closer the ratio (mass properties/material properties) approaches unity the nearer do mass properties approach those of the material.

The factors that determine the bearing capacity of a rock mass have been introduced into rock mass classification systems (Bieniawski 1989) and the Hoek-Brown empirical rock failure criterion (Hoek 1983). Serafim and Pereira (1983) have indicated correlations between *RMR* (Rock Mass Rating) and mass cohesion, friction and

Fig. 11.6. Influence of joint density on deformation modulus and bearing capacity (after Berkhout 1985)

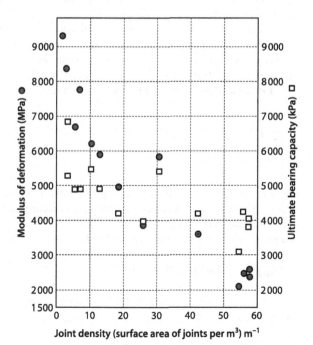

Joint density (surface area of joints per m³) m⁻¹

modulus, while Laubscher (1984), utilising a rock mass rating system arrives at a basic strength for a rock mass. It would seem clear that use of a rock mass rating system is a way forward for establishing a method of estimating rock mass bearing capacity, but much information has to be assembled in a database to allow development of a reliable system; this is being done via in situ monitoring of ground movement under foundation loads and numerical modelling of the same ground. The parameter for strength and stiffness which a model requires so as to predict the movements measured are those that can be attributed to a rockmass rating system for the ground (Marinos and Hoek 2000). Enough experience already exists to recommend guideline bearing capacities for various rock types or formations but the data required to provide a non-empirical prediction is still developing.

11.5
Foundation Settlement on Rock

It is commonly assumed that settlement of structures on rock foundations is almost always immediate or at least very rapid and may be calculated using the formula appropriate for the settlement of a rigid circular plate, namely:

$$S = \frac{Q\left(1 - v^2\right)}{r\,\pi\,E_s}$$ (11.10)

where S = settlement (m), Q = foundation load (N), r = radius of circular foundation (m), E_s = (secant) modulus of deformation (N m^{-2}), and v = Poisson's ratio.

The modulus of deformation, may be obtained by plate bearing tests on 'samples' of the rock mass. Which modulus is chosen as representative of mass behaviour depends on the way the mass behaves, but a secant modulus to either a maximum design foundation stress, or higher, may be appropriate. A useful but approximate formula for the settlement of a square foundation is:

$$S = \frac{q\,B}{E_s}$$ (11.11)

where B = foundation width (m) and q = foundation pressure (N m^{-2}).

If no in situ tests are available to measure E_s, recourse may be considered to determine the deformation modulus from the Rock Mass Rating by the Geomechanics Classification (Bieniawski 1989). Bieniawski offers the correlation for $RMR > 50$ of:

$$E_{mass} = (2 \times RMR) - 100 \text{ (GPa)}$$ (11.12)

where E_{mass} = in situ modulus of deformation in units of GPa and RMR = Rock Mass Rating.

These very simple approaches are, of course, appropriate to only rather simple problems. In the case of major structures on heterogeneous rock masses recourse may be had to finite element analysis to allow calculation of rock deformation. Here, the problem lies not only in the mathematics involved in such solutions but also in finding truly appropriate mass property parameters to put into the equations.

One difficulty with rock masses is that anisotropy of the mass will deform the shape of the bulb of stress under the foundation. Gaziev and Erlikhman (1971) have shown by model testing that the bulb of stress will be deformed depending on the inclination of stratification (Fig. 11.7). Obviously such a stress distribution will have a profound effect on deformation and perhaps also on bearing capacity.

If rock deforms unequally under a foundation then the problems of differential settlement arise. Such problems are particularly significant in the case of concrete dams for if the modulus of deformation of the rock formation is greatly different from that of the dam concrete then the concrete may deform to the extent that it cracks or that the watertight seals linking concrete monoliths are opened.

Another basic problem of concern in rock foundations is the problem of *creep*. Tests have shown that any rock has an *'instantaneous strength'*, measured in the standard unconfined compression test and a *long-term strength*, which may be assessed in a creep test. The long-term strength may be perhaps but one quarter of the instantaneous strength. Thus structures imposing a long-term load on a creep sensitive rock foundation could, perhaps after many years, bring that rock to the point of failure. What seems to reduce this potential problem to relatively slight proportions is the fact that most major structures are founded on the stronger rocks so that even their long-term strength is more than adequate to sustain the foundation load. However, some consideration might have to be given to potential creep problems if heavy structures are founded on obviously weak rock which might be creep sensitive. *Such rocks are the clayey rocks, highly porous and weakly cemented rocks and rocks which might contain gypsum, anhydrite or salt.*

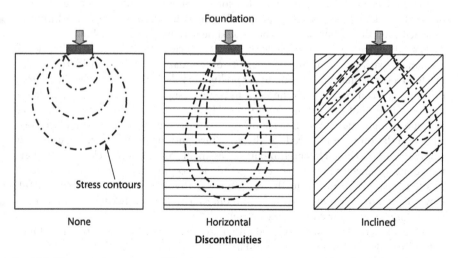

Fig. 11.7. Changes in stress pattern following discontinuity structure (Gaziev and Erlikhman 1971)

While creep phenomena might not lead to failure they might lead to settlement that is somewhat greater than that calculated on the basis of elastic parameters. Meigh (1976) has described how the settlement of a nuclear reactor founded on Keuper Marl was calculated to be about 40 mm with an estimated total settlement of about 60 mm. After application of 90% of the total load the settlement had already exceeded 60 mm and was increasing under virtually constant load. Creep characteristics of the rocks were investigated and taking these into account settlements of about 115 mm could be anticipated. By the end of 1974, eleven years after construction commenced, settlement had reached about 102 mm on one reactor and 120 mm on the other; time/settlement graphs indicated that settlement appeared to be complete.

11.6
Preliminary Estimates by the Engineering Geologist

Every one of the calculations outlined so far requires the ground to be idealised in some way so that a "*typical*" value for this or that parameter can be used. What should the engineering geologist "*do*" to contribute relevant data? It helps to know the data required, hence the introduction above, and the geology that will be relevant.

11.6.1
Estimates of Bearing Capacity

Over the years engineers have developed, as a consequence of experience, some general knowledge of the likely bearing capacity of most ground masses. This knowledge has been assembled and presented in codes of good practice in foundation engineering. The table presented in this text (Table 11.2) is that given in the 1975 edition of Blake's '*Civil Engineer's Reference Book*'; it is not repeated in more modern editions of that work. Once published, tables of this sort become too readily accepted by engineers as short cuts for design purposes, which may lead subsequently to arbitration in which even the authors of the tables could be held partly accountable.

The table gives '*presumed bearing values*' for various types of soils and rocks. A presumed bearing value is the net intensity of foundation loading considered appropriate to a particular type of ground *for preliminary design purposes*. The idea is that the engineer, confronted with the need to begin designing a structure, might determine the nature of the ground from the study of a geological map, use the table to assign bearing values and thus get some idea of the suitability of the site and the possible foundation type required for the structure and the ground investigation that would be appropriate. This preliminary assessment might lead to selecting an alternative site, to changes in site layout design, to the recognition of certain requirements in site investigation and so forth. Used this way the table can be extremely useful. However, *it must never ever be used as an alternative to undertaking either a site investigation, or a proper engineering design.*

While the table may be considered rather old-fashioned, its consideration of foundations on rock embodies certain fundamental principles, namely,

i   the stronger the rock material the greater the 'presumed bearing value',
ii the greater the number of discontinuities in the rock mass the lower the 'presumed bearing value',
iii that whatever the character of the rock mass the 'presumed bearing value' will be reduced by weathering and faulting.

Many structures founded on rock may not have the economic, industrial and social worth that merits the profound and detailed investigations given to major works. For these structures, the office blocks, the minor buildings and the like, a philosophy which may be followed is not to try to determine the true bearing capacity of the rock mass by elaborate testing and calculation, but to ask the designer what bearing capacity is necessary for reasonable foundation design, *and then to determine whether or not the rock mass is adequate for that purpose.* If this approach is undertaken then the presumed bearing values given in the table are a good starting point for investigation. The investigation may be directed towards determining whether or not there are

Table 11.2. Approximate presumed bearing values for rock and soil masses

Croup	Class	Type of soil or rock	Presumed bearing value (kPa)	Remarks
I Rocks	1	Hard igneous and gneissic rocks	10 000	These values are
	2	Hard limestones and sandstones	4 000	based on the
	3	Schists and slates	3 000	assumption that
	4	Hard shales and mudstones, soft sandstones	2 000	the foundations
	5	Soft shales and mudstones	600 to 1 000	are carried down
	6	Hard sound chalk, soft limestone	600	to unweathered
	7	Thinly bedded limestones, sandstones, shales	To be assessed	rock
	8	Heavily shattered rocks	after inspection	
II Non-cohesive soils	9	Compact gravel or compact sand and gravel	>600	Width of foundation (B) not less
	10	Medium dense gravel or medium dense sand and gravel	200 to 600	than 1 m. Ground-
	11	Loose gravel or loose sand and gravel	<200	water level assumed to be not
	12	Compact sand	>300	less than B below
	13	Medium dense sand	100 to 300	foundation
	14	Loose sand	<100	
III Cohesive soils	15	Very stiff boulder clays and hard clays	300 to 600	Group III is susceptible to long
	16	Stiff clays	150 to 300	term consolidation and settlement
	17	Firm clays	75 to 150	
	18	Soft clays and silts	<75	
	19	Very soft clays and silts	Not applicable	
IV	20	Peat and organic soils	Not applicable	
V	21	Made ground or fill	Not applicable	

The 'presumed bearing value' is the value of the safe bearing capacity of the ground that may be adopted for preliminary design purposes. This must be checked and finally established by investigation.

geological factors, such as weathering, or geomorphological factors such as disturbance of the ground by landsliding which make the presumed bearing value invalid.

11.6.2
Influence of Geological Factors on Foundation Performance

Variations in Mass Strength in Alternating Layers

It is not uncommon in bedded rocks to find a layer of strong foundation rock overlying some softer or weaker material. While the upper layer may well be strong enough to carry the foundation load the underlying layer may be overstressed and fail. Kaderabek and Reynolds (1981) have described the problems of building on Miami Limestone; a young deposit 100 000 years old and forming a layer 3 to 14 m thick. Often this layer occurs as a 'mat' or plate of rocky material resting on sand. The basic problem is whether or not the mat is able to carry a proposed building load.

The authors have recognised four possible modes of behaviour of the mat which must be considered when attempting to assess its bearing capacity. These are: punching or diagonal failure of the mat (Fig. 11.8a); local crushing of the limestone under the foundation; beam tension failure (Fig. 11.8b); and settlement of the mat into the underlying soils.

The authors suggest that the first failure mode is most likely to occur if the ratio H/B is small (H = mat thickness, B = foundation width). This might happen if foundations are sunk deep into the limestone (as for basements) as the effective thickness of the mat is thus reduced. To examine this possibility the authors suggest that a cylindrical failure surface is assumed, the surface area of this cylinder equalling the perimeter of the footing times the thickness of the limestone below the foundation. The ultimate resistance to shearing then equals the shear strength of the rock times the surface area of the cylinder.

To overcome the second problem the authors suggest that the theoretical contact pressure of the concrete foundation on rock be limited to no more than the uncon-

Fig. 11.8. Failure modes of foundations on Miami limestone (after Kaderabek and Reynolds 1981)

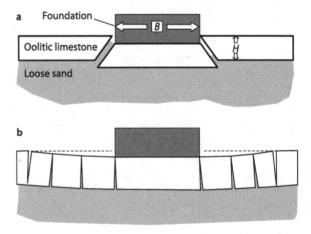

fined compressive strength of the rock material. The third type of failure may occur when the H/B ratio is large and the flexural strength of the mat small. In such a case the tensile strength of the rock is the dominating factor. To determine the settlement of a structure on a limestone mat which has been established as unlikely to fail in any of the three modes given above, the authors calculate the deformation of the mat using the elastic properties of the rock and thence calculate consolidation of the soils below using conventional soil mechanics techniques.

The properties of the Miami Limestone are not uniform and the authors have determined the shear strength of the rock mass (necessary to assess the possibility of punching failure) by undertaking anchor pull-out tests.

Influence of Weathering

All materials wheather once they are exposed to the near surface environment. Underground, chemical weathering takes place aided by percolating groundwater. Soils are less likely to weather than rocks as they are in many cases already the products of weathering. Changes can occur within them; e.g. in tropical climates the development of lateritic soils is a consequence of weathering and is of importance in engineering in such areas and the behaviour of *"quick clays"* is attributed to leaching (Kenney 1964).

The effects of weathering are much more obvious in rock. The susceptibility of a rock to weathering depends upon its composition; in general the more clayey the rock the greater its susceptibility. Thus shales could be considered to be more susceptible than sandstones. If these two rock types are found in an interlayered rock mass and if they have been subjected to exposure to weathering then the weathering in the shale would be more severe than that of the sandstone. In dipping strata this can mean that the shale would be weathered to a greater depth than the sandstone, so that weathered shale could be found *under* almost fresh sandstone (Fig. 11.9a). This has an important effect on foundation bearing capacity. While weathering, differential or otherwise, may be recognised as a potential problem it is very difficult to anticipate the style and severity of weathering in a specific rock mass because this not only depends on lithology but also on its geological history (unloading, glaciation, etc.).

In Europe and other parts of the world this is made particularly difficult because of the changes in climate and weathering agencies which occurred during the Pleistocene. Higginbottom and Fookes (1971) have described how, at the Pembroke Power Station (Fig. 11.9b) softened shales and mudstones were reduced to the consistency of a clay as a consequence of periglacial action, so that decomposed rock could be found beneath fresher materials.

It is important to recognise that certain rock types have distinct styles of weathering. Thus the coarser grained igneous rocks (such as granite, gabbro and dolerite) are susceptible to spheroidal weathering. In this, weathering eats into the rock material, beginning at the joints which conduct percolating water. The zone of weathered material widens around the joints until rounded core stones of almost fresh rock lie in a matrix of sandy residual soil. The boundary between the residual soil and the core stones is often sharp; intervening weathering grades occupying but a few millimetres of matrix, if they can be said to exist at all.

It may be very difficult to recognise this condition from borehole evidence (Fig. 11.10). Thus two boreholes, A and B in the figure, may be, by chance, so situated

Fig. 11.9. Consequences of differential weathering in dipping layered rocks; **a** the general problem; **b** at Pembroke power station (after Higginbottom and Fookes 1971)

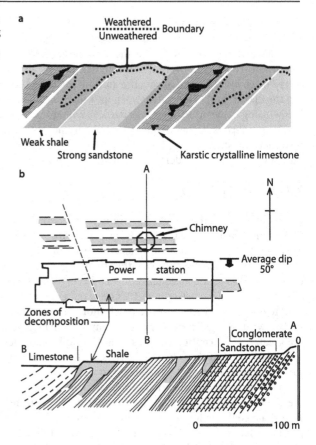

Weak shale
Strong sandstone
Karstic crystalline limestone

that borehole A shows mostly fresh rock and borehole B mostly residual soil. Unless the results of these boreholes are interpreted in the light of background knowledge of the likely weathering style, an interpretation of rock mass conditions could be erroneous. It is always a good idea, particularly if working in an area previously unknown to the investigator, *to examine natural outcrops*, quarries, cuttings etc. to obtain an impression of the style of weathering of the local rocks. In this way one is better able to interpret mass weathering from the information given by a series of boreholes drilled for the design of foundations.

Soluble rocks, the limestones, and the salt, gypsum and anhydrite bearing rocks, present especial problems. The phenomenon of the solution weathering of limestone is perhaps the most widely known. Areas of limestone in which extensive underground drainage has evolved as a result of solution weathering and which display surface swallow holes, *sinkholes*, *dolines*, and the like may be described to be '*karst*' (after the type area in Yugoslavia). The solution process, by dissolving away the calcium carbonate of which limestone is chiefly composed, leaves behind a mass of unweathered limestone containing holes. If a limestone weathering product remains it is often red ferruginous clay (*terra rossa*) but there is not a lot of it. Limestones, although full of so-

Fig. 11.10. Corestone weathering in granite. Borehole A would appear to be in almost solid rock; borehole B in granite sand. Knowledge of the style of weathering of such rocks is required to interpret correctly the nature of the mass from the results of the boreholes

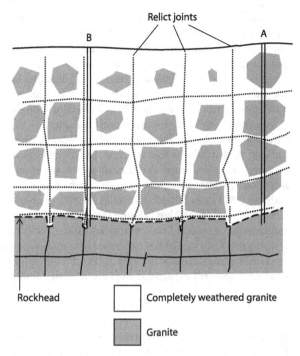

Relict joints

B   A

Rockhead   Completely weathered granite

Granite

lution caverns, tend to form ridges through which rivers may flow in narrow steep sided valleys. Such valleys are well suited topographically for dam construction but the engineer has to accept the hazard of solution cavities.

If such a site is chosen for a dam the designer faces two major problems, first, that the rock mass is weakened by the caverns within it so its bearing capacity is suspect and, second, that leakage under the dam and through abutment rock could be through karst. It is sometimes found that the problem is not the relatively simple one of caverns in otherwise solid rock; in the geological history of the limestone mass caverns may have collapsed, the debris become re-cemented, then further solution occurred and so on. Sometimes caverns are partly filled with soil which may be the weathering product of the limestone itself or have been washed into the caverns from some overlying formation. Boreholes sunk along a dam axis may reveal a nightmare melange of such features. Many papers have been published describing such conditions. For example, Moneymaker (1969) has described the conditions encountered in a number of dams built in the USA; in the case of the Kentucky Dam, solution features extended for some 70 m below the valley floor. In some cases the solution activity may be seen to be concentrated along a particular band of limestone, this band being particularly susceptible to solution as a result of texture, mineralogy, jointing or perhaps because it is bounded by impermeable layers of shale. Solution activity revealed by karst need not be of recent geological origin; Bless et al. (1980) have shown that the karst found in Devonian Limestones in Belgium was of early Carboniferous age.

A major problem that faces all engineers dealing with any form of construction on karstic limestone is whether or not the limestone solution is *continuing* or likely to be re-activated as a result of the construction work. In dams the danger is that water, slowly seeping through the foundation rock under the hydraulic head of the reservoir, will enlarge fissures and form cavities. This depends partly on water chemistry and lime-stone mineralogy and it is prudent to measure solution parameters of carbonate rocks to help assess these possibilities (James 1981). Existing limestone caverns may be filled with grout and concrete to render the rock mass impermeable, but grout cannot penetrate caverns filled with clay debris. This debris might, however, be slowly washed out by percolating water from reservoir leakage.

If, on any site, bedrock is found under a layer of younger terrestrial sediment such as glacial drift it follows that at some time in geological history the bedrock surface was an exposed landscape and may display features associated with its erosion. Features such as faults, dykes and so forth may be found eroded out to form deep gullies. Rawlings (1971) has described glacial meltwater channels which were found in the foundation rock of the Llyn Brianne Dam in Wales. In glaciated areas the bedrock may contain major dislocated rock blocks and shear zones formed from the movement of overlying ice. Knill (1968) has described the significance of such features in dam construction. Bedrock outside the area of direct ice erosion may be severely disturbed by 'periglacial' features associated with permafrost conditions.

Discontinuities

All rock masses contain discontinuities and most rock masses have at least three sets of discontinuities cutting them. Their significance to foundations is related to their frequency of occurrence, orientation and openness. Orientation is important in that, relative to the direction of the stresses imposed by the foundation and the shape of the landscape on which the foundation is placed, sliding may occur along unfavourably oriented discontinuities. Frequency is important in that the more frequent the discontinuities the more opportunities there are for them to be open and to slide.

Not all loads imposed by foundations are vertical. In the case of dams the pressure of the water in the reservoir is resisted by the mass of the dam, the connection of the dam to the ground, and the mass strength of the ground (Fig. 11.11). During construction a dam imposes vertical load on its foundation but once the reservoir begins to fill the water pressure on the dam adds a horizontal component to the foundation load. The magnitude of this horizontal component depends on the type of dam. Gentle sloped earth dams have a lesser component than steep sided concrete dams, for the force of the water is at right angles to the water retaining component of a dam slope. The ratio of dam mass vertical pressure to horizontal water pressure tends to become smaller as the dam becomes thinner and steeper, reaching its ultimate in thin arch dams, where the dam concrete transfers the water pressure to rock abutments.

If horizontal forces exist then there is always the possibility that the dam foundation might slide along a low angled weak discontinuity, such as a shear zone, fault or bedding plane, in the foundation rock (Fig. 11.12).

One obvious area of danger is the interface between the cast-concrete, or placed earth/rock fill, and the rock foundation. For this reason foundation rock is always carefully cleaned before the foundation is placed and sometimes sprayed with gunite. In concrete dams much attention is paid to a good concrete/rock bond and cored boreholes may be sunk through the interface to check this.

Fig. 11.11. Earth, gravity and arch dams. As the dam becomes thinner there is less mass to resist water thrust and the direction of thrust changes to give a greater horizontal component

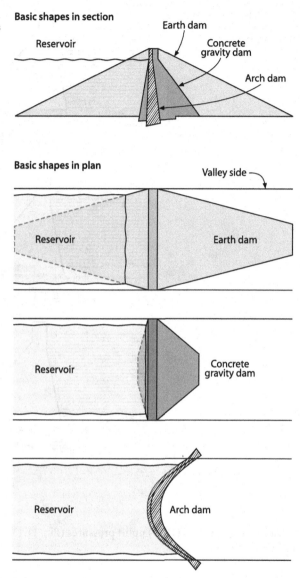

Basic shapes in section

Earth dam

Reservoir

Concrete gravity dam

Arch dam

Basic shapes in plan

Valley side

Reservoir

Earth dam

Reservoir

Concrete gravity dam

Reservoir

Arch dam

The reader should at this point recognise the similarity between the sliding dam and the rock block sliding on a slope that forms the first page in many texts on slope stability. The mechanism is the same; the difference is the origin of the forces and their orientation relative to the horizontal. Answers to the problem are also similar. Frictional forces may be increased by long cable anchors, as was done in the Les Cheurfas dam in Algeria (Walters 1971) and water uplift pressures may be reduced by drainage wells sunk into the rock mass and discharging into galleries. Grout curtains or concrete cut-offs, and blanket grouting lengthen seepage paths and so reduce leakage and uplift pressures.

Fig. 11.12. Potential planes for sliding of dam foundations under reservoir pressure

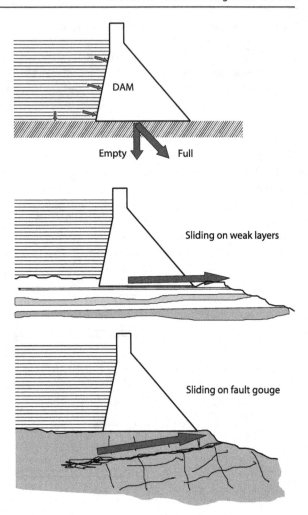

According to Jaeger (1979) uplift pressures (Fig. 11.13) may be calculated using the formula:

$$F_u = \frac{\gamma_w \, B\left(jH_r + H_t\right)}{2} \tag{11.13}$$

where F_u = uplift force (N), B = base area of dam (m^2), H_r = height reservoir (m), H_t = height tailwater (m), γ_w = density of water (N m^{-3}), and j = a factor, not more than one, which is related to the percentage of the base area of the dam on which the uplift forces act. It may be reduced to values of 0.8 or 0.9 using the judgement of the design engineer.

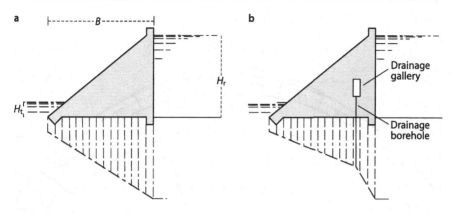

Fig. 11.13. Reduction of water uplift pressures on dams by drainage

In Fig. 11.13a uplift pressures are represented by the area of the shaded trapezium below the dam. With drainage this area is reduced (Fig. 11.13b). Drainage holes are sunk to a depth at least that of the height of the drainage gallery above the dam base. Some engineers prefer large diameter holes or shafts to boreholes, for these are less easily blocked. The water rises into the gallery and drains away. How much rises depends upon rock mass permeability.

If the dam is so designed that the reservoir water thrust is almost totally resisted by the rock foundation with little resistance given by the weight of the dam, as in arch dams, then the direction and strength of discontinuities in the rock mass is of vital importance in assessing rock mass resistance. For such projects extensive and detailed geological mapping must be undertaken to determine orientation and location of discontinuities and laboratory and field tests are required to determine their strength.

Having determined these factors then calculations must be undertaken to analyse dam stability. These calculations are again based upon the sliding block concept. The data required to allow calculation includes: (1) orientation of discontinuities, (2) orientation of dam loads, (3) magnitude of dam loads, (4) shear strength parameters of discontinuities and (5) estimates of water pressures on the dam and in discontinuities.

The investigation of the foundation conditions of the Monar Dam in Scotland provides a good example of this type of analysis; Henkel et al. (1964). The Monar dam is a double curvature gravity dam (resembling an arch dam) on the river Farrar in Scotland. It is not a large dam, being 35 m high and 161 m long. The concrete of the arch is divided into eleven blocks separated by construction joints which were sealed after the concrete had cooled (Fig. 11.14). Construction took place between 1959 and 1963. The dam is founded on metamorphic rocks which are psammitic granulites of Pre-Cambrian age. The rocks have been folded in at least two periods of deformation and are generally strong and little weathered but contain thin pelitic layers, pegmatites and occasional quartz veins. The bulk properties of the rock mass are controlled by a system of joints. These were examined in exposures and cores and were found to fall into sets, shown in Fig. 11.14. The dyke was of heavily sheared, slickensided and weathered lamprophyre, about 1.8 m thick and outcropped upstream of the dam, dipping

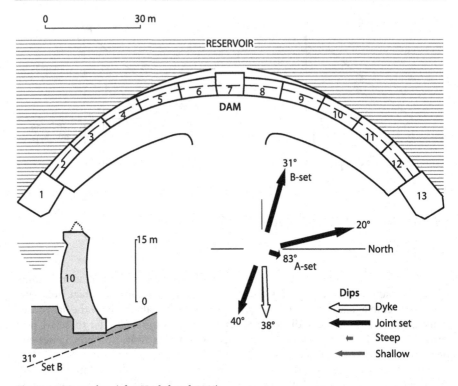

Fig. 11.14. Monar dam (after Henkel et al. 1964)

under the dam to be about 18 m below the dam base in the river section. Joints were tested for peak and residual strength and as a consequence of orientation and strength it was considered that sliding of the dam might occur along the joint set dipping westward at 31°. Accordingly the stability of six of the eleven component blocks was examined. In making the stability analysis it was assumed that resistance to movement of the block was provided only by the foundation rock.

Three types of forces were involved in the stability calculation. These were:

i forces transferred to the dam foundation by the dam structure. These forces were vertical and parallel to the dam axis;

ii forces due to the weight of the rock wedge forming the dam foundation;

iii water forces, consisting of thrust at the back of the foundation wedge and uplift on the joint B at the base of the wedge.

These forces were resolved into components acting normal to and along the plane of sliding. However, the joint A also provided some restraint to sliding and this was taken into account. It was assumed that the orientation of the B joints was 31° dip towards N284° and that the A joints were vertical, striking N285°, and that the angle of friction of the B joints was 50° and that of the A joints 25°. The factors of safety of

Table 11.3. Monar dam – increases in factors of safety as a consequence of drainage

Block No.	Factor of safety	
	With drainage	Without drainage
1	3.3	3.3
3	2.7	2.2
5	5.3	3.7
7	3.0	2.3
9	4.1	2.7
11	3.6	2.8

each block were calculated assuming no drainage and drainage measures to reduce water pressure (Table 11.3).

These results demonstrate the increase in safety achieved when water pressure is reduced by installing drainage boreholes. Later in the analysis the effects of variation in joint orientation were examined and it was found that, while the factor of safety was in some cases reduced, it was never-the-less adequate. The technique used to analyse the situation was essentially that used to analyse the stability of the sliding block on a slope but was complicated by variations in force direction relative to geological structure. Performance of the dam after construction was monitored and it behaved satisfactorily.

Other dams have been less fortunate. The 151 m high Canelles dam in Spain (Alvarez and Sancho 1974), was founded on karstic limestone which gave a problem of leakage, solved by an extensive grouting programme. The limestones dipped upstream (Fig. 11.15) and contained extensive vertical joints near parallel to the valley sides. It was feared that these joints would lessen the resistance of the rock mass to the reservoir water pressures transmitted by the arch dam to the rock mass. Samples of joints were cut out of tunnel walls and tested for shear strength.

Calculations confirmed these fears and accordingly a concrete buttress, held in place by prestressed cable anchorages (Fig. 11.15), was proposed to provide additional resistance to possible sliding on the vertical joints. However, possible problems with cable corrosion and relaxation (as later experienced at Les Cheurfas dam) recommended their replacement by tunnels, filled or lined with heavily reinforced concrete, which cross the joints and act as shear pins to reinforce them.

11.7
Foundations on Slopes

Every slope has a stability that depends on the strengths of materials and discontinuities in the slope, action of groundwater, slope angle and so forth. If a structure is built at the top of the slope then the load imposed by that structure adds to the slope weight and to the potential driving forces that can initiate failure. It is relatively easy to assess the stability of a slope in fairly uniform soil which is additionally loaded by a structure built at the top of the slope, however in many soils and in rocks this is some-

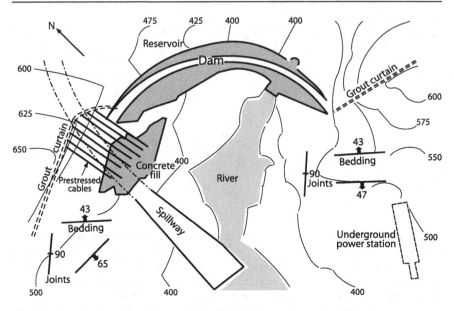

Fig. 11.15. Canelles dam (after Alvarez and Sancho 1974)

what more difficult because the slope stability may be influenced by particular horizons and by discontinuities. Larnach and Bradshaw (1974) examined the stability of slopes in limestones and mudstones on which a hotel was to be built overlooking the Avon Gorge, Bristol, England. They found that the slope had a potential for sliding on clay layers in a direction oblique to that of the dip of the beds. Their solution to the problem was to take the foundations of the proposed hotel down to a depth below the potential sliding plane using deep piles or piers.

It is common for the foundations of bridges to be placed on slopes especially as bridges tend to be as short as possible and are often located to cross narrow places in valleys. The valley side in rock will probably be both weathered and relaxed, most intensively weathered along discontinuities so that some may be weaker than others. Inevitably, bridge foundations tend to be as close to the valley edge as possible and thus in an area of most severely weathered rock; this has to be either penetrated to sound rock or supported with cables, anchors, bolts etc. or improved. In such a situation a single continuous highly weathered discontinuity could determine the stability of the foundation. Placed on one side of it, the foundation could be potentially unstable, where as if placed on its other side, perhaps only a metre or two away, it is stable (Fig. 1.2).

In the case of arch bridges (without mid-span piers) the thrust of the bridge structure is not vertical but directed into the valley sides. If there is a possibility of slope failure into the valley along discontinuities dipping out of the valley side, then the bridge thrust may improve slope stability by adding to the forces acting normal to the potential sliding discontinuities. However, it must be remembered that this condition exists only *after* completion of the bridge. Until the arch is formed abutment loads tend to be vertical and would therefore increase potential instability. Thus the design of a foundation on a rock slope should take account of stability *both during and after construction.*

11.8
Construction Problems

11.8.1
Foundation Excavations

Recommendations for foundation design based on the results of site investigation have to include comments on the level of the stratum capable of sustaining the proposed foundation pressure (without exceeding allowable settlement), the stability of the excavation sides, likely occurrence of groundwater in the excavation and the method of excavation.

The foundation material exposed at the bottom of an excavation has been relieved of the weight of the material removed, and may rebound slightly and so loosen the ground. This might result in the opening of any discontinuities, perhaps increasing mass permeability and making it easier for groundwater to flow into the excavation. If the ground is relatively weak and the excavation deep, the stresses from the side slopes may cause the foundation ground to bulge upward or heave. Some clays may take up water and swell, giving a form of ground heave. An excavation may be made in ground overlying an aquiclude which confines an aquifer containing water under pressure (Fig. 11.16). As the excavation is deepened the point must come when the weight of the material overlying the confined aquifer is insufficient to restrain the uplift pressure i.e.

$$x\gamma_e + y\gamma_a < \gamma_w h \qquad (11.14)$$

where γ_e, γ_a and γ_w are unit weights of near surface ground, aquiclude and water respectively. Excavation can only safely continue if water pressures in the aquifer are reduced by some form of dewatering around the excavation.

Fig. 11.16. Sub-artesian head gives the potential for ground heave and water inrush in a deepening foundation excavation

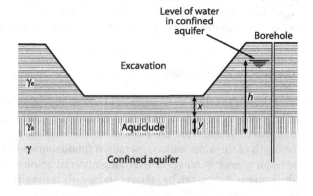

Most foundations on rock are too shallow to reach unweathered rock mass; open joints, erosion gullies and so forth may be encountered. Having established that these are not indications of a greater problem, such as karst at depth, mine collapse, landsliding, etc., they are generally first cleaned out as well as possible and then back-filled with compacted concrete or grout, to restore the integrity of the mass so that, following the principles outlined above, the ground becomes more like that of the material than its weathered mass.

The rock mass examination and any remedial works should be undertaken as soon as possible after excavation, for weathering and decay starts as soon as the ground is exposed. While in many rocks the effect is not particularly significant if the period of exposure is a few weeks, in some rocks it could be quite serious. The clayey rocks, the mudstones, shales, siltstones and similar materials can be particularly susceptible to swelling phenomena or may simply decay due to wetting and drying, freezing and thawing and temperature changes. Rocks containing readily soluble minerals such as rock salt and gypsum should also be protected. One answer to this problem is to seal newly exposed rock by covering it with a thin layer of concrete (as shotcrete, or gunite or some similar grout-like mixture). This may be removed if need be before casting foundation concrete on freshly excavated rock.

The sides of any excavation must be stable during construction of the foundation, and consideration should be given to the design of these slopes and, if necessary, the means of temporarily supporting them.

11.8.2
Stepped Foundations

A foundation may be constructed on a slope in such a way as to avoid problems of foundation stability, but difficulties could arise for particular foundation elements within the foundation excavation itself. Thus, for example, if a foundation, comprising multiple pad footings, is constructed on sloping ground, steps may have to be excavated in the slope onto which the pads will be cast, so that some will be on a higher level than others. Depending on the height of the steps, the spacing of the footings and rock discontinuity orientation and spacing, some footings could be unstable. In horizontal foundations the bulbs of stress from each pad footing will overlap to give a uniform stress field at depth. If the foundation is stepped, then the stress from the higher foundation will partly increase and distort the stress field under the next lower foundation, and so on. This might result in unequal settlement across the footing width.

11.8.3
Foundations Piled to Rock

Often rock is too deep to be reached by open excavation from the surface. This depth will depend on the character of the overburden, whether it requires support, the need to support adjacent structures, ease of soil disposal and backfilling the excavation, and the depth of the water table. Generally, if foundations for standard structures have to be taken to rock because the overlying material is too weak, the maximum depth to which an excavation can be taken economically is usually 6 to 8 m; if deeper than that then piles are normally considered. The choice between using driven or bored piles depends on a number of factors, Table 11.4, as pile design is a complex process and certain parameters have to be derived by site investigation to allow this to take place.

Bored Piles into Rock

Bored and cast-in-place piles ending in rock may penetrate into the rock to pass through any weakened upper weathered zone. In such 'rock-socketed' piles, the end-bearing load capacity may be the safe bearing capacity of the rock mass times the area of the pile base, while the skin-friction load capacity is the 'skin-friction' per unit area times the area of the pile shaft within the rock socket. A simple conservative approach is to assume that end-bearing load capacity plus skin friction load capacity is the maximum safe carrying capacity of the pile.

In soft rocks the sides of the pile hole may be smooth and values of skin-friction between pile concrete and rock may be used to calculate total skin friction load. If the rock is strong it is probable that the sides of the rock socket will be rough so that the load will be carried by the key between rock and concrete (Fig. 11.17). In such cases the rock/concrete key rather than the rock at the base of the pile may carry most of the load, and if so the critical factor in the carrying capacity of the pile is the shear strength of the weakest member of the rock/concrete key.

Table 11.4. Factors to be considered in choosing between driven and cast-in-place piles

	Bored and cast-in-place concrete piles		Driven piles	
	Advantages	Disadvantages	Advantages	Disadvantages
1	Available in very much larger diameters then driven piles (and thus in carrying capacity) and diameters can be varied in the foundation	Often difficult to emplace in heavily water bearing overburden	The carrying capacity can be estimated from the driving performance	May be prevented from reaching bedrock by boulders in the overburden
2	Can penetrate through boulders in overburden and bedrock easily recognised	Quality of concrete in pile may be suspect if emplaced in difficult conditions e.g. underwater	Installation not seriously affected by water in the overburden	Limited carrying capacity because of limited maximum diameter
3	Bedrock condition at final depth can be inspected and tested in large diameter types	Pile shaft may 'neck' in soft overburden over rock		Final performance may be uncertain in some rocks
4	Carrying capacity can be improved by forming 'bell' at toe or deepening into the rock (rock socketing)	Material excavated has to be removed from site		Ground heave may be detrimental
5		Careful disposal required if material excavated is contaminated		Public nuisance due to noise and vibration during driving

Fig. 11.17. The design of rock sockets for bored and cast-in-place piles. The graph (Cole and Stroud 1976) offers the factors to complete Eqs. 11.15 and 11.16

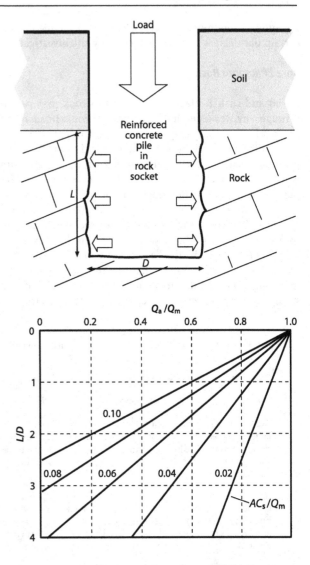

Design of such piles best follows pile tests in which skin friction factors and end bearing factors are measured for the particular rock type. However, such tests are expensive and on sites with variable rock types a pile test for each rock type would not be economic. Accordingly, certain empirical formulae have been devised which suggest the depth of rock socket required to carry the pile load. The rock socketed part of the pile under load may be viewed as a cylindrical shear box test, the shear force being that imposed by the pile load, which acts vertically, while the normal force is that imposed by the diametral expansion of the pile in response to the vertical load (Fig. 11.17). Any formula must take into account shear strengths of rock and concrete and the load on, and strength of, the rock at the base of the pile. Cole and Stroud (1976)

have offered a formula to determine the depth (L) of the rock socket to carry the required pressure (Q_m) to be imposed on the pile. This is:

$$\frac{L}{D}\left(\frac{AC_s}{Q_m}\right)=\frac{1}{4}\left(1-\frac{Q_a}{Q_m}\right)$$ (11.15)

where Q_m = maximum pressure to be carried by the pile, Q_a = allowable pressure on the rock at the base of the pile (its bearing capacity), C_s = shear strength of the rock in the socket and A = shaft adhesion factor (taken as 0.3).

The authors calculated Q_a using the formula:

$$Q_a=\frac{N_c C_b}{F}$$ (11.16)

where C_b = shear strength of rock below the pile base, N_c = bearing capacity factor, taken as 9 and F = safety factor, usually 3.

Note that this method gives the depth of the rock socket for the pile to carry the required load, but does not indicate the pile settlement. However, the method is considered conservative so that piles designed by this method will usually give settlements under the working load within the limits of tolerance of a normal structure. The method takes no account of the roughness of the socket wall and the relative strengths of concrete and rock. Most rock sockets are examined before placing concrete and the observer may decide, if the rock surface is rough and strong, to reduce socket depth. Alternatively, if the rock is weak and the surface smooth and muddy, it may be prudent to be conservative.

Driven Piles into Rock

In 'hard' rock it is unlikely that the pile can be driven for any significant depth and so this is an unlikely construction technique for use in such ground. On 'soft' rock the pile may be driven considerable depths. Piles may be heavily reinforced concrete with a steel point, or steel cylinders or 'H' beams. The piles have usually to be very strong so that they can withstand the hammering required to obtain the penetration to ensure lateral stability. Piles can pass down through weak soils to meet a rockhead that slopes steeply. As with the stepped foundation, higher piles might give additional load to lower piles; further if a group of such piles is loaded the lateral resistance of the soil may be inadequate to restrain the pile toes from slipping on the inclined rockhead surface. Steeply inclined rockhead topographies are common in glaciated areas especially on the sides of fjords, which often form the sites of jetties for shipping; such jetties would be founded on driven piles. Since many fjords are also in hard rock, the problem of the pile penetrating into rockhead to give support against slippage over rockhead, is one of importance. To overcome this, hardened steel tips may be welded on to the end of a conventional 'H' steel pile and serve to gain a few centimetres penetration.

Such piles are not readily commercially available and have to be especially manufactured. Accordingly it is very important for the engineering geologist to:

1. recognise the geomorphological history of the area, to suspect that rockhead may be steeply inclined beneath soft soils, and
2. design an investigation to determine the topography of rockhead and nature of the overlying soil, so as to allow pile toe and pile lengths to be designed.

In marine structures 'dolphins' may have 'raker' piles inclined from the vertical. In such a case recognition of rockhead topography is also clearly important.

Piles driven into soft rocks (weak sandstones, shales, mudstones, marls, chalks and the like), are largely held in place by the shearing resistance between the pile and the rock. This becomes greater as more pile length is driven and greater depth is reached because lateral pressures increase until, for that particular type of pile and hammer, the pile reaches a 'set'. In soft rocks the first 'set' may be transitory because the rock may 'relax' and the forces holding the pile diminish. Accordingly it is common to re-drive a pile in soft rock for a second time, possibly after some days. The cause of this relaxation is not fully understood but suspected to be related to the re-establishment of porewater pressures reduced by dilation of the ground during pile driving.

In soft chalk it has been possible to drive piles for many metres penetration *without* reaching a set, only to find that if the pile is left for a few days it will bear the working load. The precise nature of this phenomenon is also unclear, but evidently the rock is pulverised by driving and, if saturated, becomes a liquid with little frictional resistance to shaft penetration. However, if the ground is left to 'settle', (during which time some of the pore pressure generated by piling drains away and permits frictional resistance between the ground and the pile to be re-established) the rock resumes a proportion of its original strength, grips the pile shaft and provides end and side bearing resistance.

It is quite possible that rockhead is irregular; this may be discovered as the piling progresses (Fig. 11.18a). It is thus always better to begin piling from the lowest level of rockhead and to work upwards, for the opportunity then exists to modify pile design (Fig. 11.18b). Whether or not this is a major problem will also depend upon the lateral resistance offered by the overburden soils but, clearly, *investigations for structures that may be founded on piles over a buried valley should include defining the shape of the valley*.

It should be noted that piles, either driven or cast-in-place, may heavily and locally load rock and attention should be given to the quality of the foundation material beneath piles. The opportunity to inspect the material before pile installation may argue in favour of a bored, rather than a driven pile foundation.

11.9
Field Assessment of Exposed Foundations

A time comes when the foundation excavations are opened and the suitability of the foundation mass must be validated by inspection. The engineering geologist examining the excavation has to consider whether or not the mass exposed conforms to the

Fig. 11.18. The correct sequence of piling may help overcome problems caused by steep rockhead slopes

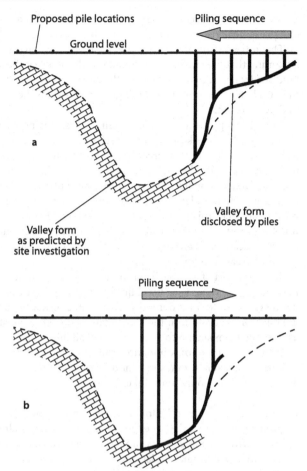

Proposed pile locations

Piling sequence

Ground level

a

Valley form
as predicted by
site investigation

Valley form
disclosed by piles

Piling sequence

b

prognosis of the site investigation and can thus adequately fulfill its function as a foundation. For ordinary structures there is usually no time available to undertake any material or mass testing of ground, so that only simple techniques may be employed. When the structure to be built is not a major work like a power station or a dam, it is likely that the foundation design proposal is based on less information than that which can be derived from viewing the excavation itself. The site investigation report should have recommended an allowable bearing pressure for the foundations based on the geotechnical properties of the foundation mass at a certain depth below surface. *The first part of any inspection is thus to check that the mass exposed corresponds to that anticipated.* This tends to be relatively easy for foundations on soil but may be much more difficult for rock. The reason for this is that in most cases the rock mass will be to some degree weathered and weathering may have significantly affected mass properties, perhaps in an irregular way. If allowable bearing pressures could be gauged by a rock mass classification then values of parameters, gathered from borehole information and utilised in that assessment, could be verified in the field, and if necessary, allowable bearing pressures adjusted. However, no such rock mass classification system presently exists; perhaps it soon will.

The author, confronted with this problem, has tried to clarify his own thinking. Assuming that the site investigation has indicated that the allowable bearing pressure required by the design engineer can be carried by the rock mass at a particular depth, the problem reduces to determining whether or not the rock mass exposed *conforms* to that expected.

Table 11.5 approaches this problem by first posing a series of questions that the investigator has to ask of the information derived from a rather ordinary site investigation undertaken using boreholes, with some testing of rock materials. With regard to bearing capacity the reader will see that, providing the rock mass is composed of adequately strong materials with closed discontinuities, the only issue is that of the comparing the foundation pressure with the compressive strength of the rock. With most fresh rocks this is unlikely to be a problem. However, as is generally recognised, if discontinuity intensity increases, discontinuities become open, and layer strength contrasts occur, then either the foundation pressures have to be reduced or the ground has to be treated to improve its quality.

The same sort of approach may be adopted to examine settlement, utilising the rock material E_{secant} value as a 'baseline' for assessing best case settlement, thence anticipating greater and /or differential settlement as other factors, as indicated in the table, are taken into account. In the investigation stage the questions are answered as well as possible from such information as is available.

If the assessment of foundation conditions is found to be correct, no further action need be taken. If not, some modification of foundation construction is required. The actions commonly undertaken include:

i   increasing the size of the foundations to reduce bearing pressure,
ii deepening the foundation either to reach a better quality rock mass or to increase the downward force operating against any upward force coming from failure of ground beneath the foundations,
iii removing weak materials and replacing by mass concrete, and
iv improving the ground by such techniques as grouting.

Table 11.5 applies to the common foundation excavation on level ground. Should the foundation be of a bridge, at the top of a sloping valley side, then the excavation should be examined to see if any discontinuities existed which would allow sliding of the bridge abutment into the valley. Clearly, each excavation examination should take account of the particular circumstances pertaining to each structure.

The ideas outlined above apply to structures of modest significance. Structures such as dams, nuclear power stations, major bridges and the like are of sufficient economic and social significance to require in situ testing to establish rock mass bearing capacity and settlement characteristics. However, such tests as the plate bearing test, test only a comparatively small volume of the mass that may be stressed, and an examination of the foundation excavation must determine whether or not the 'sample' of the mass tested is representative of the whole. To do this the readily observable rock mass characteristics of the 'sample', as given in Table 11.5, must be recorded to allow comparison to be made with the more extensive rock mass revealed in the foundation excavations.

Table 11.5. Factors contributing to bearing capacity and settlement assessments

	Questions to be answered using information derived from site investigation	Answer	Deductions to be made relative to	
			Bearing capacity	Settlement
A	Is the unconfined compressive strength of the rock material(s) adequate to carry the foundation pressure?	yes	Proposed foundation pressure can be applied	Settlement 'baseline' values deduced from E_{secant} of material Check possibilities of creep for highly porous, weakly cemented rocks
		no	Proposed foundation pressure must be changed or weak materials removed	
B	If the rock mass consists of alternating layers of different rock materials, is there a major strength contrast between the layers?	yes	May be hazards due to punching of strong layers into underlying weak layers; weaker layers could be yet weaker due to differential weathering. Will ground treatment help?	Settlement governed by most deformable material
		no	Proposed foundation pressure can be applied.	
C	Are discontinuity bounded rock blocks bigger or smaller than the likely size of the foundation?	bigger	Proposed foundation pressure can be applied.	Settlement likely to increase with discontinuity density increase, particularly if discontinuities open.
		smaller	May be a problem if discontinuities open or are infilled with weaker material.	
D	Are discontinuities open or closed?	open	May be a problem if strong/weak layer contrast and/or if dip of rocks significant, say > 30°. Can open discontinuities be filled?	Open discontinuities may give unconfined rock columns, allowing greater deformation, and generally reduce stiffness of the mass.
		closed	Proposed foundation pressure can be applied.	
E	Are open discontinuities filled?	filled	Possible problems of open discontinuities reduced depending on infill type.	Filled discontinuities reduce open discontinuity problem.
		not filled	Possible problems of open discontinuities as given in D.	
F	Will weathering profile give 'hard points' at possible foundation level?	yes	If problem of differential weathering between dipping layers, then excavation and replacement of weaker materials required. If problem due to 'corestone' weathering then general reduction of bearing capacity to that of the most weathered material may be required.	'Hard points' will give differential settlement across or between foundation elements, depending upon contrast in deformability between stronger and weaker materials.
		no	Proposed foundation pressure can be applied.	

The investigator examining an excavation can only see what is visible and conditions may change below surface. If the excavation is large enough, it may be useful to measure seismic velocities of the mass, and shear stiffness with depth, to give assurance that mass quality does not deteriorate with depth. Such work takes little time and would give no undue delay.

The examination of the foundation excavation may well reveal features which were entirely unanticipated by the investigation. Such features could be minor faults, shear zones, open joints, solution cavities, mine shafts and such like. How significant these could be will depend upon the project, but they should not be viewed lightly. They might be regarded as symptoms of a 'disease' which may have a greater influence on site security than the symptoms themselves. Thus opened joints in a hillside site could be the result of erosion, glacial shear, old mining subsidence etc. but could also indicate continuing slope movement. If this last is the cause of open joints then site stability could be in jeopardy. Waltham et al. (1986) have reported the problems revealed by the foundation excavations for the Remouchamps Viaduct in Belgium. Solution caves in Devonian limestones, sufficiently spectacular to be a tourist attraction, were known to be in the vicinity and had been, as far as possible, mapped. Extensive (by conventional standards) borehole investigations found no major cavities but foundation excavations for an abutment and a pier revealed significant caverns. Thereafter further investigations on all piers on limestone incorporated probe boreholes on a 2 m grid.

The accuracy with which site foundation conditions can be predicted depends on the complexity of the geology and *the opportunity for accurate calculation may not exist*. In such a case the foundation design must allow for an element of geological uncertainty. This uncertainty as to the nature of the foundation geology is not necessarily a consequence of inadequate investigation. It may be that, for reasons beyond the control of the investigator, access to all parts of the site is limited. It may also be that the geology is simply too complex to be unravelled by the available site investigation techniques, however extensively applied. This is particularly the case with older rock masses, those which may have suffered several phases of tectonic disturbance and perhaps also metamorphism. Tight sharp folds, multiple faults and shear zones may be so closely spaced that correlation between boreholes, even with the aid of geological mapping of surface outcrops, may be totally inadequate to reveal the true geological situation.

In such a case the investigator must be under no illusions as to the accuracy of any forecast of sub-surface geology that can be made. It is important to recognise that a range of possible foundation problems might exist in order that solutions for them may be devised in alternative foundation designs. These solutions must be allowed for in the provisions of the construction contract.

11.10
Further Reading

Simon N, Menzies B (2000) A short course in foundation engineering, 2nd edn. Thomas Telford, London
Tomlinson MJ, Boorman R (2001) Foundation design and construction, 7th edn. Prentice Hall, Harlow, Essex
Woodward J (2005) An introduction to geotechnical processes. Spon Press, London
Wyllie DC (1999) Foundations in rock, 2nd edn. E&FN Spon, London

Chapter 12

Dynamic Loading of the Ground

The ground may be subject to dynamic loading by either natural phenomenon or the activities of man. Some sources of dynamic loading are the action of heavy machinery, traffic, especially trains, explosions, mine collapse and earthquakes. Vibrations travel through the ground and may cause damage to property and, at the very least, annoyance to residents. The public expectation of the severity of the damage likely to be caused by such vibrations is often much greater than that which takes place, but it is generally desirable to record accurately their effects. In urban tunnelling projects, for example, it is prudent to monitor trial blasts and machine noise at ground level, and, if necessary, adjust blasting and tunnelling procedures to minimise disturbance.

While vibrations caused by man are of importance in engineering, the problems they pose pale into insignificance in comparison with those brought about by earthquakes. Earthquakes are a severe natural hazard faced by mankind, as witnessed by the tsunami from the Banda Aceh earthquake of 26 December 2004, which is estimated to have killed at least 270 000 and left many more homeless; few other natural hazards kill and damage on this scale.

12.1
Engineering Geology and Earthquakes

To engineer safely in a seismic area it is necessary to possess the following:

1. an estimate of the likely strength, frequency and location of future earthquakes. This may be assessed from knowledge of the *geology of the region* around the construction site and a survey of *past earthquake events.*
2. knowledge of *site geology,* in order to assess the ground response to previous earthquakes and the likely response of the ground to a future earthquake. This would determine whether any possibilities existed for phenomena such as liquefaction, land spreading and flow slides to occur, which are associated with weak saturated deposits, often of Quaternary age.
3. an assessment of the likely response of the proposed structure and any other structures in the vicinity of the proposed structure, to the anticipated tremors.
4. knowledge of the vulnerability of coastal sites (if that is where the structure is to be built) to sudden flooding by tidal waves (*tsunamis*) generated by off-shore earthquakes that can cause an abrupt vertical displacement of the ocean floor.

Attention must be given to the effects of earthquakes on infrastructure (roads, water, supply, electricity supply, etc.) as well as on the principal constructions so that, in the event that a major earthquake occurs, sufficient infrastructure remains for relief measures to be implemented.

12.2
Sources and Characteristics of Earthquakes

Many of the recent and best documented earthquakes can be attributed to movement along faults associated with either inter-plate or intra-plate adjustments to global plate movements. Such faults may be continually moving but from time to time movement may be partially or totally restricted by unevenness and lack of planarity on the fault surface, until sufficient stress builds up to break through the crustal obstruction. This release of stress and sudden relaxation of elastic stored strain energy in the ground generates the seismic shock. Many faults give rise to numerous very small and, to human senses, imperceptible earthquakes which may be described as "*seismic noise*". This may increase prior to a major shock and some earthquakes are preceded by "*foreshocks*", together with small but measurable changes in the character of the crust and shape of the ground. The main shock may be followed by "*aftershocks*", sometimes of a severity close to that of the main shock, and perhaps extending over a period of many weeks after the main event but with diminishing strength.

A mechanical shock generates mechanical waves and in an earthquake the waves are *compression* (*P*) and *shear* (*S*) (also known as *longitudinal* and *transverse* waves, respectively); the former oscillate particles in the direction of wave travel and for this have been called "push-pull" waves, the latter at right angles to it. An earthquake also generates "*surface*" waves where most of the motion is at ground level; "*Love*" waves shake the ground horizontally and "*Raleigh*" waves move it vertically and horizontally.

The *P* wave velocity is significantly greater than the *S* wave velocity. Seismographs record the time of arrival of waves at a seismic station and there will be a time difference between *P* and *S* wave arrival; that difference increases the further away the observation station is from the source. If the earthquake is recorded by a number of seismic stations and *P* and *S* wave velocities are known, the location of the source or the "*focus*" of the earthquake can be established by triangulation. Seismographs commonly record the accelerations produced by the earthquake as "*accelerograms*", which display components of the accelerations experienced by a particle in three orthogonal planes, two horizontal and one vertical. Ideally, at a particular station, the *P* wave motions are the first to be recorded, principally in the "up/down" component, thence the *S* wave motions, and thereafter diminishing surface waves. Accelerations are usually expressed as proportions of "*g*", the acceleration due to gravity, and are commonly much less than "*g*", but have been known to exceed it in very severe earthquakes. Ground shaking caused by an earthquake is usually of short duration, often less than one minute between onset and end, while the cycles of "strong motion", associated with peak accelerations, occur within a yet shorter period, perhaps less than 10 seconds. These patterns of movement tend to be quite different from those created by rock bursts in mines where the stresses are so high that their supporting pillars and tunnel walls literally burst apart, generating a very short lived local earthquake (known in the mines as a "*bump*"). Likewise, underground nuclear explosions radiate a different signal from natural earthquakes; indeed were it not for the "*cold war*" and mutual distrust of the nations our seismic network would not be as good as it is because we were all listening to each other!

12.2.1
Magnitude

The strength of an earthquake at source is described as its "*magnitude*"; seismologists recognise several types of magnitude but two measures are commonly used:

1. *Richter Magnitude* (M) where

$$M = \log_{10}\left(\frac{A}{T}\right) + q(\Delta, h) + c \tag{12.1}$$

Here, A = maximum amplitude of the wave (10^{-6} m), T = wave period (secs), q = function correcting for the distance from the source of the quake, Δ = angular distance from seismometer to epicentre, h = focal depth of the quake and c = an empirical constant related to the station and regional characters.

2. *Moment Magnitude* (M_w) where

$$M_w = \frac{2}{3}\log_{10}\mu Au - 6 \tag{12.2}$$

Here, μ = shear modulus, A = ruptured area of the fault, and u = average displacement of the fault. The multiple (μAu) is called the *seismic moment* of an earthquake Mo (Nm).

Both magnitudes give roughly similar values but (M) is more closely related to the *effects of a quake*, and thus valuable to planners, whereas (M_w) relates to the *cause of the quake* and is thus valuable to scientists. Engineers are interested in both. The Banda Aceh quake of 26 December 2004 registered 9.0 on the Richter scale and 9.5 for its Moment magnitude.

These and other magnitudes that can be calculated, are a measure of the energy (E, in ergs) released by the earthquake at source. For example, the relationship between E and M is

$$\text{Log}_{10}E = 12.24 + 1.44\,M \tag{12.3}$$

The magnitude "scale" has no upper limit and because of the logarithmic function in the equation, one unit of magnitude increase approximates to a 27 times increase in energy release.

12.2.2
Intensity

The *"focus"* of an earthquake, also called its *"hypocentre"*, is the point below ground level where the earthquake nucleates; e.g. where the snag on the fault shears. The *"epicentre"* is the point at the ground level, immediately above the focus. A schoolchild may observe, by throwing a stone in a pond, that waves passing through a medium lose energy (are *"damped"* or *"attenuated"*) with increasing distance from their source. Earthquake waves are similarly damped and the strength of the shaking experienced on the surface (the *"intensity"*) will, to some degree, depend upon the distance from the earthquake focus to any point where intensity is measured on the surface.

The amount of damping will depend not only on distance but also on the nature of the geology through which the wave travels, and to determine how geology influences intensity seismologists undertake intensity surveys after major earthquakes to produce intensity maps (Fig. 12.1); these show *"isoseismal"* lines, i.e. lines joining points where similar ground motions were experienced; they are assessed by examining the damage done. Table 12.1 shows a Japanese scale of intensity to illustrate this idea. In homogeneous ground, and on a regional scale, the isoseismals should be almost circular – like the waves from a stone thrown into water, but distortions may reveal broad geological structure.

Various isoseismal scales have been developed over the years. One commonly used in Europe and America is the Modified Mercalli (MM) scale which has twelve intensity grades. The MSK scale (after Medvedev, Sponheuer and Kárník), is a modification of this (Grünthal 1993), and there are others, each modification trying to produce more accurate criteria for intensity assessment. One problem has always been that the damage to a structure depends not only on the strength of shaking, i.e. ground movement, but also the type of structure, its age and condition. In the very detailed Herkenbosch study (Fig. 12.2) intensities (as judged from structural damage to property) varied from house to house! (Den Outer et al. 1994).

The European Macroseismic Scale of intensity (Grünthal 1993) recognises (*a*) the difference between intensity effects on humans, objects and nature, and damage to structures, (*b*) five different classes of structural damage (Grade 1 = least damage;

Fig. 12.1. Microseismic map of the 1992 Roermond earthquake. (Haak et al. 1995)

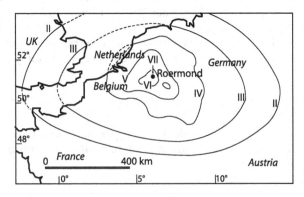

Table 12.1. Japanese (JMA) seismic intensity scale

No.	Strength	Characteristics
0	Not felt	Too weak to be felt by humans; registered only on seismographs
1	Slight	Felt only feebly by persons at rest or those who are sensitive to an earthquake
2	Weak	Felt by most persons, causing slight shaking of windows and Japanese latticed sliding doors
3	Rather strong	Shaking of houses and buildings, heavy rattling of windows and Japanese latticed sliding doors, swinging of hanging objects, sometimes stopping pendulum clocks and moving liquids in vessels. Some persons are so frightened as to run out of doors
4	Strong	Overturning of unstable objects, spilling of liquids out of vessels
5	Very strong	Causing cracks in brick and plaster walls. In many cases un-reinforced concrete block walls collapse and tombstones overturn. Many automoblies stop due to difficulty with driving. Occasionally, poorly installed vending machines fall. Landslides in steep mountains are observed
6	Disastrous	Causing demolition of more than 1% of Japanese wooden houses. Landslides, fissures on flat ground accompanied sometimes by spouting of mud and water in low fields
7	Ruinous	Causing demolition of almost all houses; large fissures and faults are observed

Grade 5 = greatest damage), and (*c*) six different classes of vulnerability of structure. Table 12.2 presents a précis of the scale.

As the strength of shaking is related to the acceleration of particles, attempts have been made to relate acceleration to intensity; Table 12.3 illustrates one example (for the original see Okamoto 1973).

12.2.3
Amplification

The damage accompanying earthquakes often depends upon the nature of the soil immediately underlying structures. In the 1944 Tonankai earthquake in Japan (Okamoto 1973), 26% of the houses that totally collapsed were founded on clay, 3.5% were on sand, 1.4% were on gravel and 0.2% were on rock. The general experience is that the softer the ground the greater the shaking. This phenomenon is known as *amplification* (Fig. 12.3).

Amplification ratios (surface shaking to bedrock shaking) may be as much as 4 to 1. The implication is that, if it is possible to do so, structures are best not built on soft ground. If they must be built on such ground precautions against damage should be taken.

12.3
Liquefaction

Deposits of loose silts and fine sands, without significant clay content to fill the voids between the particles and bind the silt and sand particles together, will attempt to compact when shaken (i.e. the volume they occupy will "*contract*", in soil mechanics

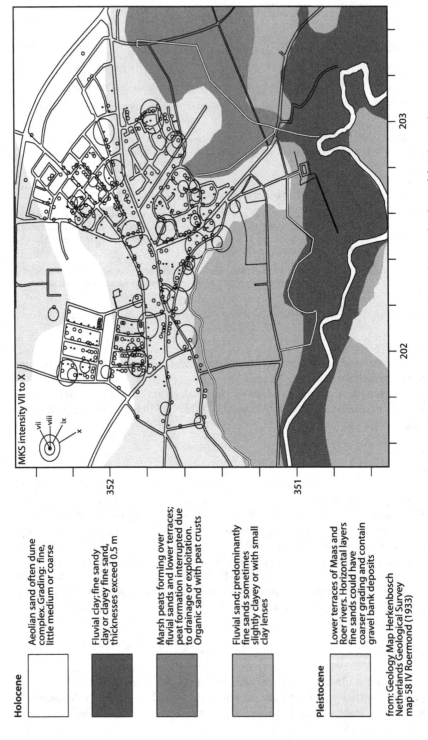

Fig. 12.2. Earthquake intensities for the 1992 Roermond earthquake (based on Den Outer et al. 1994 and Maurenbrecher and de Vries 1995)

Holocene

Aeolian sand often dune complex. Grading: fine, little medium or coarse

Fluvial clay; fine sandy clay or clayey fine sand, thicknesses exceed 0.5 m

Marsh peats forming over fluvial sands and lower terraces; peat formation interrupted due to drainage or exploitation. Organic sand with peat crusts

Fluvial sand; predominantly fine sands sometimes slightly clayey or with small clay lenses

Pleistocene

Lower terraces of Maas and Roer rivers. Horizontal layers fine sands could have coarser grading and contain gravel bank deposits

from: Geology Map Herkenbosch Netherlands Geological Survey map 58 IV Roermond (1933)

MKS intensity VII to X

vii
viii
ix
x

Table 12.2. A Précis of the European Macroseismic Scale 1992

Intensity	(a) Effects on humans	(b) Effects on objects and on nature	(c) Damage to buildings
I Not felt	Not felt	No effect	No damage
II Scarcely felt	The tremor is felt only by very few	No effect	No damage
III Weak	Felt indoors by a few	Hanging objects swing slightly	No damage
IV Largely observed	Felt indoors by many, outdoors by only very few	China, glasses, windows rattle. Hanging objects swing. Moderate vibration	No damage
V Strong	Felt indoors by most, outdoors by few. Sleepers awake. Strong shaking or rocking felt	Hanging objects swing considerably. Poorly supported objects may shift or fall	Damage grade 1 to a few buildings
VI Slightly damaging	Felt indoors by most, outdoors by many. A few persons lose their balance and fall. Many frightened and run outdoors	Objects fall and furniture shifted	Damage grade 1 to many buildings, a few suffer grade 2 damage
VII Damaging	Many frightened and run outdoors. Many find it difficult to stand	Furniture shifted, some overturned. Many objects fall from shelves. Water splashes from containers, tanks and pools	Many class B and a few class C buildings suffer grade 2 damage. Many class A and a few class B suffer grade 3 damage. A few class A buildings suffer grade 4 damage. Damage noticeable to upper parts of buildings
VIII Heavily damaging	Many find it difficult to stand, even outdoors	Furniture may be overturned. Heavy objects fall to ground. Tombstones moved or overturned. Waves may be seen on very soft ground	Many class C buildings suffer grade 2 damage. Many class B, few class C suffer grade 3 damage. Many class A, few class B suffer grade 4. A few class A buildings suffer grade 5
IX Destructive	General panic. People thrown to ground	Many monuments and columns fall or are twisted. Waves seen on soft ground	Many class C buildings suffer grade 3 damage, many class B and few class C suffer grade 4 while many class A and few class B suffer grade 5 damage
X Very destructive			Many class C buildings suffer grade 4 damage. Many class B and few class C suffer grade 5 as do most class A buildings
XI Devastating			Most building of class C suffer grade 4 damage. Most buildings of class B and many class C suffer grade 5
XII Completely devastating			Practically all structures above and below ground are destroyed

Note: Vulnerability classes A, B, and C apply to 'adobe', brick and reinforced concrete buildings respectively. Classes D, E and F apply to buildings constructed to increasing levels of anti-seismic design i.e. F is more resistant than E which is more resistant than D.

Table 12.3. Approximate relationship between Japanese seismic intensity scale (*JMA*) and maximum acceleration (1 gal = g/1 000); *MM* = Modified Mercali scale

JMA scale	Maximum acceleration (gal.)	MM scale	Maximum acceleration (gal.)
0	< 0.8	1	< 1
1	0.8 – 2.5	2	1 – 2
2	2.5 – 8	3	2 – 5
3	8 – 25	4	5 – 10
4	25 – 80	5	10 – 21
5	80 – 250	6	21 – 44
6	250 – 400	7	44 – 94
7	> 400	8	94 – 202
		9	202 – 432
		10–12	> 432

Fig. 12.3. An extract of records of seismographs showing horizontal motions at various depths in the strata about San Francisco Bay (Helley et al. 1979)

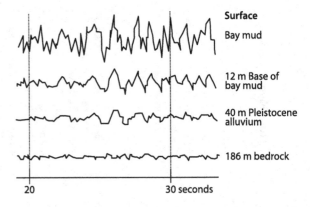

Surface
Bay mud

12 m Base of bay mud

40 m Pleistocene alluvium

186 m bedrock

20   30 seconds

terms). If these materials are saturated no compaction can take place until the water is driven from their pores. However the low permeability of fine grained soils (in comparison with coarser granular materials) prevents this taking place rapidly. Consequently, if the shaking is strong enough and continues long enough, pore pressures build up until the granular material with water in the pores becomes water with grains in suspension; i.e. it becomes a *quicksand*. This is "*liquefaction*" and it may drive fine sand in suspension upwards to appear at surface as "*sand volcanoes*"; evidence that this has occurred may be seen in excavations (Fig. 12.4). If these, and other features, can be found in association with datable deposits then the approximate date and magnitude of the earthquake that caused them can be established (Davenport and Ringrose 1985, 1987).

When liquefaction happens, and usually it occurs 6 to 8 m below ground level, structures may sink into the ground by foundation failure, often tilting severely. Buried

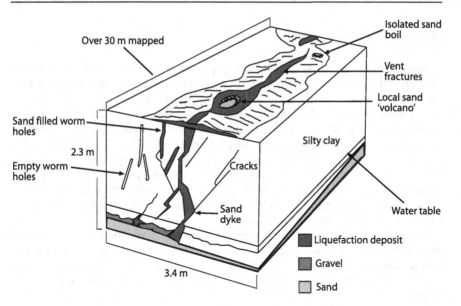

Fig. 12.4. Sub-surface evidence for liquefaction observed in trial pits at Herkenbosch after the Roermond 1992 earthquake (from Davenport et al. 1995)

artefacts, such as petrol filled tanks in garages, pipelines and sewers, may float and rise up to surface. If liquefaction occurs under sloping ground then the whole mass resting on a liquefied layer, may move down slope. A classical example of an earthquake causing liquefaction was at Niigata, Japan in 1964; there, slopes on either side of the River Shinano moved into the river by as much as 23 m (Hamada et al. 1987). Interestingly areas with SPT-N-values (Chap. 6) of 0 to 10 suffered severe damage whilst those of 20 to 40 suffered no damage and it is frequently found that the lower the relative density of the silt or fine sand the greater is its likelihood to liquefy (Seed and Idriss 1971). Great reliance is thus placed on measurements of the *in situ relative density* of sands using the Standard Penetration Test, but this is an empirical test, which requires exact adherence to standard procedures, equipment, and corrections to the results for reasonable and comparable data to be obtained. The thickness of material liquefying does not have to be great; thin lenses of saturated fine sand, if liquefied, may render any slope containing them unstable.

The reader may wonder why such attention is given to the liquefaction of fine saturated sands as an earthquake related hazard. Many earthquake prone areas in the world are associated with high mountain ranges near to the sea. Rivers from the mountains can deposit fine sediments as coastal alluvial plains at the foot of the mountains which, being flat and near water, are ideal locations for agriculture and development. Often the wide river mouths and deltas are also suitable for harbours, river transport and industrial development. Thus the areas most attractive for human settlement can readily be associated with those factors conducive to liquefaction.

12.4
Other Effects of Earthquakes

Earthquake intensity may be assessed as a proportion of "g", the strength of the earth's gravitational acceleration. During an earthquake the local "g" will change and any feature which depends upon the value of "g" for its stability may become unstable. Thus a slope whose stability is related to the weight of ground above a potential sliding surface may become unstable, because the downward forces promoting stability may be reduced relative to the horizontal forces promoting instability, which could also be increased. Most major earthquakes in mountainous areas are associated with widespread landslides arising from an imbalance of forces which promotes rapid displacements that can cause such an increase in porewater pressures within the ground that the near surface masses so affected are able to flow like a fluid, downslope, sometimes with great speed and devastating effects.

Foundations of buildings are designed assuming a given load intensity on the ground. This may increase during the earthquake, perhaps sufficiently to cause foundation failure. Ground waves can cause lateral movements in foundations which may not all be in the same direction at the same time, bringing about failure of structural columns.

Earthquakes may be caused by fault rupture and such ruptures may extend to the surface with severe effects on any structures above or below ground. Ground shaking may damage buried pipelines and cables, disrupting supplies of water, gas, electricity and the disposal of sewage. Water well casing may be damaged, cutting off supplies of groundwater. Water levels in aquifers may be temporarily or permanently disturbed. As mentioned earlier, much more widespread damage can follow a fault rupture that displaces the ocean floor and generates a tsunami

12.5
Assessing Seismic Risk and Seismic Hazard

The term seismic risk indicates the likelihood that earthquakes of given intensities or magnitudes will occur within an area or at a site. The basic data required to assess seismic risk is a catalogue of past events and an understanding of regional geology. Accurate and complete earthquake records are a recent development; seismograph networks were limited before the 1950s. Before this century accounts of seismic events are more journalistic than scientific and before the 17th century records are very sparse indeed. To extend the seismic record to earlier than about 4 000 years ago 'palaeoseismic' studies are undertaken; these seek geomorphological and soil structure information (as in Fig. 12.4) which can be used to indicate whether a major seismic event had occurred (Fig. 12.5)

Figure 12.5 illustrates a river meandering between its terraces (a) and (b). Faulting accompanying an earthquake in pre-historic times, ruptures the ground, displacing the terraces laterally, and downthrowing them towards the south. The river channel is also displaced. This geomorphological record records the presence of an active fault and to study that further a N-S trench is excavated across its likely position. The trench exposes displaced strata unconformably covered by recent alluvium. Dating the fault

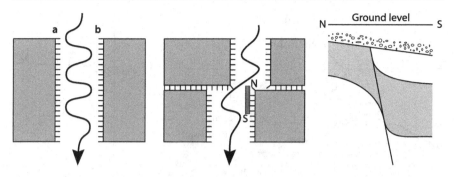

Fig. 12.5. Phases of fault movement revealed by geomorphology and geology

and thus the earthquake may be possible as it can be no younger than the alluvium and no older than the strata it displaces.

If the seismic catalogue is adequate, maps may be produced displaying, in some fashion, varying degrees of risk of earthquakes in a specific area. Figure 12.6 is a seismic zoning map for northwest Germany and the Benelux countries, showing maximum expected intensities (MM scale). A comparison between this figure and Fig. 12.1 shows that the credibility of this map was verified by the 1992 Roermond earthquake. This type of map leaves it to the user to determine the damage that may ensue from an anticipated earthquake. Other maps, such as some published in the United States, may link such predictions to building codes recommending structure strengthening against earthquake damage.

For major structures whose failure would have serious consequences for society, special studies are required to evaluate seismic risk. The first task is to establish the *"design"* earthquake; that is, the earthquake whose effects the project must resist. The characteristics to be established are its possible location, magnitude, duration, and intensity and its frequency of occurrence. Location may be established from regional and local geological studies and the seismic record. The magnitude may be estimated by assessing the length of fault rupture which could occur to generate the next earthquake, while the duration, of cycles of strong motion in particular, is related to magnitude. The intensity at a site will depend upon its distance from the focus and attenuation factors which have been derived from intensity studies of past earthquakes, such as illustrated in Fig. 12.1. Frequency of occurrence may be assessed from the seismic record.

Establishing the design earthquake requires not only the assembly of information relative to the site but a review of the latest methods of parameter assessment. Once the design earthquake has been established, the task of the engineering geologist is to assess site response. This requires a study of soils, rocks, geomorphology and geohydrology to assess how the ground may react to the design earthquake. Thus, attention would be given to assessing the characteristics of ground with a potential for liquefaction, and calculating the stability of slopes that could fail. Faults should be located to see if, should they rupture, major damage would be caused. All associated hazards should also be evaluated.

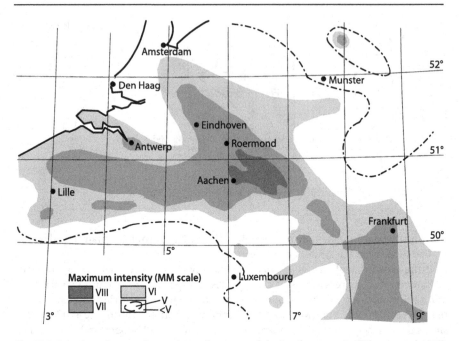

Fig. 12.6. Seismic zoning map for northwest Germany and the Benelux countries (Ahorner et al. 1976)

12.6
Ground Engineering Design against Earthquake Hazards

The surface of the earth may be divided into zones each characterised by one of three grades of seismic risk. In the first, earthquakes have been observed to occur either very rarely or never, in the second earthquakes giving some damage occasionally occur and in the third damaging earthquakes frequently occur and become a part of life. The second gives the greatest difficulty for not only is the seismic record likely to be incomplete but the public may not perceive earthquakes as a hazard requiring special, and possibly expensive, engineering precautions; only an "unexpected" and possibly damaging earthquake may sharpen that perception.

However, having established that a hazard exists, engineering design may be employed to mitigate the damage that could result. Every building rests on a foundation and static foundation designs may be modified to allow for dynamic loading. One approach in Japan (Okamoto 1973) has been to divide the country into areas with particular seismic coefficients which relate to expected intensities of shaking. In Table 12.4 the seismic coefficient k is 0 for static conditions rising to a maximum of 0.4 for maximum shaking. The Terzaghi bearing capacity factors are thence reduced as k increases to give larger area footings (and thus less bearing pressure) to compensate for increased loading under dynamic conditions. The reader can be forgiven for feeling this all sounds too simple a solution and the author admits that here only the basic ideas of 30 years ago are described as at that time the concepts were simple enough for most engineers to understand them.

φ	35°			30°			25°		
k	0	0.2	0.4	0	0.2	0.4	0	0.2	0.4
N_c	58	22	12	36	17	9	22	12	6
N_q	43	12	4	22	7	3	12	5	1
$N\gamma$	42	9	2	20	4	1	9	2	0

Table 12.4. Modification of bearing capacity factors for increasing seismic coefficients (simplified from Okamoto 1973)

Earthquake engineers can be said to design against "the last earthquake", for each new event may reveal the success or failure of some previous approach. For example, one idea to combat damage from liquefaction was to build on piles to a bearing stratum below the liquefiable layer. Hamada et al. (1987) described how a building, damaged in the 1964 Niigata earthquake, was 20 years later scheduled for reconstruction. As part of this work the old pile foundations were excavated to examine their condition. Out of a total of 304 piles, 74 were excavated and most of these were found damaged, showing fracture displacements in a direction corresponding to that of the permanent ground movements and suspected to be a consequence of liquefaction. Clearly, piled foundations alone are not the answer to liquefaction; if liquefaction induces lateral movement piled foundations may be damaged and even if this does not occur, the loss of lateral support to the pile in the liquefied layer may cause damage as piles and building shake.

Greater benefits can be obtained by *improving the ground where possible* so that it is not weakened by shaking. Paradoxically, the most commonly used technique in sands and silts is to increase their relative density by giving them a good shake, where necessary induced by dynamic loading! The objective is to attain a relative density which safeguards the deposit against liquefaction from the earthquake the structure is designed to withstand. This might be achieved by the process of *vibroflotation*, in which a vibrating cylinder, the vibroflot, is jetted into the ground, and then vibrates the ground so expelling water from its pores and enabling the material around it to consolidate and strengthen. In the 1968 Off-Tokachi earthquake, which gave MM intensity IX at the site of the Hachinohe paper mill, surrounding untreated ground cracked and liquefied, while buildings on the consolidated ground suffered only light damage (Okamoto 1973). Ground improvement may be monitored by in situ testing using the Standard Penetration Test, the static cone test or the Menard pressuremeter, and geophysical methods. The seismic design of engineered structures is continually improving to accommodate the most recent experiences.

12.7
After the Earthquake

The number of casualties caused by an earthquake is, to some degree, a matter of chance. In relatively primitive agricultural areas where the inhabitants may dwell in adobe buildings with heavy roofs, there are likely to be many more casualties if the earthquake occurs at night when all are asleep in their readily collapsible homes than if it occurs by day when most are working in the fields. The 1992 Roermond earthquake, which occurred at 3:20 A.M. on a Monday morning, caused almost no casualties although many chimneys fell into the streets. Had it occurred on the previous Saturday afternoon, when the area was densely crowded with shoppers, the consequences would have been much more serious. However, while few may be killed during the earthquake, many more may die after it due to the destruction of engineering works of social significance. Thus in past earthquakes most deaths have occurred due to fire, disease, and exposure, or to civil unrest consequent to the earthquake.

Earthquakes located off shore usually kill no one but if they initiate a tsunami their effect on coastal communities thousands of kilometres away and in different lands, can be devastating; the quake of 2004 at Banda Aceh was such an event.

Accordingly, in earthquake-prone regions attention must be given to protecting those engineering works, such as roads, bridges, airports, water reservoirs and delivery systems, which will allow relief measures to take place. As far as ground engineering is concerned this includes ensuring that roads are not blocked by landslides, runways and hospitals are located in safe places and all are designed to resist dynamic loading.

12.8
Further Reading

Jackson J (2001) Living with earthquakes: know your faults. J Earthquake Eng 5(SPl/1):5–123
Lillie RJ (1999) Whole earth geophysics: An introductory textbook for geologists and geophysicists. Prentice-Hall Inc., Upper Saddle River, New Jersey
National Research Council (U.S.), Committee on the Science of Earthquakes (2003) Living on an active earth; perspectives on earthquake science. National Academic Press, Washington, DC

So many advances can be expected that readers are advised to keep abreast with developments using the following web sites. Each site contains links to other relevant sites.

http://earthquake.usgs.gov – The home page for the Earthquake Hazards Program of United States Geological Survey. Data covering the whole world can be obtained from this site (accessed on March 2007)
http://www.emsc-csem.org – The home page for the European-Mediterranean Seismological Centre with data spreading from the mid-Atlantic ridge to the Himalayas (accessed on March 2007)
http://www.eri.u-tokyo.ac.jp – The home page for the Earthquake research Institute of the University of Tokyo (accessed on March 2007)

Chapter 13

Ground Reaction to Changes of Fluid and Gas Pressures

13.1
Subsidence due to the Pumping Out of Fluid

The fluids found naturally underground are groundwater, oil and gas. Of these the fluid most familiar to engineers and most frequently encountered in civil engineering is groundwater. Groundwater is derived from rainfall and when rain infiltrates the ground it *recharges* the groundwater reserves that are there; if this occurs at a certain rate water levels will rise and with them the water pressure in the pores and fissures of the ground. This water will be draining towards a river valley and ultimately to the sea. It is when the rate of recharge is greater than the rate of drainage (or *discharge*) that water level will rise. When the reverse is the case, water levels will fall. Any measure of natural water levels will show they rise and fall with time in a seasonal way, perhaps over a few tens of metres. Pore water pressures will likewise rise and fall in a seasonal way too.

This natural cycle has superimposed upon it an artificially stimulated set of changes when wells are sunk and fluids are pumped from the ground. Such unnatural changes in fluid levels, due to the withdrawal of water, are almost entirely due to the activities of man. Wells are sunk into aquifers and petroleum reservoirs to permit their fluids to escape up them in a controlled manner but this reduces the pore pressures in them; as fluid pressure reduces so fluid levels reduce too. This reduction in pressure brings about changes in the balance between the total stresses in the ground coming from its self weight, and the pressure of fluids in its pores; i.e. the *effective stresses* within the ground. These stress changes cause *consolidation* of the ground in the areas where fluid pressure has been reduced; i.e. its particles move more closely together; the density of the ground increases and because of that the strength of the ground usually increases as well.

Consolidation causes *settlement* in oil and gas reservoirs and groundwater aquifers, which occurs with time and can be explained as follows. The pressure of oil or gas or water in situ depends mainly on its depth. The weight of the overburden at that depth acts partially through the grain structure of the fluid bearing rock and partially through the fluid itself. During fluid withdrawal, the pressure on the fluid will decline so upsetting the balance of forces and transferring more of the overburden weight to the grain structure; in other words, *the effective stress will increase*. The rate of increase will be the same as the rate of pressure decrease, however, for reasons explained later, the response of the ground i.e. consolidation and settlement, can be either at a similar rate *or at a much slower rate* than this.

13.1.1
The Principle of Effective Stress

A porous mass such as soil, and to a certain extent, rock, consists of particles surrounded by pore spaces. These pore spaces may be filled with air, water, oil or gas. The contents of the pore spaces can be thought of as offering some support to the particles they surround. If the contents are strong, as can be the case when pores contain mineral cement, the support they offer can be considerable and much force is required to move the particles – the material will be strong; it will be able to resist deformation under load, in particular it can resist shear. If the pore contents are weak, and fluids have little or no shear strength, then the support they offer is small especially if the fluids can move – the material will be weak and offer little resistance to deformation under load.

Consider porous ground where the pore spaces are filled with air under normal atmospheric pressure. At any point below ground surface, the weight of the material above that point is almost entirely carried by the solid skeleton of particles because the air around the particles can offer only its pressure as a support. In Fig. 13.1a the small arrows indicate load carrying point contacts between grains. In such a case the total load per unit area (σ) on the base of the element of ground equals h (the height of the element) times γ (the unit weight of the dry ground) and may be described as the "*effective pressure* or *effective stress*" (σ') at that depth, the resultant of all the contacts.

Fig. 13.1. Effective stress

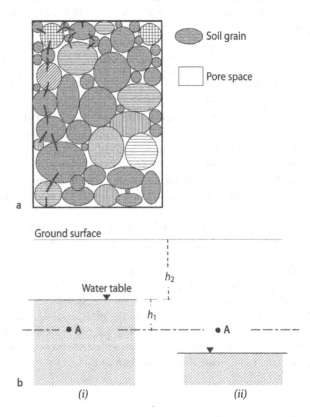

Soil grain

Pore space

Ground surface

Water table

h_2

h_1

• A

• A

a

b

(i)

(ii)

If the fluid, be it a liquid or a gas is under an additional pressure and cannot escape from the pores then its support to the grains increases; indeed it has been shown time and again that when a fluid of given pressure 'u' exists in pores then the total pressure at a depth in the sample is given by either

$$\sigma = \sigma' + u \tag{13.1}$$

or

$$\sigma' = \sigma - u \tag{13.2}$$

Figure 13.1b shows the stresses on a point A below ground level when water pressures change. In (i) the effective stress σ' at point A is given, following Archimedes' principle, by either

$$\sigma' = \gamma_g h_2 + \left(\gamma_g - \gamma_w\right)h_1 \tag{13.3}$$

or

$$\sigma' = \gamma_g\left(h_1 + h_2\right) - \gamma_w h_1 \tag{13.4}$$

where γ_g = unit weight of solids and γ_w = unit weight of water.
However, if the water table is lowered below the level of point A, as in (ii)

$$\sigma' = \gamma_g\left(h_1 + h_2\right) \tag{13.5}$$

The effective stress on point A has thus been increased by ($\gamma_w \times h_1$) and there has been a gradual increase in effective stress in the soil through the h_1 zone. This produces deformation of the soil within the zone, leading to consolidation of the zone and to surface subsidence.

The terms "*consolidation*" and "*compaction*" cause some confusion. Geotechnical engineers in soil mechanics, rock mechanics and engineering geology, use the term consolidation to mean a reduction in pore space as result of increase in effective stress and "*compaction*" to describe densification of soils by mechanical tools to increase the soil strength (i.e. for road embankments). The term compaction is used by *petroleum engineers* to define reduction in pore spaces of soils or rock produced by an hydrocarbon extraction; in other words, consolidation due to an increase in effective stress. In this context the two words are synonymous.

The equations describing either process can be stated in a simple way; total settlement at the surface, Δh, is related to the deformation properties of the ground, the reduction of the fluid pressure Δu (= the increase of the effective stress, $\Delta \sigma'$) and the thickness h of the formation in which the pressure declines.

The basic formula to calculate total consolidation when time is infinity is of the form:

$$\Delta h = m_v \times \Delta \sigma' \times h \tag{13.6}$$

where m_v is the coefficient of volume compressibility and is defined as the relative reduction in length per unit of stress increase in an axial direction under a constant loading rate in the stress range $\Delta\sigma'$.

In order to predict the compaction behaviour of a producing aquifer or reservoir the following information is required:

- the full extent of the reservoir area expected to be subject to pressure decline;
- a map of reservoir body thickness (h), including the area *beyond* the field where pumping is to be undertaken (remember these fluids will flow from outside the area being pumped towards the pumps – that is why wells "work";
- a prediction, based on future field development, of the reduction in fluid pressure that is anticipated at each point of the reservoir, i.e. the distribution of Δu or $\Delta\sigma'$ at a given time;
- the coefficient m_v, which is best measured in situ but failing that on representative samples using appropriate apparatus and procedures.

These "simple" models describe and are therefore used to predict "final settlement" however it is important to appreciate that the time taken for this settlement to occur *can be many years*, depending on the permeability of the soil or rock, and on the length of the drainage path for the pore fluid. Terzaghi (1925) presented a pioneering publication on the concept and a mathematical model for describing consolidation with time. Many subsequent workers, notably Biot (1941), Koppejan (1948), Gibson et al. (1967) and Den Haan (1992) have since refined and adapted the basic concepts of his model so that several metre high piles of papers could be assembled on the subject, each consolidating under its own weight. These papers not only describe consolidation processes but also present case histories where ground response to pumping can be explained using this model.

Helm (1984) has reviewed many of these computational techniques and their assumptions. Subsidence calculations based on formulae such as Eq. 13.6 assume *uniformity in the consolidation characteristics of the ground*. Such assumptions must be verified by field data to obtain layer thickness and by field and laboratory measurements to investigate the consolidation properties of the individual layers. The whole exercise is much helped by survey levelling and deep observation wells.

13.1.2
The Role of the Engineering Geologist

It is easy to become overwhelmed by the mathematics and mechanics of the calculations involved with predictions of settlement but these efforts are worthless if the ground model they use is incorrect. The role of the engineering geologist is thus clearly defined; it is to *supply the correct ground model*. The ground has to be divided into layers and zones that can be ascribed typical values for their mechanical and hydraulic properties. This means the vertical profile has to be known, especially the presence and thickness of potential confining layers; that requires the sensible location of boreholes based on the existing knowledge of local geology, good drilling and meticulous core logging. The lateral variation of these vertical profiles needs also to be established particularly if they are bounded by faults. This is another way of saying that the *bound-*

aries of the problem need to be defined. The next aspect to define is the fluid pressure at depth; it will not be uniform and changes can be quite marked across low permeability boundaries. Such changes are often only detected during drilling and unless the drilling equipment is able to detect and record such changes they will have to be noted by the engineering geologist on site; in other words, drilling for these studies requires on site supervision. Lithologies, levels and pressures all constitute what is described by numerical modellers as the *initial conditions*. But they are not sufficient to complete the prediction; values for compressibility are required for the layers and zones defined, and obtaining them can be a major problem. Sometimes values of horizontal stress are also requested – and obtaining them is an even greater problem.

Samples for testing need to be selected and protected from the effects of disturbance, especially from stress relief. In many cases such disturbance is impossible to prevent and the best method by far is to change fluid pressures in the ground and measure how the *ground* responds; i.e. to undertake appropriate in situ trials from which values for stiffness and compressibility which link cause (i.e. fluid pressure reduction) to effect(settlement at ground level) can be calculated. That will need instrumentation and monitoring prior to field testing, and for this the position of the instruments has to be defined both in plan *and in vertical section*; they must be related to the relevant geology and vertical settlement across the most compressible layer needs to be measured.

Finally, the engineering geologist should be consulted over the likely variability of surface settlement because it is unlikely to be uniform. Near surface geology is relevant to this assessment, especially the location and nature of Quaternary deposits; air photos and mapping of the near surface deposits can be of considerable value in this aspect of the work. Risk assessments for likely damage to property cannot be completed sensibly without such mapping.

In the following sections examples will be given of settlement and the circumstances associated with it; in each case the principles outlined above can be clearly seen.

13.2
Water

Groundwater occurs in aquifers which may be "*unconfined*" at some locations i.e. have their upper surface at ground level (you can walk on them), and "*confined*" elsewhere i.e. their upper surface is not at ground level but under a confining layer. The situation shown in Fig. 13.1b is that for an unconfined aquifer and lowering of the groundwater table within it gradually increases the effective stress with depth below the original groundwater level, in both the drained and saturated zones. In Fig. 13.2 an aquifer is confined between two low permeability layers ("*aquitards*") and the water pressure within it results from water levels *outside* the confined zone.

When water is pumped from the confining layer, fluid pressures drop and the layer and its confining layers (where pressures will also drop but more slowly) will tend to consolidate; if the aquifer and its confining layers form a mechanically stiff rock mass the consolidation is likely to be small. If the aquifer and its confining layers are more soil-like, e.g. uncemented sediments, then consolidation can be large. Such responses have been understood for almost a century and were well set out by Meinzer (1928) and Terzaghi (1929). Extraction of water from compressible ground has resulted in

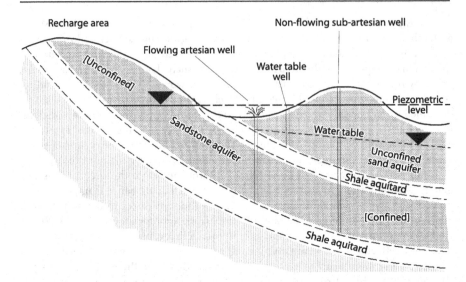

Fig. 13.2. Unconfined and confined aquifers

considerable settlement in Venice, Bangkok and in many delta systems around the world (Table 13.1). A frequent occurrence in Venice, largely as a consequence of subsidence following water extraction from aquifers at depth, is the periodic flooding of St. Marks Square when tides are high at the northern end of the Adriatic, Fig. 13.3. (The photograph shows Professor Helm walking along a cat-walk leading to the Cathedral as part of the Seventh International Symposium on Land Subsidence; 2000). The imperceptible lowering of the land associated with groundwater extraction is often only realised by the increased frequency of flooding, when people begin to realise that "something is happening"; by the time survey levelling reveals the presence of subsidence it is usually much too late to rectify the causes of the subsidence i.e. excessive pumping.

How could this situation have been allowed to happen? It is unfortunately the case that when water supplies are required it is the properties of aquifers and associated aquicludes that are gathered (aquifer yield, drawdown/pressures, water quality and recharge obtained from pumping tests and monitoring of observation wells); little if any regard seems to be given to the consolidation properties of these formations, suggesting that the consequences of extraction in relation to subsidence is not an issue for water supply even though consolidation is irreversible. *The engineering geologist must be aware of the dangers of this hydro-schizophrenia.*

13.2.1
Subsidence in Unconfined Aquifers

Subsidence from the reduction in groundwater levels in unconfined aquifers may be complicated by other aspects of geological variability in addition to layer variability, lateral variations and the like, because recent near surface deposits are capable of hav-

Table 13.1. Settlement in major cities (Dolan and Goodell 1986)	City		Maximum sub-sidence (m)	Area affected (km^2)
	Coastal			
	1	London	0.30	295
	2	Venice	0.22	150
	3	Po Delta	1.40	700
	4	Taipei	1.90	130
	5	Shanghai	2.63	121
	6	Bangkok	1.00	800
	7	Tokyo	4.50	3000
	8	Osaka	3.00	500
	9	Niigata	2.50	8300
	10	Nagoya	2.37	1300
	11	San Jose	3.90	800
	12	Houston	2.70	12100
	13	New Orleans	2.00	175
	14	Long Beach/ Los Angeles	9.00	50
	15	Savannah	0.20	35
	Inland			
	16	Mexico City	8.50	225
	17	Las Vegas	0.30	320
	18	Denver	0.66	40
	19	San Joaquin Valley	8.80	13500
	20	Baton Rouge	0.30	650

ing considerable variation in their thickness. Figure 13.4 (Foose 1968) shows how the non-uniform compaction of unconfined aquifer sediments infilling valleys in the bedrock below may give rise to differential settlement and surface shearing.

In areas where pinnacle weathered limestone is overlain unconformably by more recent soil deposits the further removal of carbonates or soil infill by groundwater movement may result in the development of soil arches spanning cavities within the rock (Fig. 13.5). If the groundwater table is lowered, say from level A to level B in Fig. 13.5, stresses in the arches can increase to the point where they cause failure. Collapse of the arch permits sudden settlement and subsidence at ground level.

Foose (1968) records how increased pumping from a mine in the Hershey Valley, Pennsylvania, lowered the groundwater table in sediments over an irregular bedrock surface in Ordovician limestones, and brought about the development of such sinkholes. When the groundwater table was allowed to recover to its former levels the development of further sinkholes ceased. Such sinkholes may be very large. In December 1962 a mine crusher station near Johannesburg, South Africa fell into a sinkhole 30 m deep and 55 m wide, with the loss of 29 lives.

Such soil bridged cavities may occur anywhere where soil covers irregular solution weathered limestone rockhead. Concentrated water inflows, as may be formed by soakaways, used to allow surface water to drain away from newly built roads and housing developments, can cause the soil bridges to fail and sinkholes to develop in ground prone to this.

Fig. 13.3. Frequent flooding of
St. Marks Square Venice at high
tide (30 September 2000)

Fig. 13.4. Ground shearing
from unequal subsidence of
recent sediments over an uneven bedrock surface

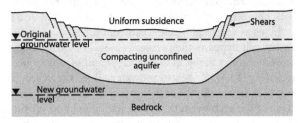

Fig. 13.5. Development of sinkholes by the collapse of soil
arches in pinnacle weathered
limestone as a consequence of
the reduction of the groundwater table (from *A* to *B*)

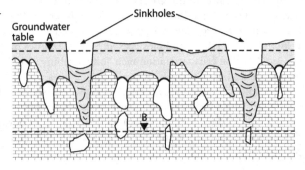

13.2.2
Subsidence in Confined Aquifers

The vulnerability of confined aquifers to consolidation comes from the fact that their confining layers can permit much higher water pressures to be developed within them than occurs in unconfined aquifers; so high that at times wells drilled into them bring water gushing to the surface as a fountain (these are called *artesian* wells). Pumping can produce considerable lowering of water pressures in such aquifers with an equally considerable increase of vertical effective stress within them, and hence consolidation. Figure 13.6 shows a typical subsidence bowl developed by the extraction of groundwater beneath the city of Hanoi, Vietnam (Nguyen and Helm 1995). Groundwater is taken from an unconfined aquifer and a lower confined aquifer, separated from it by a layer of compressible clay. Since the beginning of extraction in the early part of the century when the water levels were 2 to 5 m below ground level, the piezometric level in deep wells has sunk into the confined aquifers, and fallen to 37 to 40 m, as measured in 1993. This has not only caused the confined aquifer to consolidate *but also has initiated consolidation in the confining compressible clays.*

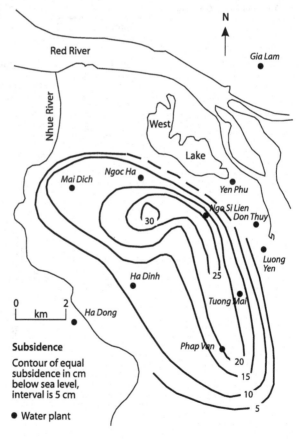

Fig. 13.6. Subsidence developed from the extraction of groundwater from beneath the city of Hanoi, Vietnam (Nguyen and Helm 1995). Wells are located in the water plants

A similar situation is to be found in the city of Bangkok, Thailand. The city is situated about 40 km from the sea on either side of the Chao Phrava River which runs
through a large flat deltaic plain. This delta fills of a broad deep basin with recent sediments and available evidence suggests that these have formed nine confined sand and
gravel aquifers, separated by clay strata of very low permeability (Young et al. 1995).
The profile of the surface of the bedrock is still undetermined, but its depth in the
Bangkok area is known to be between about 550 and 2 000 m below ground level. The
city is generally only 1 to 1.5 m above mean sea level.

Water has been extracted from these deltaic unconsolidated sediments at an ever
increasing rate since 1960 with the result that there are now serious hydrological and
geotechnical problems associated with depletion of aquifers, arising from a deterioration of water quality as saline water is drawn into the aquifers, and general land
subsidence. The large draw downs in the piezometric level of the main production
aquifers, has resulted in the installation of deeper and deeper wells, thus making the
problem worse. Some consolidation of the aquifers has occurred but much greater
consolidation of the intervening clays and overlying near surface clays has been recorded; remember, recently deposited clays tend to have porosities of >60%, much

Fig. 13.7. Subsidence damage in
Bangkok

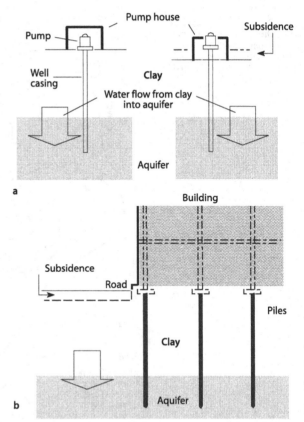

greater than those of sands and gravels (<30%) and thus will suffer greater changes in thickness on consolidation.

Maximum subsidence in 1978 was 75 cm but by 1980 the maximum *rate* of subsidence was of the order of 10 cm per year. Alarm at the settlement caused this to be reduced to between 4 and 5 cm per year by 1988, by reducing pumping rates (Phien-Wej et al. 1989). Figure 13.7 illustrates some of the effects witnessed there; consolidation of the upper soft clay causes the pump house of water wells to sink onto the pump causing the well to fail – another is then sunk, a little deeper this time. Buildings founded on piles taken to the sand aquifer are bordered by roads founded on the clay that has consolidated; this separates from the building but more dangerously can drag down the piles and cause them to fail. These are the consequence of pumping water from the shallowest aquifer. Subsidence from pumping the deeper aquifers could damage any structure whose foundations were taken down to below the upper aquifer. All structures, perhaps already damaged from shallow aquifer subsidence, will suffer a change in elevation as a consequence of deeper aquifer and aquitard compaction. A similar situation exists in Mexico City.

Significant subsidence problems are not necessarily associated with the extraction of large quantities of groundwater; small quantities can also have dramatic effects. In Stockholm, hollows in the bedrock scoured by glaciers, sometimes over crush zones, may be filled with clay to give isolated *"clay lakes"*, each with its own water table (Morfeldt 1970). Tunnel boring under these areas may experience *slight* seepages through crush zones, of little significance to either tunnel construction or stability. However, such drainage lowers the water table in the clay lake, producing consolidation, subsidence and marked differential settlement, with serious damage to structures at ground level. Here uneven bedrock and changes in thickness of the compressible layer combine to magnify the effects of consolidation

13.3
Subsidence in Rock Aquifers and Reservoirs

13.3.1
Extraction of Water

Rock aquifers, such as limestone and sandstone, are much stronger than uncemented and weakly bonded sediments, such as sands and silts, and suffer very little compression from the changes in effective stress that extraction of water within them creates. Nevertheless if the aquifer is deep enough for the loads upon it to be high compared with its strength, it will deform when the water pressures within it are lowered. This hardly ever causes problems of settlement at ground level because the movements are small and their origins are at such depths that the effects of any differential movements are spread over a large area by the time they reach ground level – they become imperceptible. However, the lithologies of a sedimentary sequence can easily place a sediment that would be described as rock against one that would not be described as rock. A typical example of this occurs beneath London where the major confined aquifer, the chalk (of Cretaceous age) is unconformably overlain and confined by the London clay (of Tertiary age). The chalk beneath London was severely pumped for water supply in the 19th and early part of the 20th century, and the question whether London is sinking as a result, has been asked. Wilson and Grace (1942) used Terzaghi's equations for consolidation to examine the long-term pumping and coupled subsidence effects in the London Basin from 1850 to 1940. They concluded that settlement should be expected from the consolidation of the London clay – this being a result of the reduction of pore water pressure in the strata beneath it. They did not discuss the possibility of the chalk compressing as the effective stresses increased but subsequent analyses shows that the changes in effective stress would probably have been too small to make a significant contribution to the settlement that could reasonably be expected from the London clay alone. This is not the case in the oil fields where the levels of stress and stress change produced by pumping can easily match the strength of the sedimentary reservoirs from which oil and gas is being extracted causing the reservoir rock in the vicinity of the well to fail.

13.3.2
Extraction of Oil and Gas

Contrary to experience from the extraction of water from the chalk under London or elsewhere on land in Great Britain, subsidence associated with the extraction of oil and gas from the chalk of the Ekofisk reservoirs in the North Sea caused more than 3.5 m settlement. In this instance, lowering oil pressure in the chalk around the wells increased effective stresses at depth so much the yield point of the chalk was passed, making it highly compressible and causing it to fail. The reasons for this compressibility under the change in effective stresses that were imposed on the chalk reservoir, are complex and reflect the geological history of the ground. In studying this, Sulak and Danielsen (1989) had to differentiate between mechanical and chemical mechanisms. Mechanical compaction (in petroleum engineering language) is comparable to consolidation: the expulsion of water under gradually increasing loads from the accumulation of overburden sediments. Chemical compaction describes pressure solution processes that continue to influence the rock when mechanical compaction is largely complete – a process that would probably be recognised as "*creep*" i.e. slow and long-term deformation under constant effective stress. To appreciate how these mechanisms can operate in chalk it is necessary to understand the material as well as the mass. The chalk is largely made from microscopic biogenic platelets of silt to clay size particles, that are the fragmentary remains of larger rings of platelets known as *coccoliths*, which are themselves broken down from planktonic unicellular algae known as *coccospheres*. The platelets and less frequent coccoliths, and coccospheres, accumulate on the sea floor to create an open and porous delicate structure that is preserved in an early cementation phase of sedimentation on the floor of the chalk sea; the bonds between these particles are likened to "spot welding". Once effective stresses increase sufficiently, due to pumping, this light cementation breaks down and the chalk begins to disaggregate; when this happens completely the chalk becomes a slurry, with the consistency of toothpaste. At this point compaction is rapid and dramatic. Table 13.2 (Iingersoll and Ulstein 1987) summarizes the settlements that have taken place in major oil fields.

The Goose Creek oil field in Texas provides another example of a subsidence bowl (Fig. 13.8) and a typical example of seismogenic faulting associated with fluid extrac-

Table 13.2. Documented cases of oil and gas field subsidence, from Ingersoll and Ulstein (1989) with updates location *1:* Xu (2002); *18:* Hirono (1970) and Takeuchi et al. (1970); *22:* Gambardella and Bortolotto (1991)

City		Subsidence (m)	Area (km^2)	Depth (m)
1	Lake Maracaibo: Bachaquero and Langunillas, Venezuela	4.32	450	1 200
2	Buena Vista, CA	0.27	48	1 130
3	Clinton, TX		12	820
4	Dominguez, CA	>0.07	7	1 430
5	Edison, CA	>0.09	6	1 100
6	Ekofisk, North Sea	3.4	35	3 125
7	Eureka Heights, TX		6	2 540
8	Fruitvale, CA	>0.04	14	1 370
9	Goose Creek, TX	>1.0	6	600
10	Groningen, Netherlands	0.65		3 000
11	Huntigdon Beach, CA	1.22	16	930
12	Inglewood, CA	1.73	5	900
13	Kern Front, CA	>0.34	19	745
14	Long Beach, CA	>0.61	7	1 690
15	McKittrick, CA		6	360
16	Midway-Sunset, CA			
	Central	>0.49	65	555
	Globe Area	>0.43	15	1 020
	Sunset Area	>0.18	20	590
17	Mykawa, TX	>0.10	6	900
18	Nigata, Japan (land area)	>1.5	15	380–610
	1959–1968	>0.75	245	
		>0.05	3 000	
19	Orange Co., CA	>0.05	5	1 480
20	Paloma, CA	>0.07	23	3 800
21	Playa del Rey, CA	>0.29	2	1 520
22	Po Delta, Italy (land and sea)	3.5		up to 600
	1951–1960	>2	3	
		>1.5	100	
		>0.75	1 400	
23	Rio Vista, CA	>0.30	98	1 300
24	River Island, CA	>0.23	18	1 250
25	San Emido Nose, CA	>0.06	4	3 900
26	Santa Fe Springs, CA	0.66	6	1 300
27	Saxet, TX	>0.93	25	1 800
28	South Houston, TX	0.09	7	1 200
29	Tejon-North, CA	>0.09	10	2 800
30	Torrence, CA	>0.10	27	1 230
31	Valhall, North Sea	0.6		
32	Webster, TX		11	1 670
33	West Ekofisk, North Sea	0.6		
34	Wilmington, CA	>8.84	29	1 000

tion. The Goose Greek displacements were accompanied by minor seismicity within the marginal zone of the subsidence bowl of the oil field, and characterised by high angle normal faulting in which the interior blocks dropped down towards the centre of the field. The faults were located along the margins of the subsidence bowl, within the zone of radially orientated horizontal extensional strain; the maximum subsidence was about 1 m.

The Wilmington oilfield in California produces oil from depths of between 600 and 2 000 m and has been pumped since 1932; it has much greater subsidence than Goose Creek. Subsidence is believed to have begun in 1937 but was first recognised by levelling in 1941. At its greatest point of depression the subsidence was 2.4 m in 1947, 4 m

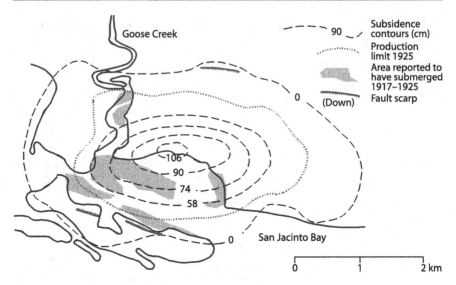

Fig. 13.8. Goose Creek oil field subsidence. The subsidence shown occurred in the period 1917–1925. The 'production limit' encloses the producing wells during that period

in 1951, 6.1 m in 1954, 7.9 m in 1958 and 8.8 m in 1965. Because the subsidence bowl is deep, horizontal displacements associated with the vertical movements were large, reaching a maximum of about 3.6 m. Small earthquakes, with hypocentres at 450 to 550 m, occasionally occurred.

Although the subsidence at Wilmington is high for on-shore oil and gas fields, it may well be exceeded by that in the Ekofisk oil and gas field, mentioned earlier. There, subsidence, first noted as significant round about 1984, had reached 6 m in 1994 and may reach as much as 18 m at the centre of a subsidence bowl, which itself is likely to be about 12 km in diameter in 2028. This will bring about changes of significance to platform foundation performance and design (Broughton et al. 1997). The practical aspects of determining the parameters for modelling compaction are recorded in Richard Goodman's recent biography of Karl Terzaghi (1999). Key amongst these is either obtaining sufficient "*undisturbed*" samples (in this instance rock core) from depths of over 1 000 m without them being unduly influenced by considerable stress relief or measuring stiffness in situ at those depths.

13.3.3
Extraction of Gas

The Groningen gas field was discovered in 1959 and is now known to extend over a net thickness ranging from 70 m in the south to 200 m in the north, and covering an area of some 900 km². The gas reservoir belongs to the Permian Rotliegendes Formation, is confined by the salt deposits of the Permian Zechstein Formation and overlies the Carboniferous (Fig. 13.9). Its depth of burial is approximately 2 900 metres. When the true size of the field was recognized, it was realised that subsidence would accom-

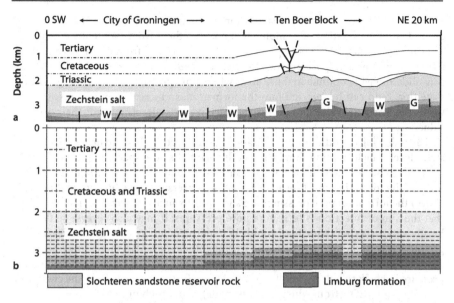

Fig. 13.9. A profile through the Groningen gas field (Van Hasselt 1992); (**a**) a simplified geological profile showing fault bounded reservoirs; G = gas, W = water; (**b**) the model for finite element study

pany extraction of the gas because the total drop in reservoir pressure (when the reserves have been exhausted) was expected to be approximately 30×10^6 N m^{-2}.

A geological model of the entire area likely to become affected by compaction and subsidence was designed on the basis of the available seismic stratigraphy and well logs. Subsidence, calculated on the basis of Eq. 13.6, was presented in the form of maps for the years 1975, 1990 and 2005, indicating maximum subsidence of 160 mm, 600 mm and 850 mm respectively. While the development of the gas field has been somewhat different from that anticipated at the time of making these initial estimates, those predictions now appear to be rather conservative, probably due to the difficulty of establishing representative values for compressibility. Actual maximum subsidence measured in 1990 was 180 mm. Modern calculations employ finite element techniques (Fig. 13.9) and benefit from back calculating compressibility from known gas pressure changes and measured ground movements; not surprisingly they now appear to give reasonable forecasts when compared to measured values. A maximum subsidence of about 360 mm is predicted for the year 2050. The reservoirs are fault bounded and over recent years small earthquakes, of Richter magnitudes 2 to 3 and with shallow hypocentres, have occurred. These are thought to arise from differential compaction at fault boundaries

13.4
Reduction and Monitoring of Subsidence from Fluid Withdrawal

Subsidence can be halted and in some cases stopped by no longer reducing pore fluid pressures. Sometimes this requires more than the cessation of pumping out because extraction will have created a distribution of different pore pressures through the ground and the cessation of pumping will not stop them from eventually equalising. Thus the areas with lowest pore pressures will be recharged from the areas of highest pore pressures and all pore pressures in between will be affected too. In other words consolidation will continue in those areas where pore pressures are declining. To alleviate this, fluid can be injected into the formation, so increasing pore pressures towards their original value and decreasing effective stress levels towards their original value.

The following conditions must be satisfied for injection fluids to decrease effective stress levels: the formation should accept fluids into its pore spaces but its permeability needs to be low enough for pore pressure to build up. Further, the injection of fluids into the formation must be at such rates and pressures that (in conjunction with above) the formation pore pressures are significantly increased over a wide area. Note that the aim of injection is to restore pore pressures; it may also reverse settlement locally and for a short time but is unlikely to permanently restore ground level – the ground has consolidated and that is an irreversible change. Also, increasing pore pressure can promote chemical effects. Cook et al. (2001) describe how increasing pore pressure in chalk to halt pore collapse, resulted in *further* collapse from the change in solubility of $CaCO_3$ (from which chalk is made) under high pressures.

Fluid injection has been applied successfully in the Wilmington Oil field; large scale injection was initiated in the upper four reservoirs and a response was noted almost immediately, both in decreasing and halting subsidence; it also increased oil production (Leps 1987). As the water injection programmes proceeded, surface uplift (called rebound) was noted and measured over wide spread areas as the elevations of over 900 bench marks were surveyed every 3 months. As a result of injection rates as high as 84×10^3 m^3 per day, the rebounded area spread to over twenty or more square kilometres between 1960 and 1967. During this period the rate of ground level lowering at the centre of the subsiding area, which was about 87 mm yr^{-1} before injection commenced, reduced to approximately 78 mm yr^{-1}.

Injection methods employed 27 km north west of Wilmington, at the Inglewood Oil field resulted in differential movements along a fault that passed beneath the embankments of the Baldwin Reservoir (Jansen 1987); the embankment displaced allowing reservoir water to escape and the embankment to be eroded, culminating in its failure.

13.4.1
Measuring Consolidation and Compaction

Periodic measurements of the ground surface elevation usually are the only means for recognising and calculating the initial stages of surface subsidence. In the Netherlands, a country where large tracts of land surface are close to or below sea level, changes in land elevation are closely tied to the maintenance costs of drainage systems. Parts of the Netherlands settle at rates of between 1 and 2 mm per year as a result of consolidation of the deeper Tertiary clays and sands. Such measurements, on a yearly basis, are within the error of the survey methods especially if older data is used and survey points are not protected from accidental movement; measurements have to be made over a span of several years to be able to determine most settlement rates.

In the northern provinces of Groningen and Friesland extraction of hydrocarbons has accelerated settlement. Here a number of factors combine to produce the settlement experienced: natural gravitational consolidation of the near surface sediments, extraction of near surface and deeper groundwater, to suit the needs of agriculture and water supply, local hydraulic mining of salt and extraction of hydrocarbons from deeper reservoirs. Claims for damage resulting from land subsidence cannot be easily resolved in the courts as proper measurements are scarce so making it impossible to distinguishing the contribution of each of these factors. Various approaches have been tried to resolve this problem; one uses stochastic methods based on movements at depth to determine settlements at the surface. Another has examined sedimentation thicknesses either side of the faults of the Roer graben in the Netherlands to determine rates of subsidence (Zijerveld et al. 1992). Others use down-hole geophysical logs correlated with geotechnical parameters to model subsidence rates in the Cainozoic deposits (Kooi 2000). Despite these approaches for determining subsidence the data for use in predictions remains both inaccurate and insufficient. The following techniques describe direct measurement techniques which, if installed, give by far the most satisfactory data.

Cable Measurements

The principle is to measure the length of a cable which is held under constant tension by means of an anchor weight at the bottom of a deep well and counterweights at the surface (Fig. 13.10). This method was used in the San Joaquin and Santa Clara Valleys, California during the early 1900s, where water is withdrawn from confined aquifers. A series of holes to various depths will identify the layers that are compacting

These techniques can also be used when settlement is occurring in sediments overlying karstic limestone and/or mined ground and be the only means of determining whether settlement is the result of either consolidation (which at least makes the ground stronger) or subsidence from a collapsing void at depth (which definitely makes the ground weaker). With the former the length of the upper holes will decrease relative to the deeper ones where as in the latter the length of the deeper holes will increase relative to the shallower ones.

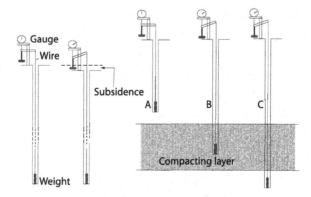

Fig. 13.10. Subsidence measurements by cable and weight. A comparison of the relative movements recorded by A, B and C serves to identify approximately the location of the layer undergoing compaction

Collar Counting

In general, when drilling an oil or gas well, casing is installed in the hole to the depth where the productive formation has been reached. Subsequently the annulus between the casing and the ground is sealed with cement grout bonding the casing to the ground. If the ground settles it drags the casing down with it, which in turn shortens due to its elasticity. A simple method of measuring formation compaction is therefore to periodically measure the vertical distance between the well casing joints; this is done using a geophysical tool which can de lowered down a well and is designed to measure aspects of its construction, such as the condition of the casing. The shortening of the distance between joints will give a good indication of the vertical compaction or stretching (if the distance lengthens) in the formations around the individual wells.

Marker Bullets

In a later stage of consolidation when casing shortening exceeds the maximum elastic deformation of the casing, collar movements may no longer be representative of formation movements, and fixed points (markers) placed in the formation itself may provide a better means of measuring compaction. This can be accomplished by measuring the relative displacement of marker bullets shot into the formation at regular distances. The bullets can be traced periodically by means of borehole logging using a Geiger counter if they are radioactive, or some other appropriate detector.

13.5
Further Reading

Barends FBJ, Brouwer FJJ, Schroeder FH (eds) (1995) Land subsidence; natural causes, measuring techniques in the Groningen gas fields. 5th International Symposium on Land Subsidence, The Hague. Balkema, Rotterdam
Cripps JC, Bell FG, Culshaw MG (eds) (1986) Ground water in engineering geology. The Geological Society of London (Engineering Geology Special Publication 3)
Poland JF, Davis GH (1969) Land subsidence due to the withdrawal of fluids. In: Varnes DJ, Kiersch G (eds) Reviews in engineering geology, vol II. Geological Society of America, Boulder, Co. USA (pp 187–269)

Epilogue

So, what is engineering geology about and what does an engineering geologist do?

I hope these pages written by David Price and completed by his colleagues have made the reader aware that *geology* is fundamental to engineering geology. It used to be said, and David would say it, that *"the best geologist is the one who has seen the most rocks"* and that *"no one can truly think of themselves as a geologist unless they have also visited a volcano, a glacier, and a coral reef"*. That is true for the best engineering geologist too but in addition the engineering geologist needs to have seen the most soils and the greatest variety of tectonic, and geomorphic settings. It also helps immensely to have witnessed failure of the ground, be it the natural failure of a slope as a landslide or an avalanche, the natural failure of the crust as an earthquake, or the failure of ground in a mine or below a foundation. However, it is the experience of catastrophe that matures the engineering geologist; the devastation caused by a tsunami, a mudflow, strong ground motion, the collapse of strata underground, the flooding of mines and tunnels; the area covered by these and their speed of occurrence is awesome and powerfully illustrates there is nothing steady state about geological processes, not even the slow and silent contamination of groundwater. This is the real world in which engineering geology is practiced and it requires what used to be studied under the name of *Practical Geology*.

So, what should an engineering geologist be able to do?

The answer is *"Practical Geology"*; it is important to be able to identify the common and especially the dangerous minerals and rocks, to be able to read geological maps and to be able to map at any scale, not only rock types but their structure too, together with the superficial evidence of events during the Quaternary; in short, to know how to address a volume of ground so as to reveal its relevant aspects for geotechnical engineering. It is important to be able to keep a good field notebook, to be able to sketch the salient features of site and landscape, and to be able to interpret from these geological products the natural processes that formed them, for it is these processes that the engineering geologist must reveal. Engineers analyse processes, in particular stability, and it is here, in the description and quantification of materials and processes, that the engineering geologist usually needs most help, in the form of further education.

Most young geologists have received little if any proper education in mechanics and its application, especially to the natural world, and that is where they need further training at an appropriate stage in their career. It is extremely frustrating to operate in geotechnical engineering without having a workmanlike grasp of forces and their moments, stress and strain, deformation and failure, failure criteria, hydraulic head and groundwater flow. Without these, landscapes cannot be read in a meaningful way and the products of natural processes cannot be used to quantify the process that formed them. These are the tools which facilitate the interface between geology and engineering, that link field to laboratory and design to analyses; they take time to assimilate and most geologists need help to acquire them.

But will all this be relevant to the future?

Absolutely; this book describes the core of the subject. The ground will deform and fail in the future in the same way as it has done in the past; the geological controls on strength, deformation, conductivity and durability have been placed there by nature and will not go away! There is no doubt that numerical modelling will change the way geology is used in geotechnical engineering. Limit equilibrium analyses that require propositions for relevant boundaries, material properties and mass processes, will give way to forms of analyses that will indicate the failure and deformation to expect with time. Far from removing the need for geology these analyses, when used properly, increase the demand for good quality geology; they require information about the geological history of the site so as to set initial stresses and a basis for separating the lateral and vertical variations needed to make model output equal measured ground response. Second rate geology will be of no value to numerical modelling. Likewise, the use of sophisticated methods for data analyses is fatally flawed if the data is not founded on a solid understanding of basic site geology. The chemistry of the ground will play an increasingly important role in predictive ground engineering and here a major problem exists with no obvious solution to it at present; viz, that most geologists, chemists and engineers are unable to talk to each other in meaningful terms. However, from the many attempts made to solve this problem, it is certain that the solution will require geologists to have a good grasp of their subject and practical experience in its use.

Geologists understand that the Present is the key to the Past and for engineering geologists the past is an important key to the future; the value of case histories to predictions of likely ground response bears witness to this. David Price reminded his students that "...there is no substitute for geological knowledge; you either acquire it before the engineering work starts or suffer the painful, and often expensive, lesson of being taught it as the work proceeds." It was his wish that this book would promote the former and prevent the latter.

The Editor

References

Ahorner L, van Gils JM, Flick J, Houtgast G, Ritsema AR (1976) First draft of an earthquake zoning-map of NW Germany, Belgium, Luxemburg and the Netherlands. Kon. Ned. Meteor. Inst., De Bilt (Publication 153)

Aikas K, Loven P, Sarkka P (1983) Determination of mass modulus of deformation by hammer seismograph. International Association of Engineering Geology, Paris, France (International Symposium on Soil and Rock Investigations by In-situ Testing, vol 1, pp 131-134)

AlpTransit (2002) St. Gotthard excavation and support drawing (*http://www.alptransit.ch/index.htm*)

Alvarez A, Sancho J (1974) Study of a rock mass supporting an arch dam. Proceedings 3rd Congress on Rock Mechanics, Denver, 2

Anon (1970) The logging of rock cores for engineering purposes. Report of a Working Party of the Engineering Group. Q J Eng Geol 3:1-24

Anon (1995) The description and classification of weathered rocks for engineering purposes. Report of a Working Party of the Engineering Group. Q J Eng Geol 28:207-242

Baoshan Z, Chopin L (1983) Shear wave velocity and geotechnical properties of tailings deposits. Bulletin of the International Association of Engineering Geology 26/27:347-355

Barton NR (1971) A relationship between joint roughness and joint shear strength. Proceedings International Symposium on Rock Fracture, Nancy, France (Paper I.8)

Barton NR (1988) Rock mass classification and tunnel reinforcement selection using the Q-system. In: Kirkaldie L (ed) Proceedings Symposium on Rock Classification Systems for Engineering Purposes. American Society for Testing and Materials, Philadelphia, (ASTM Special Technical Publication 984:59-88)

Barton NR (2000) TBM tunnelling in jointed and faulted rock. Balkema, Rotterdam

Barton NR (2005) Comments on a critique of QTBM. Tunnels & Tunnelling 37(7):16-19

Barton NR, Stephenson O (eds) (1990) Rock Joints. Balkema, Rotterdam (Proceedings International Symposium on Rock Joints)

Barton NR, Lien R, Lunde J (1974) Engineering classification of rock masses for the design of tunnel support. Rock Mech Rock Eng 6:183-236

Bekendam R, Pottgens JJE (1995) Ground movements over the coal mines of southern Limburg, The Netherlands, and their relation to rising mine waters. In: International Association of Hydrological Sciences (ed) Proceedings 5th International Symposium on Land Subsidence (Publication 234, pp 3-12)

Bekendam R, Price DG (1993) The evaluation of the stability of abandoned calcarenite mines in South Limburg, Netherlands. In: Sousa R, Grossmann F (eds) Eurock'93. Balkema, Rotterdam (pp 771-778)

Berkhout TJGM (1985) Model tests to assess the deformational characteristics of jointed rock foundations. Department of Mining Engineering, Delft University of Technology, Delft (Memoirs of the Centre for Engineering Geology in the Netherlands 32)

Bichara MM, Lakshmanan J (1983) Determination direct des densités du sol et de reablais a partir de mesures gravimetriques. Bulletin of the International Association of Engineering Geology 26/27: 171-174

Bieniawski ZT (1984) The design process in rock engineering. Rock Mech Rock Eng 17:183-190

Bieniawski ZT (1989) Engineering rock mass classifications. Wiley, New York

Bight GE (1997) Destructive mudflows as a consequence of tailings dam failures. Geotechnical Engineering 125(Jan):9-18 (Proceedings of the Institution of Civil Engineers)

Binns A (1998) Rotary coring in soils and soft rocks for geotechnical engineering. Geotechnical Engineering 131(Apr):63-74 (Proceedings of the Institution of Civil Engineers)

Biot MA (1941) General theory of three dimensional consolidation. J Appl Phys 12:155-164

Bishop AW (1948) A new sampling tool for use in cohesionless sands below groundwater level. Geotechnique 1:125-131

Bjelm L, Follin S, Svensson C (1983) A radar in geological sub-surface investigation. Bulletin of the International Association of Engineering Geology 26/27:175–180

Blake LS (ed) (1975) Civil engineer's reference book. Butterworths, London

Bless MJM, Bouckaert J, Paproth E (1980) Environmental aspects of some Pre-Permian deposits in NW Europe. Medelingen Rijks Geologische Dienst (The Netherlands) 12:1–14

Bowden AJ, Lamont-Black J, Ullyott S (1998) Point load testing of weak rocks with particular reference to chalk. Q J Eng Geol 31:95–104

Broch E, Franklin JA (1972) The point load strength test. Int J Rock Mech Min 9:669–696

Brook N (1985) The equivalent core diameter method of size and shape correction in point load testing. Int J Rock Mech Min 22:61–70

Broughton P, Aldridge TR, Nagel NB (1997) Geotechnical aspects of subsidence related to the foundation design of Ekofisk platforms. Geotechnical Engineering 125(Jul) (Proceedings of the Institution of Civil Engineers)

Brown ET (ed) (1981) ISRM suggested methods: Rock characterization, testing and monitoring. International Society for Rock Mechanics, Commission on Testing Methods, Pergamon Press

BS1377 (1990) British Standard Methods of test for soils for civil engineering purposes. Part 2: Classification tests. British Standards Institution, London

BS4019 (1974) Rotary core drilling equipment. British Standards Institution, London

BS5930 (1999) Code of practice for site investigations. British Standards Institution, London

BS EN ISO 14688-1 (2002) Geotechnical investigation and testing. Identification and classification of soil. Identification and description. British Standards Institution, London

BS EN ISO 14688-2 (2002) Geotechnical investigation and testing. Identification and classification of soil. Principles for a classification. British Standards Institution, London

BS EN ISO 14689-1 (2003) Geotechnical investigation and testing. Identification and classification of rock. Identification and description. British Standards Institution, London

Cai JG, Zhao J, Hudson JA (1998) Computerization of rock engineering systems using neural networks with an expert system. Rock Mech Rock Eng 31:135–152

Carranza-Torres C, Fairhurst C (2000) Analysis of tunnel support requirements using the convergence-confinement method and the Hoek-Brown rock failure criterion. Technomic Publishing Company, Inc. (Proceedings GeoEng2000, Melbourne)

Cole KW, Stroud MA (1976) Rock socket piles at Coventry Point, Market Way, Coventry. Géotechnique 26:470–462

Cook CC, Andersen MA, Halle G, Gislefoss R, Bowen GR (2001) An approach to simulating the effect of water-induced compaction in a North Sea reservoir. Society of Petroleum Engineers, Reservoir Evaluation and Engineering (Paper SPE 71301:121–127)

Cosma C (1983) Determination of rock mass quality by the crosshole seismic method. Bulletin of the International Association of Engineering Geology 26/27:219–226

Darracott BW, Lake MI (1981)An initial appraisal of ground probing radar for site investigation in Britain. Ground Eng (April)

Davenport CA, Ringrose PS (1985) Fault activity and palaeoseismicity during Quaternary time in Scotland-preliminary studies. In: Earthquake engineering in Britain. Thomas Telford, London (Proceedings Conference organized by the Institution of Civil Engineers and the Society of Earthquake and Civil Engineering Dynamics, pp 143–155)

Davenport CA, Ringrose PS (1987) Deformation of Scottish Quaternary sediment sequences by strong earthquake motions. In: Geological Society London (ed) Deformation of sediments and sedimentary rocks (Special Publication 29:299–314)

Davenport CA, Lap JMJ, Maurenbrecher PM, Price DG (1995) Liquefaction potential and dewatering injection structures at Herkenbosch: Field investigations of the effects of the 1992 Roermond earthquake, the Netherlands. Geologie en Mijnhouw 73:365–374

Deere DU (1968) Geological considerations. In: Stagg KG, Zienkiewicz OC (eds) Rock mechanics in engineering practice. John Wiley and Sons, pp 1–20

Deketh HJR (1995) Wear of rock cutting tools. In: Laboratory experiments on the abrasivity of rock. Balkema, Rotterdam

Den Haan EJ (1992) The formulation of virgin compression of soils. Geotechnique 42:465–483

Dennis JAN (1978) Offshore structures Quart. J Eng Geol 11:79–90

Den Outer A, Maurenbrecher PM, de Vries GT (1994) Earthquake-hazard mapping with questionnaires in the Roermond Area (The Netherlands). In: International Association Engineering Geology (ed) Procedings 7th International Congress. Balkema, Rotterdam (vol 01.3, pp 2087–2094)

De Ruyter TFM (1983) Engineering geological properties of Maastrichtian chalk. ENCI Quarry Maastricht, Engineering Geology Section, TU Delft (Internal Publication)

Dolan R, Goodell HG (1986) Sinking cities. Am Sci 74:38–47

Early KR, Skempton AW (1972) Investigations of the landslide at Walton's Wood, Staffordshire. Q J Eng Geol 5:19–41

Easton WH (1973) Earthquakes, rain and tides at Portuguese Bend Landslide. Bulletin of the Association of Engineering Geologists 10:173–194

Eddleston M, Walthall S, Cripps J, Culshaw MG (eds) (1995) Engineering geology of construction. Eng Geol Special Publication 10

Edmond JM, Graham JD (1977) Peterhead power station cooling water intake tunnel: An engineering case study. Q J Eng Geol 10:281–301

Fecker E, Rengers N (1971) Measurement of large scale roughnesses of rock planes by means of profilograph and geological compass. In: International Society of Rock Mechanics (ed) Proceedings International Symposium on Rock Fracture. Rubrecht, Nancy, France (vol 1, paper 18)

Fookes PG (1978) Middle East – Inherent ground problems. Q J Eng Geol 11:33–49

Fookes PG (1997) Geology for engineers: The geological model, prediction and performance. Q J Eng Geol 30:293–424

Fookes PG, Hawkins AB (1988) Limestone weathering: its engineering significance and a proposed classification scheme. Q J Eng Geol 21:7–31

Foose RM (1968) Surface subsidence and collapse caused by ground water withdrawal in carbonate rock areas. Proceedings XXIII International Geological Congress, Prague (Section 12, Engineering Geology in Country Planning, pp 155–166)

Franklin JA, Chandra A (1972) The slake durability index. Int J Rock Mech Min 16:141–156

Franklin JA, Broch E, Walton G (1971) Logging the mechanical character of rock. T I Min Metall A 80:1–9

Gambardella F, Bortolotto S (1991) The positioning system GPS for subsidence control of the terminal reach of the Po River. IAHS Publ 200:433–441

Gates WCB (1997) The hydro-potential value: A rock classification technique for examination of groundwater potential in fractured bedrock. Environ Eng Geosci 3:231–267

Gaziev E, Erlikhman (1971) Stresses and strains in anisotropic foundations. In: International Society of Rock Mechanics (ed) Proceedings International Symposium on Rock Fracture. Rubrecht, Nancy, France

Geological Society Engineering Group Working Party Report (1972) The preparation of maps and plans in terms of engineering geology. Q J Eng Geol 5:297–367

Geological Society Engineering Group Working Party Report (1988). Engineering geophysics. Q J Eng Geol 21:207–271

Geological Survey of Western Australia (1974) Geology of western Australia. Geological Survey Memoir 2

Gibbs HJ, Holtz WG (1957) Research on determining the density of sand by spoon penetration testing. Proceedings 4th International Conference on Soil Mechanics and Foundation Engineering, vol 1, p35– 39

Gibson RE, England GL, Hussey MJL (1967) The theory of one dimensional consolidation of saturated clays. Geotechnique 17:261–273

Goodman RE (1980) Introduction to rock mechanics. John Wiley and Sons

Goodman RE (1995) Block theory and its application. Geotechnique 45:383–423

Goodman RE (1999) Karl Terzaghi. The engineer as an artist. American Society of Civil Engineers Press

Goodman RE, Moye DG, van Schalkwyk A, Javandel I (1965) Ground water in-flows during tunnel driving. Eng Geol 2:39–56

Goodman RE, Korbay S, Buchignani A (1980) Evaluation of collapse potential over abandoned room and pillar mines. Bulletin of the Association of Engineering Geologists 17(1):27–37

Greenfield RJ (1979) Review of geophysical approaches to the detection of karst. Bulletin of the International Association of Engineering Geologists 16:393–408

Grima AM, Bruines PA, Verhoef PNW (2000) Modelling tunnel boring machines performance by neuro-fuzzy methods. Tunn Undergr Sp Tech 15:259–269

Grunthal G (ed) (1993) European macroseismic scale 1992. Conseil de l'Europe. Cahiers du Centre Européen de Géodynamique et de Seismologie 7:1–79

Haak HW, van Bodegraven JA, Sleeman R, Verbeiren R, Ahorner L, Meidow H, Grunthal G, Hoang-Trong P, Musson RMW, Henni P, Schenkova Z, Zimova R (1995) The macroseismic map of the 1992 Roermond earthquake, the Netherlands., Geologie en Mijnbouw 73:265-270

Hack HRGK (1998) Slope stability probability classification – SSPC. International Institute for Aerospace Survey and Earth Sciences (ITC), The Netherlands (Publication 43, 258 pp)

Hack HRGK, Price DG (1990) A refraction seismic study to determine discontinuity properties in rock masses. International Association of Engineering Geology (ed) Proceedings 6th International Congress. Balkema, Rotterdam (vol 2, pp 935–941)

Hack HRGK, Price DG (1995) Determination of discontinuity friction by rock mass classification. International Society of Rock Mechanics (ed) Proceedings 8th Rock Mechanics Congress, Tokyo, Japan. Balkema, Rotterdam, pp 23–27

Hack HRGK, Hingera E, Verwaal W (1993) Determination of discontinuity wall strength by Equotip and ball rebound tests. Int J Rock Mech Min 30:151–155

Hack HRGK, Price DG, Rengers N (2002) A new approach to rock slope stability – a probability classification (SSPC). Bulletin of the International Association of Engineering Geology and the Environment 62:167–184

Haines A, Terbrugge PJ (1991) Preliminary estimation of rock slope stability using rock mass classification systems. In: International Society of Rock Mechanics (ed) Proceedings 7th Rock Mechanics Congress, Aachen, Germany. Balkema, Rotterdam, pp 887–892

Hamada M, Yasuda S, Andlsoyama R (1987) Permanent ground displacement induced by soil liquefaction during 1983 Nihonkia-Chubu and the 1964 Niigata earthquakes. Proceedings 5th Canadian Conference on Earthquake Engineering, Ottawa, pp 533–542

Harroun DT (1940) Stability of cohesive earth masses in vertical embankments. National Academy of Sciences, Washington DC (Proceedings Highway Research Board)

Helley EJ, Harwood DS, Doukas MP, Irwin WP, Herd DG, Hanna WF, Brabb EE, Griscom (1979) Northern California. Reston, Virginia (U.S. Geological Survey Professional Paper)

Helm DC (1984) Field based computational techniques for predicting subsidence due to fluid withdrawal. Man-induced land subsidence. Geological Society of America (Reviews in Engineering Geology VI)

Henkel DJ, Knill JL, Lloyd DG, Skempton AW (1964) Study of the foundations of the Monar Dam. In: International Commission on Large Dams (ed) Transactioins 8th Congress on Large Dams. Edinburgh (vol 1, pp 425–441)

Herrenknecht (2002). TBM Botlek tunnel, The Netherlands (*http://www.herrenknecht.com*)

Higginbottom IE (1976) The use of geophysical methods in engineering geology. Part 2: Electrical resistivity, magnetic and gravity methods. Ground Eng 9(2)

Higginbottom IE, Fookes PG (1971) Engineering aspects of periglacial features in Britain. Q J Eng Geol 3:85–117

Hirono T (1970) niigata ground subsidence and ground water change. IAHS Publ 88:144–161

Hobbs DW (1964) The tensile strength of rocks. Int J Rock Mech Min 1

Hoek E (1983) Strength of jointed rock masses. Géotechnique 33:187–223

Hoek E, Bray J (1981) Rock slope engineering. Institution of Mining and Metallurgy, London

Hoek E, Brown ET (1994) Underground excavations in rock. Institution of Mining and Metallurgy, London

Hoek E, Franklin JA (1968) A simple triaxial cell for field and laboratory testing of rock. T I Min Metall A 77:22–26

Hoek E, Wood D, Shab S (1992) A modified Hoek-Brown criterion for jointed rock masses. In: Hudson JA (ed) EUROCK'92. Thomas Telford, London, pp 209–214

Hongey C, Rongher C, Tsan-Hwei H (1994) The application of an analytical model to a slope subject to rock falls. Bulletin of the Association of Engineering Geologists 31(4)

Hustrulid W (1976) A review of coal pillar strength formulas. Rock Mech Rock Eng (Supplement) 8:115–145

Hustrulid W (1982) Underground mining methods handbook. American Institute of Mining, Metallurgical, and Petroleum Engineers, New York

Hudson JA (1992) Rock engineering systems. Ellis Horwood Ltd., England

Hunt PH, Moskowitz BM, Banerjee SK (1995) Magnetic properties of rocks and minerals. In: Ahrens TJ (ed) Rock physics and phase relations: A handbook of physical constants. American Geophysical Union (AGU Reference Shelf 3, pp 189–204)

Hutchinson JN (1988) General report: Morphological and geotechnical parameters of landslides in relation to geology and hydrogeology. In: Proceedings 5th International Symposium on Landslides (Lausanne). Balkema, Rotterdam (vol 1, pp 3–35)

Ingersoll RH, Ulstein G (1989) Case history of a subsidence investigation for Stratfjord Field. OTC Paper 5406:407–415

International Association of Engineering Geology (1976) Engineering geological maps: A guide to their preparation. IAEG Commission on Engineering Geological Maps. The Unesco Press, Paris

Jaeger C (1979) Rock mechanics and engineering. Cambridge University Press

James AN (1981) Solution parameters of carbonate rocks. Bulletin of the International Association of Engineering Geology 24:19–26

Jansen B (1987) A review of the Baldwin Hills Reservoir failure. In: Leonards GA (ed) Dam failures. Eng Geol 24:1–81

Kaderabek TJ, Reynolds RT (1981) Miami limestone foundation design and construction. Proceedings American Society for Civil Engineers. J Geotech Eng 1(07), GT7 (July)

Kastner H (1971) Statik des Tunnel- und Stollenbaues: Auf der Grundlage geomechanischer Erkenntnisse (in German). Springer-Verlag, Berlin

Kellaway GA, Taylor JH (1968) The influence of landslipping on the development of the city of Bath, England. Proceedings XXXIII International Geological Congress, 12, Prague

Kenney TC (1964) Sea level movements and the geologic histories of post glacial marine soil at Boston, Ottowa and Oslo. Geotechnique 14:203–230

King LH (1979) Aspects of regional glacial geology related to site investigation requirements – Eastern Canadian Shelf. In: Offshore Site Investigation. Graham & Trotman, London

Kitano K (ed) (1992) Rock mass classification in Japan. Eng Geol (Special Issue)

Knill JL (1968) Geotechnical significance of some glacially induced rock discontinuities. Bulletin of the Association of Engineering Geologists 5(1)

Knill JL (1970) The application of seismic methods in the prediction of grout take in rock. In: Institution Civil Engineers (ed) Proceedings Conference on In Situ Investigations in Soils and Rocks. British Geotechnical Society, London, pp 93–100

Knill JL (1973) Rock conditions in the Tyne Tunnels, Northeastern England. Bulletin of the Association of Engineering Geologists 10:1–19

Knill JL (1978) Cow Green revisited. Inaugural lecture, Imperial College, London (see also Kennard MF, Knill JL (1969) Reservoirs on limestone with particular reference to the Cow Green scheme. J Inst Water Eng 23:87–136

Knill JL (2003) Core values: The first Hans-Cloos lecture. Bulletin of Engineering Geology and the Environment 62:1–34

Knill JL, Price DG (1972) Seismic evaluation of rock masses. XXIV International Geological Congress, Montreal (Section 13)

Koerner RM (1970) Effect of particle characteristics on soil strength. American Society of Civil Engineers, Journal of Soil Mechanics and Foundations Division, pp 1221–1234

Kooi H (2000) Land subsidence due to compaction in the coastal area of the Netherlands: The role of lateral fluid flow and constraints from well-log data. Global Planet Change 27:207–222

Koppejan AW (1948) A formula combining the Terzaghi load-compression relationship and the Buisman secular time effect. 2nd International Conference on Soil Mechanics, Rotterdam (vol 3, pp 32–37)

Kovári K (1993). Gibt es eine NÖT (in German)? Geomechanik-Kolloquium, Salzburg (42:17)

Kovári K (1994). Gibt es eine NÖT? Fehlkonzepte der Neuen Österreichischen Tunnelbauweise (in German). Tunnel 1

Krank KD, Watters RJ (1983) Geotechnical properties of weathered Sierra Nevada granidiorite. Bulletin of the Association of Engineering Geologists 20:173–184

Larnach WJ, Bradshaw R (1974) An investigation of the stability of a proposed hotel in Avon Gorge, Bristol. Q J Eng Geol 7:27–41

Laubscher DH (1984) Design aspects and effectiveness of support systems in different mining situations. T I Min Metall 93

Laubscher DH (1990) A geomechanics classification system for rating of rock mass in mine design. J S Afr I Min Metall 90:257–273

Laughton C, Nelson PP (1996) The development of rock mass parameters for use in the prediction of tunnel boring machine performance. In: Barla J (ed) Eurock'96. Balkema, Rotterdam, pp 727–733

Leca E, Leblais Y, Kuhnhenn K (2000) Underground works in soils and soft rock tunneling. Technomic Publishing Company, Inc., Melbourne (Procedings GeoEng2000, vol 1, pp 220–268)

Legget RF (1939) Geology and engineering. MacGraw-Hill, New York

Legget RF, Karrow P (1982) Handbook of geology in civil engineering. McGraw-Hill, New York

Leps TM (1987) Ground subsidence analysis prior to the Baldwin Hills reservoir failure. In: Leonards GA (ed) Dam failures. Eng Geol 24:143–154

Lopez A, Pirris PH (1997) Informe Geolgico-Geotecnico de Factibilidad del Sitio de Presa N2, San Jos-Costa Rica, Engineering Geology Area. Design Service Center, UEN PySA, ICE (Technical Report)

Marinos P, Hoek E (2000) GSI: A geologically friendly tool for rock mass strength evaluation. In: Technomic Publishing Co.Inc., Melbourne, Australia (Proceedings GeoEng 2000, pp 1422–1440

Maris PC (1982) On the assessment of relative density by the measurement of longitudinal wave velocity. Doctoral thesis, Delft University of Technology

Mathews MC, Hope VS, Clayton CRI (1997) The geotechnical value of ground stiffness determined using seismic methods. In: McCann DM, Eddleston M, Fenning PJ, Reeves GM (eds) Modern geophysics in engineering geology. Geological Society of London, Engineering Group, London (Special Publication 12:113–123)

Maurenbrecher PM, de Vries G (1995) Assessing damage for Herkenbosch, The Netherlands, due to the Roermond earthquake of April 13, 1992. Soil Dyn Earthq Eng 49:397–404

Maurenbrecher PM, Wever T, Degen BTAI (1998) 'H-Sense': Harbour sediment mapping using Chirp reflection surveys in Norway and Sweden. Environmental and Engineering Geophysical Society (European Section), Barcelona, Spain (Proceedings 4th Meeting, pp 245–248)

McDonald AM, Allen D (2001) Aquifer properties of chalk of England. Q J Eng Geol Hydroge 34:371–384

McDowell PW, Barker RD, Butcher AP, Culshaw MG, Jackson PD, McCann DM, Skipp BO, Mathews SL, Arthur JCR (eds) (2002) Geophysics in engineering investigations. Joint publication of the Geological Society of London (Engineering Group Special Publication 12) and CIRIA, London

Meigh AC (1976) The Triassic rocks with particular reference to predicted and observed performance of some major foundations. Géotechnique 26:391–452

Meinzer OE (1928) Compressibility and elasticity of artesian aquifers. Econ Geol 23:263–291

Mellor M, Hawkes I (1971) Measurement of tensile strength by diametral compression of disks and annuli. Eng Geol 5:173–225

Merriam R (1960). Portuguese Bend landslide, Palos Verdes Hills, California. J Geol 68:140–153

Mitani S (1998) The state of art of TBM excavation and probing ahead technique. International Association of Engineering Geology, Vancouver (Proceedings 8th Congress, vol 5, pp 3501–3512)

Moneymaker BC (1969) Reservoir leakage in limestone terrains. Bulletin of the Association of Engineering Geologists 6:83–94

Morfeldt CO (1970) Significance of groundwater at rock constructions of different types. Proceedings Symposium on Large Permanent Underground Openings, Oslo, pp 305–317

Müller L (1978) Removing misconceptions on the New Austrian Tunnelling method. Tunnels Tunnelling 10(Feb):29–32

National Coal Board (1975a) Disturbances of coal measure strata due to mining. National Coal Board, London (NCB Information Bulletin 51/49:4–14)

National Coal Board (1975b) NCB subsidence engineers handbook, 2nd edn. National Coal Board, London

Nguyen TQ, Helm DC (1995) Land subsidence due to ground water withdrawal in Hanoi, Vietnam. International Association of Hydrological Sciences (Proceedings 5th International Symposium on Land Subsidence, Publication 234)

Norbury DN, Child GH, Spink TN (1986) A critical review of Section 8 (BS5930). Soil and rock description in site investigation practice. Eng Geol Special Publication 2:331–342

Okamoto S (1973) Introduction to earthquake engineering. University of Tokyo Press

Orlic (1997) Predicting subsurface conditions for geotechnical modelling. International Institution for Aerospace Survey and Earth Sciences (ITC), Delft, The Netherlands (Publication 55)

Ozmutlu S (2002) Design of intelligent decision support systems for geotechnical purposes (DIGDSS). International Institution for Geoinformation Sciences and Earth Observation (ITC), Delft, The Netherlands

Pacher F, Rabcewicz L, Golser J (1974) Zum derzeitigen Stand der Gebirgsklassifizierung in Stollen- und Tunnelbau (in German). Proceedings XXII Geomechanical Colloquium, Salzburg, pp 51–58

Paige S (ed) (1950) Application of geology in engineering practice. Geological Society of America (Berkey Volume)

Patton FD (1966) Multiple modes of shear failure in rock. Proceedings 1st International Congress on Rock Mechanics, Lisbon (vol 1, pp 509–513)

Peck RB (1969) Deep excavations and tunnelling in soft ground. Proceedings 7th International Conference on Soil Mechanics and Foundation Engineering, Mexico, pp 225–290

Phien-Wej N, Nutalaya P, Sophonsakulrat W (1989) Current land subsidence of Bangkok and contemplated remedial measures. Balkema, Rotterdam (International Symposium on Land Subsidence)

Phillips FC (1968) The use of stereographic projection in structural geology. Edward Arnold Ltd.

PIANC (1972) Report of the International Commission for the classification of soils to be dredged. Permanent International Association of Navigation Congresses, Bulletin 11, 1:13–27

PIANC (1984) Classification of soils to be dredged, report of the Permanent Technical Committee II. Permanent International Association of Navigation Congresses, Bulletin 47 (Supplement)

Ploessel MR, Campbell KJ (1979) Northwestern Gulf of Mexico – engineering implications of regional geology. In: Offshore site investigation. Graham & Trotman, London

Poschi I, Kleberger J (2004) Geotechnical risk in rock mass characterization – a concept. Tunnels and Tunnelling 36(Sept/Oct)

Price DG (1983) Subsurface geophysical methods used to define geometrical or structural characteristics. General report on theme 2. Revue Française de Geotechnique 23 (International Association of Engineering Geology, Symposium on Rock and Soil Investigations by In Situ Testing)

Price DG (1992) On the stability of existing natural and artificial underground openings for use as underground space. Delft University Press (Proceedings 5th International Conference on Underground Space and Earth Sheltered Structures, ICUSESS'92)

Price DG (1993) A suggested method for the classification of rock mass weathering by a ratings system. Q J Eng Geol 26:69–76

Price DG (1995) Weathering and weathering processes. Q J Eng Geol 28:243–252

Price DG, Knill JL (1966) A study of the tensile strength of isotropic rocks. In: International Society of Rock Mechanics (ed) Proceedings 1st International Congress, Lisbon (vol 1, pp 439–442)

Price DG, Knill JL (1967) The engineering geology of Edinburgh Castle rock. Geotechnique 17:411–432

Price DG, Malone AW, Knill JL (1970) The application of seismic methods to the design of rock bolt systems. International Association of Engineering Geology (Proceedings 1st International Conference, Paris)

Price DG, de Goeje C, Pool M (1978) Field instruments for engineering geology mapping. International Association of Engineering Geology (Proceedings 3rd International Congress, Madrid)

Price DG, Hollingbery JW, Maxwell I (1988) Rock stabilisation works to preserve the castles at Edinburgh and Stirling, Scotland. Balkema, Rotterdam (Engineering Geology of Ancient Works, Monuments and Historical Sites, Athens, vol 1,pp 37–44)

Price DG, Rengers N, Hack HRGK, Brouwer T, Kouokam E (1996) Problem recognition index applied to the Falset Area. I Congrés del Priorat, Ajuntament de Gratallops, Catalunya, Spain

Rabcewicz L (1964) The new Austrian tunneling method. Water Power (Nov):453–457

Rabcewicz L, Golser T (1972) Application of the NATM to the underground works at Tarbela. Water Power (Mar):88–93

Rawlings GB (1971) The role of the engineering geologist during construction. Q J Eng Geol 4

Raybould DRL, Price DG (1966) The use of the proton magnetometer in engineering geological investigations. In: International Society of Rock Mechanics (ed) Proceedings 1st Congress, Lisbon (vol 1, pp 11–14)

Rengers N, Soeters R, van Riet PALM, Vlasblom E (1990) Large-scale engineering geological mapping in the Spanish Pyrenees. International Association of Engineering Geology, Balkema, Rotterdam (Proceedings 6th International Congress, pp 235–243)

Romana M (1991) SMR classification. In: International Society of Rock Mechanics (ed) Proceedings 7th International Congress on Rock Mechanics, Aachen, Germany. Balkema, Rotterdam (vol 2, pp 955–960)

Rowe PW (1972) The relevance of soil fabric to site investigation practice Geotechnique 27:195–300

Schimazek J, Knatz H (1970) Der Einflulss des Gesteinaufbaus auf die Schnittgeschwindigkeit und den Meisselverschleiss von Streckenvortriebsmaschinen (The influence of rock composition on cutting velocity and chisel wear of tunnelling machines). Glückauf 106:113–119

Schokking F (1998) Anisotropic geotechnical properties of a glacially over consolidated and fissured clay. Thesis, Delft University of Technology

Schrier JS van der (1988) The block punch index test. Bulletin of the International Association of Engineering Geology 38:121–126

Schrier JS van der (1990) A comparison between statically and dynamically determined properties of rocks: a case study. In: International Association of Engineering Geology (ed) Proceedings 6th International Congress, Balkema, Rotterdam (vol 2, pp 1061–1066)

Schumm SA, Chorley RJ (1964) The fall of threatening rock. Am J Sci 262(Nov)

Seed HB, Idriss IM (1971) Simplified procedure for evaluating soil liquefaction potential. American Society of Civil Engineers, Journal of Soil Mechanics and Foundations Division 97:SM9

Serafim JL, Pereira JP (1983) Considerations of the Geomechanics Classification of Bieniawski. In: LNEC (ed) Proceedings International Symposium on Engineering Geology and Underground Construction, Lisbon (vol 1)

Serrano A, Olalla C (1996) Allowable bearing capacity of rock foundations using nonlinear failure criterion. Int J Rock Mech Min 33:327–345

Sherrell FW (1971) The Nag's Head landslips, Collumpton By-Pass, Devon. Q J Eng Geol 4:37–73

Skempton AW, Northey RD (1952) The sensitivity of clays. Geotechnique 3:30–53

Slob S, Hack HRGK, Turner AK (2002) An approach to automate discontinuity measurements of rock faces using laser scanning techniques. In: Stephanson (ed) Eurock 2002. Madeira, Balkema, Rotterdam

Sowers GF (1978) Introductory soil mechanics and foundations: Geotechnical engineering. MacMillan Publishing Co., New York

Soydemir C (1987) Liquefaction criteria for New England considering local SPT practice and fines content. 5th Canadian Conference on Earthquake Engineering, Ottawa, pp 519–525

Stagg KG, Zienkiewicz (1968) Rock mechanics in engineering practice. J Wiley & Sons,

Stacey TR (1976) Seismic assessment of rock masses. Proceedings Symposium on Exploration for Rock Engineering ,Johannesburg, South Africa

Steward HE, Cripps JC (1983) Some engineering implications of the chemical weathering of pyritic shale. Q J Eng Geol 16

Sulak RM, Danielsen J (1989) Reservoir aspects of Ekofisk subsidence, J Petrol Technol (July):709–716

Swart PD (1987) An engineering geological classification of limestone material. Centre for Engineering Geology, The Netherlands (Memoirs 49)

Takeuchi S, Kimoto S, Wada M, Shiina H, Mukai K (1970) Geological and geohydological properties of the land subsided areas. IAHS Publ 88:232–241

Tan SB, Yang KSBE, Loy WC (1983) A seismic refraction survey for an expressway project in Singapore. Bulletin of the International Association of Engineering Geology 26/27:321–326

Telford WM, Geldart LP, Sheriff RE (1990) Applied geophysics. Cambridge University Press, UK

Terratec (2002) Rock cutting nomograph (*http://www.terratec.com.au*)

Terzaghi K (1925) Erdbaumechanik auf bodenphysikalischer Grundlage. Deuticke, Vienna

Terzaghi K (1929) Discussion on compressibilty and elasticity of artesian aquifers. Econ Geol 24: 211–213

Terzaghi K, Peck RB (1948) Soil mechanics in engineering practice. J Wiley & Sons

Tomlinson MJ, Boorman R (1995) Foundation design and construction, 6th edn. Longman Scientific and Technical, Harlow Essex

Tonouchi K, Sakyama T, Imai T (1983) S-wave velocity in the ground and the damping factor. Bulletin of the International Association of Engineering Geology 26/27:327–334

Van Hasselt JP (1992) Reservoir compaction and surface subsidence resulting from oil and gas production. A review of theoretical and experimental research procedures. International Journal of the Royal Geological and Mining Society of the Netherlands 71(2)

Varnes DJ (1958) Landslide types and processes. National Academy of Science Washington, D.C. (Highway Research Board, Special Report 29)

Verhoef PNW (1997) Wear of rock cutting tools. In: Implications for the investigation of rock dredging projects. Balkema, Rotterdam

Verhoef PNW, Kuipers T, Verwaal W (1984) The use of the sandblast test to determine rock durability. Bulletin of the International Association of Engineering Geologists 29

Verwaal W, Mulder A (1993) Estimating rock strength with the Equotip hardness tester. Int J Rock Mech Min 30:659–662

Voest Alpine Bergtechnik (2002) Excavation equipment (*http://www.vab.sandvik.com/*)

Walters RCS (1971) Dam geology. Butterworths, London

Waltham AC, Vandenven G, Ek CM (1986) Site investigations on cavernous limestone for the Remouchamps viaduct, Belgium. Ground Eng (Nov)

Weaver JM (1975) Geological factors significant in the assessment of rippability. Die Siviele Ingenieur in Suid-Africa 313–316

Whittaker BN, Reddish DJ (1989) Subsidence: occurrence, prediction and control. Elsevier, Amsterdam (Developments in Geotechnical Engineering 56)

Wilson G, Grace H (1942) The settlement of London due to the underdrainage of the London clay Journal of the Institution of Civil Engineers 104:156–160 (re-printed in Cooling LF, Skempton AW, Little AL (eds) (1969) A century of soil mechanics. Institution of Civil Engineers, pp 302–317

Wyllie DC (1999) Foundations in rock, 2nd edn. E & FN Spon, London

Wyllie MR, Gregory AR, Gardner GHF (1958) An experimental investigation of factors affecting elastic wave velocities in porous media. Geophysics 23:459–493

Xu H (2002) Production induced reservoir compaction and surface subsidence, with applications to 4d Seismics. PhD dissertation, Stanford University, California

Young N, Turcott E, Maathuis H (1995. Groundwater abstraction induced land subsidence prediction: Bangkok and Jakarta case studies. In: International Association of Hydrological Sciences (ed) Land Subsidence. Proceedings 5th International Symposium on Land Subsidence, Publication 234

Zaruba Q, Mencl V (1976) Engineering geology. Elsevier, Amsterdam

Zhu J, Xiao Z (1982) The method of evaluation for the stability of karst caverns. In: International Society of Rock Mechanics (ed) International Symposium on Rock Mechanics related to caverns and pressure shafts, Aachen (vol 1)

Zijerveld L, Stephenson R, Cloetingh S, Duin E, van den Berg MW (1992) Subsidence analysis and modelling of the Roer Valley Graben (SE Netherlands). In: Ziegler PA (ed) Geodynamics of rifting; case history studies on rifts; Europe and Asia. Tectonophysics 20:159–171

Index